Technological Foundations of Cyclical Economic Growth

Technological Foundations of Cyclical Economic Growth

The Case of the United States Economy

NATHAN EDMONSON

Routledge
Taylor & Francis Group

LONDON AND NEW YORK

First published 2009 by Transaction Publishers

2 Park Square, Milton Park, Abingdon, Oxfordshire OX14 4RN
711 Third Avenue, New York, NY 10017

Routledge is an imprint of the Taylor & Francis Group, an informa business

First issued in paperback 2017

Library of Congress Catalog Number: 2008050613

Library of Congress Cataloging-in-Publication Data

Edmonson, Nathan.
   Technological foundations of cyclical economic growth : the case of the United States economy / Nathan Edmonson.
         p. cm.
   Includes bibliographical references and index.
   ISBN 978-1-4128-1012-8
      1. Technological innovations--United States. 2. Business cycles--United States. 3. Industrial policy--United States. I. Title.

HC110.T4E36 2008
338'.0640973--dc22

                                                          2008050613

ISBN 13: 978-1-4128-1012-8 (hbk)
ISBN 13: 978-1-138-51525-3 (pbk)

Contents

Preface

Several non-economist friends expressed interest in this book as it was forming, and the most frequent question that resulted from these readings was: "Why did you write this?" The shorthand answer to this question is that after I retired from full-time professional activities in 1990, several fragmentary ideas started coalescing in my mind. The book is the result of a number of years of turning over of these ideas. They just would not let go of me, and so I had to do something. The book did not emerge like Aphrodite, full-blown, and many of the ideas expressed therein grew from the original set as I organized the presentation.

What were the original ideas behind this project? Perhaps the most important one was that one-time events that lead to shifts in demand curves and supply curves are not instantaneous in their effects. I have the impression that this fact has been somewhat neglected by economists who were educated after 1965. One curve-shifter that is included in almost all such lists is changed technology. The market models to which first-year economics students are exposed, such as that of perfect competition, rest on assumption of universal adoption of the most efficient technology by all economic actors. My experience as an economist in business and industry suggests serious questioning of this assumption. To some degree, technology flux is always changing the answer to the question of what is the most efficient organization of productive assets. Wide adoption of even the most powerful of new technologies is a time-consuming process. Universal understanding of the full possibilities of a new technology simply does not exist until much ingenuity has been applied to its use. The same is true for such events as a radical change in a regulatory regime. It takes time for the affected economic actors to get used to the possibilities of the new situation.

The result of all this can be prolonged periods of uncertainty that affect senior corporate and government managements. Two questions arise. First, what of economic interest can happen while economic actors adapt to radically altered technological or regulatory conditions? Second, what can be said of the episodes of high uncertainty that occur during adaptation? A market bubble is usually defined as some sort of disconnect between how a market is perceived and how it actually is. To find such a disconnect, it could be useful to investigate some of these episodes of high uncertainty that occur in a time of technological and/or regulatory flux.

I believe that I have shed light on some of these and other questions—but far from all of them. This book is a first attempt to get my mind around an enormous subject. I am not through with these topics, and ideas still come to me.

While this project has largely been a solo effort, I do owe some recognition. My daughter, Sarah, a young physician, challenged me to get off dead center and start working beyond having ideas. She has wide-ranging interests beyond strictly professional ones, including the advancement of a number of technologies. This interest accounts for a very fascinating correspondence between us that has gone on for years. My wife, Susan, proofed some early manuscript material, but her big contribution has been in the form of general support.

Introduction

Since its beginnings in the eighteenth century, modern economic thought has recognized some sort of connection between technological advance and economic growth. Despite this general recognition, twentieth-century economics never really got to the heart of the technology-growth relationship. Some writers on economic cycles, such as Joseph Schumpeter and John Hicks, were very cognizant of the role of technology in cyclical variability, but their lines of thought were not extensively elaborated after they passed from the scene. The neoclassical growth model, which was very influential after the 1950s, dealt with technological change by treating it as external to economics, and the "new growth theory" of the 1980s and early 1990s did not add much to the neoclassical model in this regard.

If one looks away from formal economic theory and reviews the actual record of growth, it is difficult to ignore the fingerprints of technological change on growth events—both positive and negative. If new technology can foment growth, then understanding of growth—past and present—is aided by knowing what tools people had and when. Moreover, there are at least some hints that invention has itself been influenced by economic growth events. Technology appears to be intimately associated with the cyclical pattern of growth, and therefore should be of intense interest as to its origins, development, limitations, and impact. If technology is born in invention, what can be said regarding the circumstances surrounding prolific periods of technology-induced economic growth, which alternate with periods of low growth?

A basic premise of this study is that not all inventions are created equal. The focus is on a small number of inventions from which much widely familiar technology ultimately developed, these are referred to as *prototype* inventions. *Innovation* is the process by which invention is transformed into economically useful forms. Innovation of prototype inventions shapes the nature of further inventive activity. Additionally, innovation of a prototype invention foments the investment activity that underpins growth. A slate of new technologies that enables substantially lowered cost of production attracts investment because returns to such investment are high; however, once the full range of investment possibilities based on new technology has been exploited, returns to investment in it fall off and there is a decline in the rate of investment that can have deleterious consequences for general growth. Unusually high growth will not resume until technology developments create a new slate of investment opportunities.

What is the nature of the progression from invention to general economic growth? For one thing, the duration of the process is enormously variable. Consider an actual example in lead-time between invention and widely useful technology. In 1938, Chester Carlson was awarded a patent for the process that later became known as Xerography, but not until 1960 was the first successful electro-photographic copying machine—the Xerox 914—introduced. The years between those two dates saw the development of Carlson's copying machine from a crude assembly, created in a makeshift laboratory, to a commercially acceptable office tool. This development came as a result of much tinkering and considerable financial risk by the inventor, and later, by the one business organization that dared to see a future to Carlson's invention. Following the unveiling of the 914 Copier, the use of it and successor models in tens of thousands of offices called for a massive capital investment.

Why did it take so long to bring the copier, which hindsight tells us was a useful invention, to commercial reality? A prototype invention or the equivalent, such as a major change in a regulatory regime, is a disrupter of an existing economic order because it implies the obsolescence of some established technologies. Carlson shopped his invention to a number of corporations that his logic suggested should be interested. Many of these turned him down for reasons apparently related to the competitive threat to established products that photoelectric copying appeared to pose. Other reasons why some technologies languish in a nascent state include the fact that they do not yield readily to innovation efforts. This could be because alternative technologies have always accomplished the same thing with less cost. An on-the-shelf technology may be high cost because of physical laws that no one has yet discovered how to circumvent.

At the present time, there are a number of nascent technologies that have attracted interest in relation to some of the perceived problems of today. Historically and currently, choices among technologies have been largely determined by cost considerations. Revised perceptions regarding which costs should be counted, such as brought on by consciousness of environmental degradation, bring historical decisions regarding technology adoption into question. One of the main reasons for revived interest in old technologies (such as solar power, dating from 1883) as well as nascent technologies that are more recent is concern about the environmental side effects of extant technologies. No group of problems appears more likely to call for technology solutions than that of future sources of energy. Therefore, questions regarding the circumstances bearing on how quickly technologies in early stages of innovation might become widely useful are at the center of interest because many technologies of present-day interest have not been developed to the point of being cost competitive with technologies they would have to replace.

This book presents an examination of cyclical economic growth and its how it was influenced by the historically uneven progress of technology. Chapters 12 to 14 attempt no specific forecast of how future technology will evolve but concentrate on forces that are likely to influence this future evolution. These chapters call on the chronology of technology development and impact on growth presented in chapters 2 to 11. Chapter 15 concludes.

1

Basic Concepts

Invention and Innovation

A typical economics textbook of the last several decades at some point introduces the concept of *innovation*, defined as the broad process by which an *invention* is transformed from a crude prototype to a commonly used product. *Invent*, according to Webster's Dictionary, is "to think out or produce (a new device, process, etc.); originate, as by experiment; devise for the first time." An invention, then, is the result of the inventing activity. Inventions, in their earliest manifestations, often have no immediate economic significance. Economic significance commences only when an invention is first applied to a useful purpose; and indeed, the usual economics text presents the initial application of an invention as the first step in the innovation process. What the typical economics text fails to convey is the enormous variability in timing and uncertainty inherent in the transition from invention to its possible ultimate economic implications.

In fact, invention and innovation can be thought of as successive events on an ongoing "innovation stream," whose path is determined by complex creative, economic, technological, and social influences. The purpose of this book is to explore, through examples and discussion, the genesis and nature of this innovation stream and how it relates to some major movements in economic development. As suggested, the process starts with invention. However, even a superficial reading of technology history suggests that not all inventions are of equal importance. There have been a relatively small number of inventions that have changed the way people live in fundamental ways. These are the inventions on which the study focuses and to which it refers as "prototype inventions." This chapter introduces many of the key concepts which will be explored more thoroughly in later chapters.

The Prototype Invention

Not all inventions are created equal as to their potential impact on economic growth. In looking back in economic and technology history, a handful of in-

1

ventions stand out as having been particularly strong influences on subsequent economic growth. Such inventions can be thought of as *prototype inventions.* Faraday's electric motor is an example of a prototype invention. In its earliest form, the electric motor was a demonstration of the scientific principles underlying its operation but little else. What could be made of it was limited at the time of invention by the state of *enabling technology*, the result of prior inventive activity. Earlier inventions available to Faraday included such devices as galvanic cells and other products of pioneer experiments with electricity and can be thought of as *enabling inventions*. A number of enablers that later went into making the motor a useful power source, such as the ability to machine ferrous metals to fine tolerances and an ample source of electric power were still in the future in Faraday's time.

What are the characteristics of the prototype invention? First, a prototype is an original hardware manifestation of the scientific principles that underlie its operation. Faraday's electric motor was a clear example. Second, a prototype invention triggers a series of applications that make possible significant change in the efficiency with which people do things that they had already been doing. Obvious examples are the internal combustion engine, with its applicability to powering factory line shafts, and the transistor, with applications toward amplification of radio signals. Third, not only will a prototype invention lead to the more efficient achieving of what was achieved before, it will open opportunities for doing things that were never done previously. The transistor not only became a more reliable replacement for the vacuum tube in a radio signal amplifier but later became the central enabling basis for the huge world of digital technology. The internal combustion engine went from being simply a better line shaft prime mover to becoming the power basis for land transportation in the twentieth century.

In order to understand the prototype invention concept and its implications better, it is useful to reflect that improvements in the productive potential of capital equipment take place more or less continuously through time. Each year presents the opportunity to replace an incumbent unit of equipment with equipment that offers lower operating cost characteristics. As these year-by-year improvements accumulate through time, so does the level of obsolescence of the extant stock of productive capital, and at some point, the cost implications of the accumulated obsolescence become important elements in the capital-using firm's capital replacement decisions. The decision to extend or not extend the productive life of productive capital clearly depends on other factors as well, such as level of demand for consumer goods. From time to time, however, there appears an improvement in the productive potential of newly available capital of such magnitude as to create mass obsolescence in a comparatively short period of time. This is one impact of the prototype invention.[1]

Important as they were, three nineteenth-century inventions that customarily are included in accounts of the history of technology—the cotton gin, the sewing machine, and the reaper—were not prototypes by this standard. True, these increased the efficiency of performing established tasks—and the sewing machine's primary application of automating the labor-intensive process of joining one piece of cloth with another was later adapted to new tasks such as saddle making and shoemaking. However, none of the three technologies can be said to have grown directly from basic underlying physical principles.

Innovation

The period of innovation following a prototype invention can be divided roughly into two stages (Figure 1.1). In the first of these, the prototype is improved to the point where it becomes economically attractive to private investors. This standard for completion of stage-one innovation recognizes that a technology can be usable in an engineering sense well before it becomes commercially attractive. The first stage of innovation can be lengthy if development depends on an advance of some related, or enabling, technology. This was the case with some of the earlier prototypes for which the initial enabling technology was limited. The long delay between Faraday's demonstration of principle and the electric motor's widespread application can be explained, at least in part, by lagging development of other technology. Among the enabling developments that made development of the electric motor possible were improvement in the ability to machine ferrous metals to close tolerances and the availability of commercial electric power, which grew after about 1885. Prior to then, there had been several successful attempts to build a usable electric motor, but until electric power became generally available, these could not go very far. As will

Figure 1.1
The Invention-Innovation Sequence

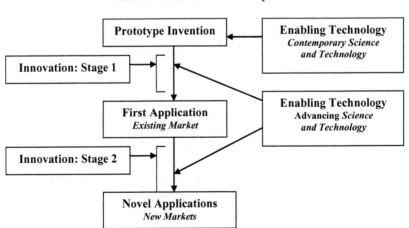

be seen, there are examples of lengthy stage-one innovation processes even in recent times.

The textbook juxtaposition of invention and innovation seems strange in light of the experience of Chester Carlson and others, for inventing is different from stage-one innovating; the latter is far more risky. A lack of enabling technology is sometimes not the sole, or even a decisive, reason for a lengthy first stage period of innovation. The textbook sequence of invention and then innovation can fail because of the lack of any person or entity willing to assume the risk inherent in innovation, especially in its first stage. Individual inventors who succeed often do so in spite of highly adverse personal financial conditions. Following his initial patent of 1938, Chester Carlson personally had to bear all the risk and expense of developing his invention into usable form as his efforts to interest companies such as Eastman Kodak met with frustration. In order for a corporation to take an interest, it was and is necessary for someone in that organization to buy into the inventor's idea of the ultimate profit potential. Failure to foresee a future for an invention such as Carlson's can come from simple lack of imagination, or from apprehension regarding the implications of the invention. Prototype inventions can be threatening to established interests; they are potential disturbers of an existing order. Eastman Kodak eventually did enter the copier business but only after Xerox had demonstrated the demand.

Historically, governments have augmented private entrepreneurs by assuming a powerful role as innovation risk takers. Government success at fostering innovation has largely, but not entirely, sprung from national defense initiatives. The effectiveness of government for fostering innovation is most potent for carrying the innovation through its risky first stage and far less successful when it tries to go beyond this point; indeed, this seems to be a globally common experience. As will be shown, some of most successful innovation streams have started with a military-supported stage one and been followed by a private sector-driven stage two.

In the second stage, the improved prototype is marketed among commercial, industrial, and individual potential users. The speed with which this stage progresses depends on how profitable the applications are anticipated to be. In an industrial setting, the decision to apply the device/system rests on some form of rate of return analysis,[2] which can be done only in the second stage, for such analysis requires that a marketable product be in sight, which is true only after the first stage. The second stage closes when all the markets for the technology that eventually appear attain maturity, a state in which the technology in question has exhausted its potential for improvement and/or its potential for being applied in novel ways. It is in the second stage of innovation that the new technology finds its full range of applicability. At the outset of second-stage innovation, a new technology attracts investment with high rates of return because it enables replacement of existing capital with more productive capital. As the new technology is applied widely, it becomes increasingly commonplace, and the returns to investment in it decline because the increment of productivity resulting from each application gets smaller.

The length of time necessary for a stage-two innovation process to work itself out to its potential is of great interest for one easily stated reason: it is the stage-two part of innovation that generates the surge of economic investment that is at the heart of a major growth episode. Generally speaking, the longer a stage-two process is active, the longer there will be a period of economic strength. Historically, there have been instances in which a number of new technologies have emerged from their stage-one innovations in a comparatively short period of time. One such example occurred in the last two decades of the nineteenth century and another in the 1960-1980 period. Sheer number of new technologies is one factor determining length of a stage-two episode, for one may suppose that the greater number of new technologies with which to work, the greater is the complexity of the stage-two process, as it can take considerable time for all the possible combinations of hardware and organization to be tried in the marketplace in the process of determining which are the most successful economically.

Apart from the raw number of new technologies in these episodes, the quality of the new technologies themselves is a factor influencing the duration of a stage-two process. Some prototypes turn out to be very powerful in the sense that their technical progeny is the basis for a major growth episode. A cluster of such powerful new technologies can underpin a foundation to economic growth for a substantial period of time. Mass (low-cost) production of major consumer goods and widespread electrification of the economy grew out of the technology legacy of the late nineteenth century. However, the growth potential of these stage two processes reached a limit before 1930. The new technologies of the late twentieth century led to low-cost computing power and ingenious applications of integrated circuitry and the technology of light, and this expansion of ideas appears to have considerable remaining life as of this writing.

Prototype Invention and the Direction of Subsequent Inventive Activity

The innovation process itself leads to substantial inventive activity. Thus, a prototype invention massively influences the nature and direction of subsequent inventive activity. The development of the machine tool industry required the invention of a number of specialized devices for finishing metals. In another example, a number of inventions in the area of diesel engine fuel injection and control in the 1920s and 1930s greatly improved the engine's usefulness as a power source in the overland transportation market. The telephone called forth inventions such as switchgear. An invention that is conceived of and applied in the direct context of a particular stream of innovation can be thought of as an *innovation invention*. The concept of innovation invention illustrates a characteristic property of the prototype invention—that of setting the general direction of follow-on inventive effort.

For any invention, prototype, innovation, or other, to be created there must be something from which to make it: materials. Thus, materials are enablers of invention and innovation such that the appearance of new materials often has the effect of making possible the emergence of earlier inventions from stage-one innovation. The technology of solid-state electronics, which has been so much at the center of technological advance in the second half of the twentieth century, rested on the properties of semiconductor materials. High-quality steels became available for general use after 1890, and it would be difficult to exaggerate the importance of these. From about 1790 until that time, there had been considerable pioneering effort to develop a system of manufacture in which parts of larger assemblies were machined to the point of not needing further work in order to be assembled into larger devices. While progress was made, achievement of universal parts interchangeability did not approach full reality until steel, with its superior ability to hold its shape under the stress of machining, became available and abundant, and tools based on very hard ferroalloys became available. New materials and the possibilities that they enable have been a central element in technological innovation of the twentieth century. The list of such materials is very long and includes such familiar names as nylon, Teflon, titanium, and Kevlar.[3]

Can a Prototype Invention be Recognized?

The question of whether or not a prototype invention, as here defined, can be identified is worth asking because there is a previous economic literature that seems to point to a negative answer, and this study clearly takes the position that the answer is positive but only after the fact. A number of years ago there was a controversy among economic historians regarding the importance of the railroad in the development of the US economy in the nineteenth century. The controversy revolved around the question of how important the railroads were in reducing the cost of overland transportation. Could some technology other than railroad have produced the cost reduction that actually transpired? Robert Fogel attacked the notion that railroads were essential. His study indicated that a growth result little short of what actually happened based on inland waterways could have transpired even if the railroads had never come into existence. His method of attacking the question was by assuming that railroad technology had never appeared and by developing a virtual US economy using this basic assumption. This method (*the contrafactual hypothesis*) naturally became highly controversial in and of itself, but the result nevertheless questioned the ability to identify an essential invention, even in retrospect.[4] In another study, Nathan Rosenberg proposed evaluating the impact of an invention/ensuing innovation by a microexamination of the impacts on other industries. Actually doing this would be extremely difficult without a real-time table of input-output coefficients, something that has never been available. Such input-output tables as do exist are tied to a fixed slate of well-defined industries, and this alone poses a problem inasmuch as what has been defined here as a prototype invention often leads to the creation of entirely new industries. The overall thrust of which this discussion is a part seems to reject the prototype invention on the grounds that there is really no such thing as an invention that comes into being in the absence of what he terms "complementary technology," which this study has termed "enabling technology."[5]

The present study takes a much different approach. The cellular telephone, with which today's world is well familiar, is built around one or more radio frequency micro-devices, which are applications of the integrated circuit (IC). The heart of the cell phone is a much-improved lineal descendant of the original invention by Kilby and Noyce around 1960. The technical details of this descent are known. Other descendants of the IC include the microprocessors that are the essential feature of today's computers of all forms. Other descendants could easily be listed. This approach avoids the problems of navigating the changing morass of detailed industry impacts that would be necessary in assessing the impact of an invention when it first emerges from stage-one innovation. Of course, the prototype invention is a retrospective concept, but it has to be. Generally speaking, there is no way to predict the future of technology as it evolves.

Multiple Innovation Streams

It is not uncommon for the innovation streams stemming from two or more prototype inventions to be moving forward at the same time, and this situation can provide the market with a choice of technologies to apply to a given problem. In some instances, two or more technologies can compete to fill the same application. Development of the electric motor was getting started in the 1890s when the diesel engine appeared in practically usable form (1897). At the time, the typical factory's machinery was driven by a rotating shaft (the line-shaft) that was powered by a large stationary steam engine at one end of the building. Rudolph Diesel's idea, shared widely in his time, was simply to replace line-shaft power and marine engines with a unit that was far more thermally efficient.[6] What is interesting is to take the view available to someone living in 1900 and contemplating the future. The observer would first see a variety of power units just coming into stage-two development. Besides the electric motor and the diesel, there would have been the Otto-cycle engine and the steam turbine. Reflection on the known characteristics of these power sources could have provided some insight but would just as likely have misled. For example, in 1900, the number of automobiles was small, and most of these were operated locally where their owners resided. The choice of automotive power consisted of electric motor, steam engine, and the Otto-cycle (gasoline) engine. Of these three options, the electric car was in a slight plurality. With the relatively strong presence of electric power, an onlooker might have been tempted to bet on that option for the future. The late twentieth and early twenty-first centuries provide outstanding examples of multiple technology development: for example, the ways of producing and distributing electric power. A more complete discussion of choice among technologies is deferred until Chapter 14.

Technological progress periodically presents the economy with a choice of options for the future. One purpose of this book is to show that when this has happened, no person or entity has demonstrated any particular ability to recognize the technological option that will succeed best in a given sphere of activity in the future. The presence of many technological options is like a roulette wheel, and viewed in this manner, the situation of 2005 is similar to that of 1900. Official efforts of pick future winners have come under the heading of "industrial policy," which after many trials in many places now has a deservedly bad reputation. Chapter 6 includes an elaboration of this statement.

Cyclical Economic Development

While a prolonged stage-two innovation episode can give the economy a foundation growth push, it will not protect from shorter-term cyclical events. There is nothing here that implies that technology underpins a uniform expansion of an economy. Indeed, experience quickly dispels that idea.[7] Technology-generated progress can be thought of as being a basement on top of which other conditions that lead to alternating periods of prosperity and recession operate.[8] The presence or absence of strong active underlying technical innovation streams can make a large difference in the severity of cyclical downturns. Innovation streams based on the internal combustion engine, the electric motor, the parsons turbine, and several others had reached states of relative maturity by 1930, and it is not obvious that any other innovation stream was present to give a growth push to the economy of the 1930s. Chapter 3 highlights the rapid economic expansion that grew out of the legacy of innovation activity that came out of the nineteenth century, and Chapter 4 develops how the end of the investment boom was a contributor to the severity of the great depression. Chapter 5 describes the development of modern materials as enablers of technological development, especially of the second half of the twentieth century.

During the 1920s and 1930s there was considerable effort by the economics profession to study and understand the economic cycle. A group of scholars, remembered as the Austrian School, made substantial contributions to the study of cycles, and writers in this group were well aware of a long cycle whose upswing phase was triggered by heavy investment activity growing out of entrepreneurial activity. Indeed, one of the observations from history emphasized by this school was that investments tend to "bunch up" in certain periods of history.[9] The last two decades of the nineteenth century was pointed out as such a period. As defined here, there was not so much a bunching of prototype inventions in the 1880s and 1890s as bunching of the onset of stage-two innovation streams from different prototype inventions, some of which had been around for a long time. Chapter 3 develops this.

The importance of various streams of innovation are the side effects in the form of economic development. Again borrowing from the economic cycle theorists, we have the idea that the cyclic episode (including upturn, crisis, recession, and recovery) typically ends with a larger economy than existed when it started.[10] An important stream of innovation imparts an upward-trend element in economic activity. A corollary of this idea is that a trend generated by an innovation stream leaves the economy with a richer fund of enabling technology than existed at the outset of the innovation period. A richer fund of enabling technology expands the possibilities for innovation on the basis of prototype inventions that appear. One can speculate that the enrichment of the fund of potentially enabling technology over time leads to an acceleration of an economy's ability to realize the full potential of any prototype invention.

Stage-Two Innovation: The Statistical Footprint

It is reasonable to ask what sort of statistical footprint does an especially fecund group of prototype inventions leave by which its impact may be recognized. The path from innovation to growth leads through stage-two innovation, wherein the application of the new technology implies investment in which firms expend their depreciation flows on capital with productivity characteristics superior to that which is replaced. This immediately suggests statistics on productivity. There is a database in this area, but an imperfect one. Official productivity statistics are of the form of a ratio between an output measure and the input of labor factor (e.g. output per worker-hour). It should be recognized that the impact of given investments on labor productivity varies considerably from investment to investment.[11] Not all investments are for the immediate purpose of displacing labor. However, technologies descended from major prototype inventions may tend to be rich in implications for the output-labor relation, such that a period of accelerated growth in the output-labor ratio can be interpreted as evidence of intense application of such technologies.

Even if the short-term objectives justifying a given investment decision do not include displacement of labor, many investments seeking other forms of cost reduction may have eventual implications for that ratio. Therefore, the statistics of aggregated private domestic investment, which reflect the characteristics of many individual investment decisions, have to impact the overall output-labor ratios over a period of years. Indeed, the lack of an immediate impact of private domestic investment on the published productivity ratio is apparent in a regression analysis whose implication was that less than 5 percent of the year-to-year variation in the national ratio of output to worker-hour input could be explained by *contemporaneous* variation in private domestic investment. An increase in the rate of growth in a national productivity measure, such as occurred in the 1910s and in the 1990s, therefore reflects an unusual accumulation of current and delayed effects of investments on the output-labor ratio. Productivity growth in the 1910s was heavily influenced by the emergence of mass production and reduced unit costs in autos and other industries. The surge in productivity growth in the 1990s can be attributed to an explosion of information technology application. In both cases, the surge was the result of a large accumulation of output-labor impacts of investments of previous years.

An expected return is the common justification for capital investment by a business organization. What are some of the reasons for investment that do not contemplate immediate displacement of labor by capital? The following describes a few of these.

Response to Environmental Regulation

This kind of investment often results in no increase in output capacity and at least a small increase in labor input.[12] Substantial investment sums are frequently involved. Examples include electric generating plants being retrofitted with stack gas scrubbing equipment for the removal of sulfur dioxide from exhaust gasses from combustion. Another example is the modification of nonferrous smelters with facilities to capture sulfur dioxide and process it into sulfuric acid. The smelting industry has made a series of such investments since the Clean Air Act, with each succeeding investment involving either major modification of existing technology or application of completely new technology. The immediate impact of this kind of investment is not to increase labor productivity, but sometimes to decrease it.

Improve Workplace Safety

The anticipated payoff to this kind of investment is to diminish worker compensation claims (or the equivalent). This was an important driver, for example, in the replacement of the line-shaft facilities before 1920. It was during the 1911-1920 period that most states enacted worker compensation laws, and line-shaft plants, with all the belting, were notoriously dangerous places to work. The standard pattern for state worker compensation law is to require employers to take out accident insurance sold by private insurance companies, and the compensation is on a no-fault basis. Prior to these laws, an injured worker had to sue the employer, and would receive no compensation if the employer could prove negligence on the employee's part. Interestingly, the worker compensation movement was strongly backed by employers as well as labor organizations. For employers, the payoff was relief from the uncertainty of the previous adversarial system wherein there was the risk of a big adverse judgment; the worker compensation laws provided an effective cap on compensation. One result of the worker compensation laws is that the number of injured workers who received compensation increased, but the average size of the award diminished.

Reduction of Production or Distribution Bottlenecks

The implication of such investments may eventually be an increase in the output possible from a given labor force, but the immediate impact on measured labor productivity may be minimal. An example is a railroad's construction of a second main track for the purpose of speeding train movement. The expected payoffs would be in the form of accelerated capital turnover (less rolling stock needed for a given service level) and greater customer satisfaction. The latter payoff might contribute to increased market share and greater pricing power. The result may also enable traffic level growth, which could impact labor productivity. Substantial investments of this kind have been made since 1995, and in many cases, these were for the purpose of accommodating growth in traffic levels that already had taken place.

Energy Saving

The saving of energy has been a major goal inside the chemical and petroleum refining industries for many decades. This history dates to well before the energy crises of the 1970s. The process of saving energy has involved many replacements of older processes with more energy-efficient, newer ones. The replacement of thermal cracking with modern fluid catalytic cracking in the manufacture of gasoline blending stocks in refineries is a prime example wherein the saving was mostly in energy. There is an almost constant pressure to reduce the energy consumption involved in the production of commodity fuel and chemical products. Indeed, usually the only strategy for a commodity producer of any kind to increase profitability is to find ways to reduce costs. The immediate impact on output-labor ratios tends to minimal in such cases.

Replacing Capital with Capital of Superior Productivity

One example is the replacement of flotation cells in copper concentrators with larger units. The technology is about the same, but the payoff relies on strong economies of scale. There are no immediate implications for labor productivity. The railroad industry's replacement of steam locomotives with diesel-electric units is another example of capital replacing capital. In this case, there was a fairly early impact on output-labor ratios, for the labor force needed to keep a fleet of steam locomotives in production is much larger than that for diesels.

The Role of Government

The impact of government on the innovation process has many dimensions. These include various forms of fomenting invention—including direct financial support of inventive activity—and in influencing the general direction of innovation. Two of the most important tools by which government influences innovation are *regulation* and *taxation*. The whole purpose of regulation is to influence some area of economic activity. The imposition of regulation over a field of economic activity provokes adaptation to the regulatory pattern, and such adaptation may have disruptive effects reminiscent of the appearance of a disruptive prototype technology. The same can be said regarding a significant change in the regulatory atmosphere, such as a deregulation. The same can also be said regarding taxation. In addition to regulation and taxation, the government grants patents that confer a term-limited monopoly on products resulting from invention. By this means, private capital is encouraged to assume the risks inherent in stage one innovation. Chapter 6 details the government's twentieth-century role in fostering invention and innovation. Chapter 7 explores the relations between regulation and taxation on cyclical economic growth.

It is difficult to talk of how inventors and innovators support themselves without recognizing the role of government. At the outset of the nineteenth century, the government was the largest potential source of financing in the economic landscape and the one market potentially large enough to absorb the output that would be possible if parts of complex assemblies were truly interchangeable. Eli Whitney's proposal to manufacture (what was then) a large quantity of muskets for the army with interchangeable parts in 1798 resulted in a government contract.[13] Scholars have pointed out that Whitney never succeeded in truly standardizing parts of muskets for reasons to be developed below, but the ideal was pursued in government arsenals (Lee at Springfield and Hall at Harpers Ferry, Samuel Colt, and others), and considerable progress toward the goal of standardization was achieved.[14] This progress, which had obvious enabling effects for industries other than small arms, was financed either by government contracts awarded to private individuals such as Whitney, or to factories under government ownership and operation.

US Government support of invention and innovation diminished in relative importance in the latter half of the nineteenth century, but it revived strongly in the second half of the twentieth. In both centuries, much of the foundation of this interest appears to have been national defense related, and government-supported invention and innovation for defense has had profound implications for non-defense economic activity. An outstanding example is the KC-135 tanker aircraft that was developed under an Air Force contract and resulted in the airframe by which much US and foreign jet commercial aviation developed—the Boeing 707. Even more dramatic was the Internet, in whose development DARPA (Defense Advanced Projects Agency) was intimately involved. At least two other US Government agencies, the National Science Foundation (NSF) and the National Institutes of Health (NIH), have been very active in supporting research in universities that is not always defense related.

The Private Sector

While the role of government as a direct supporter of invention is important, it has not been sufficient to explain all the innovation processes that have from time to time given general economic growth a powerful push. The other part of the picture has been the expansion of consumption and the profitability of developing products based on prototype inventions—some of which came into existence as a result of deliberate government activity. The internet, a government creation, has gravitated to private sector stewardship since its inception in the early 1970s with government forbearance and some encouragement. Government-private interplay is an extremely important aspect of technology development in the US and elsewhere in what used to be termed the "free world." What happens when innovation is left to government exclusively is illustrated by some of the centrally planned systems, notably that of the old Soviet Union. The USSR had no problem of creating prototype inventions through a network of research institutes and the efforts of highly educated and intelligent people. Its military spy system was effective enough to have brought in inventions originating in the West to supplement the inventive production of its own scientists and engineers. However, it was notoriously inept in realizing the potential of these inventions in its general economy. The reason was that day-to-day production of all manner of goods was incentivized almost entirely by production goals. Applying new technology has a cost in the form of temporary loss of current production, and in the Soviet scheme of things, the reward for incurring such cost was practically nonexistent and often negative. In the planned economy, there was no room for the idea of unplanned after-effects of turning an invention over to a profit-seeking private sector, for such did not exist.

As large as the role of government has been, especially as supporter of stage-one innovation in effective partnership with the private sector as supporter of stage-two innovation, the private sector itself has played a large role in the entire process, from invention through stage-two innovation. A number of industries, such as pharmaceuticals, computer software, and various electronic hardware industries, have made copious use of the patent system but otherwise have financed the total process with private capital. Private funding of innovation goes back to the nineteenth-century, independent inventor and has continued with the corporate research and development shop.

Chapter 8 describes the development of the computer, and Chapter 9 covers the technology of light in the late twentieth century. Chapter 10 covers medical technology, and Chapter 11 investigates the relationship between innovation and economic development in the late twentieth century and looks at certain similarities and dissimilarities between technology development patterns in the early and late twentieth century. Chapter 12 describes a number of technologies that have been slow to develop and that threaten to hold back other technological development in the future. All of these technology areas are commonly looked upon as private sector developments, but in many of the instances, the initial development had some relation to government, especially defense activity.

Production Function

One of the useful concepts regarding how innovation relates to production of goods and services is what economists call the *production function,* which conceptually relates the inputs of the factors of production—land, labor, and capital—to resulting production. While there have been attempts to specify production functions mathematically, the concept need not depend on an explicit specification to be useful. Each production function corresponds to a technology of production, and the availability of alternative technologies is closely related to invention and innovation. Combinations of productive factors going into a given production function depend on relations among factor costs. An external condition, such as the appearance of a new prototype invention or a significant change in government regulation, can lead to abandonment of one production function and the substitution of another. It should be noted that this concept of production function applies most realistically to the individual plant. The production function is basically a static concept and is useful for comparing one static equilibrium condition with another. Once a plant design becomes incorporated in permanent construction, the idea that the operation can attain an optimal equilibrium between capital and labor employed makes sense. The idea of a static equilibrium makes less sense as the notion of production function is expanded to more aggregate forms, such as entire industries or even entire economies. The reason is that at any time, an aggregate economy includes many plant production functions, and the one most recently adopted presumably incorporates the most efficient technology available. Thus, at any time the aggregate economy includes a mixture of technologies ranging from the newest and most efficient to older and less efficient technologies. An economy, taken in its entirety, always has components striving toward equilibrium with the totality never achieving it.

An outstanding example of how factor price relations affect the mix of factors going into a given production function can be found in a comparison of how the textile industries in the late eighteenth and early nineteenth centuries developed in the United States and in Great Britain. As compared with English conditions, both skilled labor and capital were scarcer and therefore relatively more expensive in the US, with the result that development in the US was heavily predicated on economization of skilled labor. This led to a proportionately greater interest in mechanization in the US. As an example of how a new prototype invention or a major change in government regulation can change the basic technology of production—i.e., change the production function itself—US environmental regulation since 1970 has changed the basic nature of technology in a number of industries. For example, the processes in the smelting and refining of copper have changed radically because applying the new technology (different production function) was less costly than adapting the existing technology to more stringent air quality requirements.[15]

Figure 1.2 illustrates these ideas. Suppose a certain productive activity, such as electric power production (it could be any activity), has historically used Technology A, which employs factors of production in the combination labeled as No. 1. Normally one thinks of the output as so many kilowatt hours, but there is actually another output in the form of several hundreds of tons of sulfur dioxide (SO_2), an undesirable pollutant, which has been exhausted into the atmosphere. This is what economists call a *negative externality*.[16] Suppose that the government one day makes the practice of exhausting the SO_2 illegal. Then the power producer has a problem with a definite technological aspect.

In real life dealing with this problem is highly complex, but in this illustrative example, the solution exists in the form of Technology B, which requires another combination of productive factors, labeled Combination No. 2. The product, in this example power, may be more costly to the consumer, but the entire decision to change technologies rests on externality B's being in compliance with the law.

Market forces may have been insufficient to eliminate the pollutant because its ill effects are thinly distributed and appear to be of little harm. As long as emitting the pollutant is costless for the emitter, the power company has little incentive to incur the cost of cleanup and the ratepayers have little incentive to pay this cost in their power bills. When the solution to the problem comes in the form of new regulation under a public law, the externality becomes a decisive factor in the cost and pricing of the total product. The onset of clean air regula-

Figure 1.2
Production Function

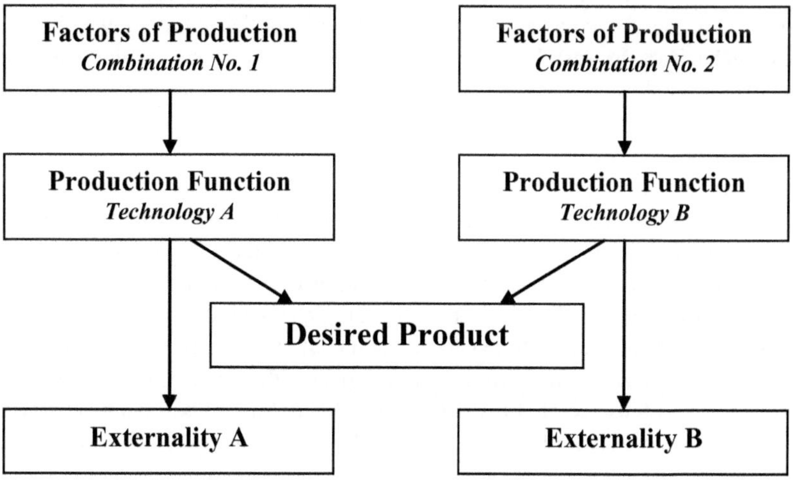

tion can be viewed as the government's effective outlawing of at least part of a currently used technology (current production function), forcing the industries using that technology either to modify the existing technology to comply with the regulation or, in some cases, to replace it with another technology (alternative production function) depending on which course complies with the new regulation with least cost.

Inventors and Discoverers

Given the importance of the prototype invention and the extensive engineering and invention work involved in innovation, one is entitled to ask what conditions foster such activity. Given that the basis for it all is intelligent and curious people, how do they support themselves while inventing and innovating? In the nineteenth century, inventing was often an individual pursuit, with the inventors supported by their own resources, investor capital, a university faculty income, or other means. Cyrus McCormick, credited with the harvester, was essentially a farmer working in his spare time. The big change that came with the twentieth century was the growth of the corporate research and development division, and the tendency was for inventors to be employees of these organizations. Stephanie Kwolek, who is the recognized inventor of Kevlar (1964), was a chemist working in the research laboratories of the DuPont Corporation. Kevlar has achieved wide usage as lightweight body armor for military and police forces and is used in a number of other applications, including music (it is used in drumheads).

Thomas Edison, whose name is attached to invention of the incandescent light bulb, was a pioneer in what later became familiar as the independent research and development laboratory, an organization that concentrated on inventing. Some of the early independent invention shops were later folded into the R & D divisions of large corporations. Thus, the independent invention shop could be viewed as a transitional form, except that it has shown signs of a comeback in the very late twentieth century. The reasons are developed in Chapter 13 below.

The Future

What does the technology future hold? One clue may come from a consideration of where past prototype inventions that that have fomented massive innovation have come from. The private sector is often accused of limiting its research effort to bringing products to market in the short term and, indeed, there is much of this. However, several examples come to mind. Bell Labs has provided the laser and the transistor, to name but two prototype inventions, and it is not clear that the innovation springing from either is mature. Government, through its defense-related research, has delivered a number of inventions to the private sector, which has developed these commercially. Moreover, it is difficult to find a stronger goad to invention and innovation than that provided by environmental regulation. Universities are very much in the picture with fields such as nanotechnology. These are but a few of the technological roulette wheels that we face at the outset of the twenty-first century. Chapter 14 explores the ways in which the economy chooses among competing technologies in the face of great uncertainty.

Notes

1. There is an extensive literature on the theory of capital replacement of the routine incremental kind. For an excellent synthesis of this body of theory, see Smith, Vernon L., *Investment and Production.* (Cambridge, MA: Harvard University Press, 1966), 128-161.
2. See Edwin Mansfield, *Technological Change,* (New York: W. W. Norton & Company, 1971), pp.73ff.
3. While the availability of steel and other materials in the twentieth century made parts standardization possible in a number of fields, full interchangeability of parts remained a problem as recently as the 1960s, when Xerox experienced difficulty in getting its contractors to provide parts to the required standards for its mechanically complex copiers.
4. Fogel, Robert, *Railroads and American Economic Growth,* (Baltimore, MD: The Johns Hopkins University Press, 1964), p. 10.
5. Rosenberg, Nathan, *Inside the Black Box: Technology and Economics,* (Cambridge, MA: Cambridge University Press, 1982), pp. 55-80.
6. Rudolph Diesel's thinking, and that of the people and organizations that backed him in his early work, has been well documented. See C. Lyle Cummins, *Diesel's Engine: Volume one, From Conception to 1918,* (Wilsonville, OR: Carnot Press, 1993).
7. Appreciation of the invention-innovation role in the economic cycle was not confined to the Austrians. J.R. Hicks expressed the idea thus: "It is only possible to make sense of the theory of the trade cycle to which we appear to have been led, to reconcile it, that is, with the most obvious facts of history, if we lay great stress upon the supply of investment opportunities which is provided by invention and innovation." J.R. Hicks, *Value and Capital (Second Edition),* (Oxford: Clarendon Press, 1965), p. 299.
8. Schumpeter expressed this idea as follows: "For our purposes it is both necessary and sufficient to list innovation, however much it may be linked to the other two [influences on the business cycle], as a third and logically distinct factor in economic change, and to submit the propositions: The kind of wave-like movement, which we call the business cycle, is incident to industrial change and would be impossible in an economic world displaying nothing except unchanging repetition of the productive and consumptive process." Joseph A. Schumpeter, "The Analysis of Economic Change," *The Review of Economic Statistics* (XVII, No. 4), May 1935. As reprinted in *Readings in Business Cycle Theory* (Homewood, IL: Richard D. Irwin, Inc., 1951), p. 7.
9. This idea was common among members of the Austrian School, notably with Joseph A. Schumpeter. See *The Theory of Economic Development,* (New Brunswick, NJ: Transaction Publishers, 1934) (Translated from the German), p.215.
10. The interest in the business cycle was a logical reaction to experience of the nineteenth and early twentieth centuries, especially the great depression that affected the entire industrial world. Much of the work on economic cycles was summarized

in Gottfried Haberler's *Prosperity and Depression*, (Lake Success, NY: United Nations). The concern with cycles immediately after World War II reflected a widely held fear that the peace would see a reversion to depression conditions. Inasmuch as events did not bear out this fear, interest in studying the economic cycle waned to the extent that many university economics departments were questioning the utility of their cycles courses in the mid-1960s.

11. The problem with representing productivity as a ratio of output to labor inputs has long been recognized: all change is attributed to the labor factor when another factor of production, say capital, could be the source of the change—as when the effect of an investment is to displace capital with superior capital. In an attempt to deal with this problem, the Bureau of Economic Analysis of the Dept. of Commerce has provided estimates of "multifactor productivity," which attempts to represent the impact of the capital factor in overall productivity change. Multifactor productivity is not used in this study for two reasons. First, the historical coverage of this statistic is insufficient for the length of history examined. Second, the capital factor is based on several assumptions that are counter to the assumptions of this study. The most important of these is that capital wastage (depreciation) is a constant proportion of the existing capital stock. Capital is subject to obsolescence, and in a period of high technology flux, the wastage of capital via obsolescence can be very high. The reality is that capital wastage as a proportion of existing capital stock is highly variable through time, and the official capital measurements make no allowance for this variability.

12. Where substantial capital facilities are required to comply with regulation, the increase in labor force is from the staffing to operate and maintain the new facilities.

13. Whitney's idea met with considerable resistance that he overcame in part by means of several demonstrations in which he assembled the locks of several muskets from a jumbled pile of parts, using parts as they came to hand. Whitney also benefited from the sympathy of a politically powerful champion, Thomas Jefferson, who had witnessed an earlier effort at parts standardization during his years in France and who was enthusiastically receptive to Whitney's proposal.

14. See David Freeman Hawke, *Nuts and Bolts of the Past: a Hisotry of American Technology, 1776-1860,* (New York: Harper & Row, 1988), pp. 97ff and 154ff.

15. Prior to the US Clean Air Act (1970), copper smelting technology did not differ essentially from the processes described in Agricola's *De Re Metallica,* a sixteenth-century treatise. Since 1970, the US industry has utilized other technologies such as flash smelting and, more recently, flash converting. By contrast, the Japanese smelting industry approached the air quality problem by tinkering with the established technology.

16. The producer that generates an externality effectively imposes the costs of dealing with the externality on third parties external to itself, such as populations living in the area. Elimination of the externality means that a producer has assumed the costs of dealing with it. In the example of an electric utility, the costs of dealing with the pollutant are passed to the ratepayers. Before, these ratepayers had paid anyway in the form of impaired health and other problems.

2

The Nineteenth Century's Legacy:
Innovation and Discovery to 1910

Introduction

Between 1885 and 1910 there emerged a number of innovation streams, each based on the completion of stage-one innovation of a prototype invention. The sheer number of technologies entering stage-two innovation was historically unprecedented for so short a time period. This bunching resulted from a combination of progress in the sciences and in the availability of new materials along with superior ability to form them into precisely-dimensioned shapes. Innovation from recent prototype inventions was augmented by innovation of much earlier prototypes brought to usefulness by the appearance of enabling technologies that had not existed at the time of invention. It was this backlog of older inventions emerging from stage one innovation that added materially to the richness of the field of new technologies available in the first decade of the twentieth century.

The importance of this unprecedented episode of technological flux and its relevance to economic development into the twenty-first century can be appreciated if one performs the mental experiment of taking the view of a person living in 1900 and trying to envision the future. Taking all that was technologically new at that time and combining the various pieces into visions of the future quickly yields a large number of scenarios, including, perhaps, some that eventually transpired. Today's technology is vastly different in its detail, but the problem of seeing the future is still one of guessing which of many possible combinations of technical elements will succeed. The visionary of 1900 had at least one advantage over that of 2000, for the technologies of 1900 had mostly emerged from their stage-one innovation. That is not true of many of the futuristic possibilities of very recent times, as will be developed in later chapters. This chapter describes the most important of the new technologies of the early twentiety century, and the following chapter develops how these were used and the general prosperity that they underwrote.

Table 2.1 presents a list of prototype inventions before 1910, dates of invention, and approximate dates of first application.[1] The difference between these two sets of dates is a measure of the length of the first stage of innovation. The list includes a number of inventions from much earlier times whose first stage innovation periods ended between 1850 and 1900. Examples include the hydraulic press, the electric motor/dynamo, the photoelectric cell, and the gyroscope. In addition, the list shows a number of prototype inventions that appeared between 1890 and 1910 which had comparatively short first-stage innovation periods. These include aniline coal tar dye, dynamite, celluloid, the telephone, and the Parsons turbine. In all these cases, commercialization (second-stage innovation) affected industrial development in the first several decades of the twentieth century.

Enabling Technology: Materials and Inventions

Machine Tool Development

The ability to machine to close tolerances made significant progress after about 1780. The Lathe was brought to a high state of capability through the efforts of Henry Maudslay and Joseph Whitworth, and Eli Whitney is credited with the milling machine. As the industrial revolution in England unfolded, a typical machining procedure consisted of bringing a rough forging close to final dimension with a lathe or, as often as not, with a hand file. The results could be finely finished to a close tolerance specification, but the process was skilled labor intensive, and large-volume production presented serious problems, especially in the United States where there was a shortage of people skilled in this work. The limitations of machining capability were glaringly apparent in the 1820s in Charles Babbage's "difference engine," an early computer. The difference engine went beyond what later became known as the calculator in having a memory, the ability to perform a series of computations according to a program, and an integrated printer. What eventually prevented its completion was extreme difficulty in obtaining gears machined to the needed precision *in the required quantities.*

Another chapter in machine tool development opened in 1797 in the United States with Eli Whitney's successful bid on a US Army contract to manufacture four thousand muskets. Whitney proposed to build the weapons by making all the parts sufficiently alike as to be interchangeable. He kept the skeptical army officials and a sympathetic Thomas Jefferson interested by means of several demonstrations wherein he presented a jumbled pile of musket parts from which he would draw parts at random and assemble a musket before his onlookers' eyes. Ideally, what Whitney was proposing was to use power machinery to make the parts to a level of finish such that little or no hand filing would be necessary.[2] This strategy recognized the shortage of skilled hand filers in the

Table 2.1
Selected Prototype Inventions to 1910

Invention	Inventor	Date of Prototype	Date of First Application	Physical Principle
Hydraulic Press	Blaise Pascal	1646	1909	Pascal's Law of pressure on confined liquids
Electric Motor/Dynamo	Michael Faraday	1832	1879	Magnetism and induction
Photography	Daguerre	1838	*	Light-sensitive chemistry
Fuel cell	William Grove	1845	1950	Electrochemistry
Gyroscope	Jean Foucault	1852	1908	Newtonian mechanics: inertia
Low-cost steel	H. Bessemer, W. Kelley	1855	1866	Basic steel chemistry
Aniline coal tar dye	William H. Perkin	1856	*	Growing knowledge of organic chemistry
Dynamite	Alfred Nobel	1866	*	Safe-to-handle nitrogen-based explosive
Celluliod	Alexander Parkes	1866	*	Growing knowledge of organic chemistry
Otto Cycle engine	Nicklaus August Otto	1876	*	Compression of air-fuel mixture
Telephone	Alexander Graham Bell	1876	*	Amplification of an electric current
Incandescent Lamp	Thomas A. Edison	1878	**	Heat applied to a filament in partial vacuum
Mechanical refrigeration	Carl von Linde, Ferdinand Carré	1878	**	Joule-Thomson effect
Solar cell	Charles Fritts	1883	1953	Anticipated quantum mechanics
Induction Motor (AC)	Nicola Tesla	1883	1895	Alternating current theory
Aluminum	Hall and Héroult	1886	*	Metallurgical chemical properties
Rayon		1892	*	Growing knowledge of organic chemistry
Radio	Guglielmo Marconi	1894	*	Radio spectrum
Parsons turbine	Charles A. Parsons	1894	*	Thermodynamics laws
X-ray	Wilhelm Roentgen	1895	*	Experiments with radiation spectrum
Diesel engine	Rudolph Diesel	1897	*	Relation of compression ratio to engine efficiency
Airplane	Wilbur and Orville Wright	1903	*	Aerodynamic principles
Synthetic ammonia	Fritz Haber	1908	*	Joule-Thomson effect
Bakelite	Leo Baekeland	1908	*	Growing knowledge of organic chemistry

Notes: *Within five years of working prototype.

** Immediate application in the brewing and meat packing industries. Cryogenic air separation was achieved in 1895.

+ Prototype discoveries leading to the germ theory of medicine are discussed separately below and refer to Table 2.2.

United States and was designed to be operable by workers fresh off farms who had undergone only a short period of training.

Whitney failed to fulfill his contract in the specified time, and he never achieved true parts interchangeability for reasons detailed below. However, the army was sufficiently intrigued with the idea to continue its support at its two national arsenals, at Springfield, Massachusetts and Harpers Ferry, Virginia and at the latter location came close to the ideal of interchangeability under the direction of John Hall, a semi-private arms maker also located at Harpers Ferry. Parts interchangeability was also the basis for Samuel Colt's arms manufacture at Hartford, Connecticut. Colt's contribution came after he become a successful army contractor about the time of the Mexican War.

The interchangeable parts approach to machining gave rise to something called the "American system" of manufacture by way of distinguishing it from the approach in England, the major world manufacturing center of the early and mid-nineteenth century. For a time, historians saw the American system as a response to the shortage of skilled machinists in the US as compared with England. This relation appears to have been overblown for several reasons according to historians since 1960. First, as already noted, true interchangeability was never achieved in the nineteenth century, although much progress was made toward this goal.[3] Second, using machinery to finish parts to final specification required substantial investment in highly specialized machines, and, therefore, was too costly to interest manufacturers other than those that could anticipate very large markets for standardized goods, such as army contractors. The private sector mass market, which could justify heavy investment in tooling, was not really to appear until the opening decades of the twentieth century.[4]

Machining parts to final specification in the nineteenth century was physically difficult for several reasons. First, tool steels then available were not up to demands imposed by parts interchangeability.[5] Tungsten steel, the first real improvement over carbon steel as tool steel, first appeared only in 1855. The steels that had previously been available were crucible steels, and the crucible process never had been adapted to high-volume production. Hence, steel was very expensive and its markets were confined to high-end products, such as fine cutlery and components of high-end clock movements. Second, the iron from which parts were made often did not hold shape under the stress of machining. This made final finishing to tolerance a matter of hand filing. The progress of machine tooling at the end of the American Civil War had been sufficient to deliver tolerances within .001 inch, but not in quantity. Quantity production of finely machined parts was a necessary condition for commercial development of such inventions as the electric motor, the hydraulic press, and others, and this problem would not be solved until low-cost quality steel became available in substantial quantity.

Abundant steel. The availability of steel at low cost was a crucial enabler of the machine tools industry, and, by extension, of all other innovation that depended on close-tolerances. In the mid-1850s, Henry Bessemer in the UK and (independently) William Kelley in the US invented the process of blowing air through a bath of molten iron by which the impurities in the iron were oxidized and the carbon content reduced to make carbon steel. The Siemens-Martin (open-hearth) process was introduced in the mid-1850s, and demonstrated important advantages over the Bessemer process, especially in the matter of product quality control. Siemens-Martin and Bessemer were competing technologies initially, but the former eventually became dominant, only to be itself replaced by the basic oxygen process, mostly after World War II.

In the US, steel's impact on machine tool performance did not really become evident until after 1890 for several reasons. The first and foremost of these was that virtually the entire capacity of the new steel industry was subscribed to by one product: rail. The nineteenth century saw the first development and massive expansion of the railway network, and the decade of the 1880s saw the high point in the expansion of railway route miles. Not only did the domestic steel industry plus imports have to keep up with demand from railway expansion, but also that from rebuilding the prior-existing railway system with steel rails.[6] A second reason related to the quality of tool steel, as understanding of the metallurgy of steel alloys had its beginnings only in the 1860s. From the 1860s onward, there was steady progress in the understanding of steel metallurgy, especially the properties of alloys. After 1890, steels became available for other markets. These included structural steel, which enabled the remarkable change in office building architecture embodied in the ability to build tall buildings.[7] Improvement in the ability to machine steels to accurate tolerances enabled the start of innovation streams based on such prototypes as the electric motor, the steam turbine, the gyroscope, the hydraulic press, and others.

Aluminum. The modern electrolytic process for producing aluminum was discovered independently in 1886 by Charles M. Hall of the United States and Paul-Louis-Toussaint Héroult of France. Hall applied for and received a US patent and, with a group of businessmen, established the Pittsburgh Reduction Company, which later became Alcoa (Aluminum Company of America). Inasmuch as aluminum had been a very expensive metal, initial demand for it really did not exist. Alcoa and its predecessor, from the start, actively sought to create markets for the metal and were successful in the endeavor.[8] Early products included kitchen cookware, but aluminum really grew in the inter-war period with the growth of aviation. The Hall-Héroult process is a large consumer of electric power, and this profoundly influenced the location of the early aluminum industry. American production initially became concentrated in the Niagara Falls area, where it became the first customer of the Niagara Falls power development in 1895. In the 1930s, production facilities were located in Tennessee and the Pacific Northwest, taking advantage of low-cost hydropower. The European industry, based on Héroult's patents, tended to locate in Alpine regions, also to take advantage of low-cost waterpower.

Vulcanized rubber. Raw natural rubber is sticky and tends to deteriorate rapidly when exposed to air. *Vulcanization* was originally the process for curing natural rubber. It consists of admixing sulfur with the raw rubber and subjecting the mixture to heat. The result of this process is a smooth and springy material with high tolerance of extremes of temperature. Discovery of the vulcanization process is usually credited to Charles Goodyear, an American inventor, who made his initial discovery in the year 1839 and secured the US patent for the process in 1843.[9] Due to some serious problems with enforcing his patent rights in the US and with maintaining such rights in the United Kingdom and France, Goodyear never made much money on the basis of his discovery. Vulcanization is an irreversible chemical process. While it is the result of heating the rubber-sulfur compound, it holds its integrity when reheated. It is thus a *thermosetting* material.

The vulcanization process was improved in subsequent years by several developments. These included the use of accelerators, or chemical additives to the rubber-sulfur compound, whose purpose was to speed up the curing process. Another improvement in the vulcanized rubber product itself was the admixture of carbon black in the rubber compound. Carbon black does several things. First, it strengthens the vulcanized product by contributing additional chemical bonds. Second, it gives the product an abrasive surface, making possible such applications as automobile tires, and therefore enabled the pioneer automobile industry at the end of the nineteenth century. The pencil eraser is a common use of vulcanized rubber; and without carbon black, a tire would have the durability of a pencil eraser.

Origins of the Modern Chemical Industry

Early Products

The invention of aniline coal tar dye (William Perkin, an Englishman, 1856) was the foundation of the modern chemical industry, which had its nineteenth-century growth on the basis of dyestuffs and other products of coal tar distillation. Synthetic dyestuffs were subsequently developed extensively in Germany. Among the many later products stemming from coal tar distillation were the sulfa drugs that were discovered in the 1930s and became a first-line defense against infection from wounds during the Second World War. The period under consideration also saw the invention of celluloid (Alexander Perkes, 1866), the pioneer plastic flexible wrap; of rayon (1892), the pioneer synthetic fiber; and of bakelite (L.H. Baekeland, 1908), the pioneer plastic. These new products collectively reflect growing understanding of organic chemistry.

Synthetic Ammonia

Synthetic ammonia, attributed to Fritz Haber in 1908, became extremely important in the manufacture of fertilizers and explosives, and was enabled by the ability to separate air into its components. In the Haber process, nitrogen from the air, is combined with hydrogen under high temperature and pressure conditions to make ammonia, NH_2. The hydrogen comes from synthesis gas, a mixture of hydrogen and carbon monoxide, which is made by either steam reforming or partial oxidation of methane. Both processes require precisely controlled quantities of oxygen; hence, the need for cryogenic air separation capability.[10] The nitrogen produced in the air separation goes into the ammonia reactor.

Early Plastics and Fibers

The oldest of the plastics was cellophane, which dates to before 1870. It is produced by extruding regenerated cellulose into thin, transparent sheets. An alkaline solution of cellulose fibers known as viscose is extruded through a narrow slit into an acid bath. The acid regenerates the cellulose, forming a thin transparent film. Cellophane has survived to the present time mainly as a packaging material, due to its resistance to dry gases, grease, and bacteria. Rayon originated in 1892 in attempts to make artificial silk from cellulose. Its most common form, viscose rayon, is made by forcing viscose through a tiny nozzle and then treating the resultant fiber chemically to harden it. Viscose rayon is almost as strong as ordinary grades of nylon, but suffers the disadvantage of losing strength through water absorption and flammability. Up to a point, the ability to absorb water served a useful purpose, for it helped rayon fabrics' comfort level, as it emulated the water-absorbing qualities of cotton. Known for some years as imitation silk, rayon was used in hosiery. Another market was as an automobile tire reinforcing material. Some motorists actually preferred rayon tire cord to nylon, for the former, when dry, is not as elastic as nylon. When a nylon-reinforced tire sat in cold temperatures without being operated, the tire would have a flat spot that would not go away until the tire became warmed in operation. Rayon tire cord avoided this problem. The pioneer non-cellulosic plastic was Bakelite, invented in 1908 by L.H. Baekeland, an American.[11] Bakelite was a trade name for the synthetic resin formed from the reaction of phenol and formaldehyde. Its first market was all the markets previously held by celluloid. However, its strong electrical insulating qualities made it useful as the structural material of a huge number of electro-mechanical devices. It is also used in molding and casting operations, in paints and baked enamel coatings, and as an adhesive.

Dynamite and Modern Explosives

No one who has studied the building of the first transcontinental railroad in the United States will fail to appreciate the invention of dynamite (Alfred Nobel, 1866), the first of the truly modern explosives that did not pose high risk to its users. The transcontinental railroad was a huge construction project for its time, and was probably the last large construction project in the world to be built with black powder and some nitroglycerine. The latter explosive is notoriously sensitive to shock, and its use bought increased productivity at the cost of tragic accidents for its handlers.

Photography

Photography is described in *Britannica* as "[a] method of recording the image of an object by the action of light, or related radiation, on a sensitive material."[12] The original invention was by two Frenchmen: Nicéfore Niépce and Louis-Jacques-Mandé Daguerre. Niépce actually made the first photographic image in 1826, on a polished pewter plate covered with a petroleum derivative called bitumen of Judea. This approach turned out to be a dead end, and Niépce began experimenting with silver compounds as light-sensitive materials. He was killed by a stroke in 1833 in the midst of his experiments, and bequeathed his notes to Daguerre, with whom he had been working. Daguerre developed a process in which a silver salt deposited on a copper plate, which he named the Daguerrotype. Daguerre provided a showing of his work to a highly interested Paris audience in 1838-1839, and for a while this process enjoyed substantial popularity with the middle classes of the time. Even though a Daguerrotype was inexpensive compared with an oil painting, it was still costly, and because it was on a copper plate, the only way to reproduce it was to take another exposure. The next step, the collodion wet-plate process appeared in the 1850s. In this, the emulsion deposited on a glass plate was a *negative* of the original scene (blacks and whites interchanged). A positive could be obtained by exposing print material to light through the negative imprinted on a glass plate. The collodion process was the means by which the American Civil War was recorded.

While the quality of the early photographic work was high in terms of razor sharpness of images and details recorded, the pictorial possibilities were almost entirely confined to landscapes and portraits. The reasons were the bulkiness of the equipment and the slowness of the light-sensitive emulsions. Bulkiness resulted from the fact that if an image of 8" x 10" were desired, the plate-holding camera back had to be of that dimension. The slowness of the emulsions required exposure times counting in minutes, even under good light conditions. This shows up in major events that were recorded by photography, such as the Crimean War, the American Civil War, and the building of the transcontinental railroad: there are no action photos from this period.

In 1880, George Eastman invented the dry emulsion, which eventually took photography from the professional's studio into popular use. Dry emulsion was backed by paper at first, and was sold already loaded into a simple box camera with up to one hundred exposures. When the user used up his roll, he sent the camera to a Kodak (Eastman's coined trade name) processing facility, where the exposures were developed and prints, plus a reloaded camera, were returned to the customer. Later, celluloid was adapted as backing for the dry emulsion, and the box camera design was modified such that the user could reload with a purchased roll of film. At the turn of the century, photographic chemistry had made sufficiently rapid strides that action pictures were beginning to appear.

Electricity and Pioneering Electronics

The Battery

The battery is a device that holds electric charge. Under this definition, it is closely related to the capacitor, but *battery* in general has come to mean a device that produces an electric current as a result of some kind of chemical reaction. Therefore, the concept includes the fuel cell. The term *battery* was apparently coined by Benjamin Franklin to describe an array of charged plates, or a simple capacitor, with which he was experimenting about 1748. The word battery came from the legal term meaning a beating, of which Franklin was reminded by the effects of touching the electrodes of his device. The modern chemical battery appears to trace its roots to the work of Alessandro Volta, the Italian physicist, who investigated the electrical results of immersing plates of different metals in salt water in about 1800.

Volta's and others' early batteries supported significant scientific research in the nineteenth century. It was in 1800 that William Nicholson used a battery to decompose water into hydrogen and oxygen. Sir Humphrey Davy studied electrolysis, and later, Michael Faraday used an improved voltaic cell as a power source in his researches in induction that led to the electric motor/dynamo. There were a number of batteries devised during the nineteenth century, and of these, the two most interesting from today's standpoint were the lead-acid battery and the fuel cell. The former appeared in the 1860s and is credited to Raymond Gaston Plant. The first lead-acid batteries suffered from short shelf life, but the concept has been improved incrementally to become the common standard power source for automotive engine starting systems. The fuel cell derived current from the combination of hydrogen and oxygen and is credited to William Robert Grove in 1839. To the present day, the fuel cell has not progressed much beyond an early stage-one innovation, but is the subject of much research effort as a future automotive and stationary power source.

Electric Motor: Stage-One Development

Development of Faraday's first prototype proceeded slowly and resulted in a commercially significant motor only in 1873.[13] The early motors operated on direct current. The synchronous alternating-current (AC) motor was invented by the Serbian-American inventor Nicola Tesla only in 1888. Tesla produced his motors in cooperation with the Westinghouse Electric Company, founded by industrialist George Westinghouse.[14] Improvement of the electric motor has since not involved many radical ideas, but has mostly consisted of better armature design and improved bearings and contact materials—the collective effect being the reduction in size relative to power output or the increase in the unit's power density.[15]

Interestingly, in light of later developments, many of the earliest applications of the electric motor were in the transportation field. This is not surprising, for in its earliest days of stage-two development, the motor was large and bulky and therefore suited to industrial applications in enterprises large enough to be able to afford their own power production facilities. These include electric trains in Germany and Ireland, based on direct-current (DC) traction motors. The pioneer US railway electrification was by the Baltimore and Ohio Railway in its tunnels under Baltimore Harbor, and involved the electrification of 3.75 miles of track in 1895. Prior to this tunnel/electrification project, the B&O had been obliged to ferry its trains across Baltimore Harbor. In operation, the fires of steam locomotives approaching the harbor tunnel would be dropped, and the entire train, including the locomotive, would be pulled through the tunnel by an electric locomotive.

At about the turn of the century, the electric motor was just beginning to penetrate the individual household market. Railway applications were examples of heavy industrial applications in which the motor user dealt directly with the manufacturer. In general, applications of the electric motor before 1895 were of this nature.[16] Reduction in the size of motors (increasing the power density) was an extremely important development in this period, for it opened not only the possibility for powering small appliances, but also created the possibility of powering machine tools and other industrial machinery with dedicated motors of appropriate small horsepower. Ultimately, the increase in motor power density led to the replacement of the awkward and inefficient line-shaft. This is developed in the following chapter. In 1890, there was a serious effort to reduce the motor to suitability as power for sewing machines.[17] The downsizing effort was well under way by 1900, when a German comment was to the effect that US practice was to use motors of minimum of five horsepower for driving machines when use of smaller motors might be economical, even though these were less efficient. This observation continued to point out the substantial economic advantages of the system of operating all machines with their own dedicated motors with horsepower as low as two.[18]

Incandescent Lamp

The application of the electric motor was substantially enabled, indirectly, by the invention of the incandescent lamp (1878). Although other inventors were working on the incandescent lamp at the time Thomas A. Edison was working on it,[19] he is remembered as the inventor because of his vision of the entire problem of marketing the lamp. He had thought through the requirement for power generators and power distribution facilities, and turned his inventive, organizing, and financial resources to providing these. The early growth of power generation and distribution facilities opened the opportunity to use the motor as well as the light bulb.[20]

There was an early controversy in the new electric power industry over the issue of direct-current versus alternating-current development. The latter was recognized as having superior ability to avoid line loss over longer distances, but the direct-current advocates, including Thomas Edison, were able to present an effective argument as long as transmission distances were small, or as long as power production and distribution were confined to densely populated urban areas, such as New York City, with production physically close to consumers. Alternating current's advantage came to be of decisive importance with problems of distributing low-cost waterpower from the Niagara Falls complex, developed in 1895.

X-rays

Wilhelm Roentgen, a German physicist, discovered X-rays in 1895 during the course of experimenting with electric current flow in an evacuated glass tube (cathode ray tube), when he observed that a nearby piece of barium platinocyanide glowed when the tube was in operation. He hypothesized that when electrons struck the wall of the tube, that a radiation ensued that traveled across the room and struck the barium platinocyanide, causing it to phosphoresce. Further experimentation revealed that these rays penetrated various solids to varying degrees and effected photographic plates. Finding that the rays did not apparently act like light, he concluded erroneously that they had no relation to light and named them X-rays. They also became known as Roentgen Rays. He took photographs of the interiors of metal objects and the bones in his wife's hand. His discovery earned him the first Nobel Prize in physics in 1901. X-rays revolutionized diagnostic medicine. The important innovation invention in the general adoption of X-ray technology was the invention of the X-ray tube by an American engineer, William D. Coolidge, in 1913. The tube consists of a cathode (source of electrons) and a tungsten anode enclosed in an evacuated glass tube. It works on the principle that when high-speed electrons strike matter of any kind, X-rays are produced.

New Power for Industry and Transportation

Otto Cycle Engine

The Otto Four-Stroke Engine (1876), credited to Nicolaus Otto, a German Engineer, was the first really successful internal combustion engine.[21] Because of its light weight, reliability, efficiency, and relative quietness, the four-stroke engine was an immediate success. More than thirty thousand engines were built in the ten years following the prototype. There had been a number of attempts to build an internal combustion engine prior to Otto, but the four-stroke engine was successful because of its novel use of the energy in the flywheel to compress the air-fuel mixture prior to ignition; earlier attempts generally overlooked the advantage of compression.

Like earlier attempts at internal combustion (IC), the Otto Engine was initially thought of as a candidate for the stationary engine market. The fuel envisioned was town gas (or coal gas) which was produced by municipal gas companies for lighting and cooking. However, in 1885, Karl Benz took advantage of the Otto Engine's light weight to power a carriage, and thereby invented the automobile. The Otto design had a far more favorable power-to-weight ratio than earlier power sources, and this is clearly what made the Otto Engine of high interest. Benz took advantage of the adaptability of the Otto Engine to burn liquid fuel in the gasoline boiling range. Petroleum fractions in the gasoline boiling range had previously had no large market and were regarded as a waste product by early petroleum refiners, whose main product was kerosene for the illumination market.[22]

Increasing power density was one of the most important characteristics of late nineteenth-century technology. Some progress in this direction had been made with the steam engine in the course of its history since the time of Newcommen, but the stationary steam engine of the 1890s could still be legitimately described as large and awkward. The appearance of the automobile, which for a number of years after Benz' initial invention remained little more than of interest to well-to-do hobbyists, began to encourage some dramatic increases in the power density of the steam engine. These were sufficiently successful that the steam engine for a while competed successfully with the IC engine for the automobile market. Other power sources that satisfied the growing needs for compact power included the electric motor and the Parsons Turbine.

Parsons Turbine

The Parsons (inv. Charles Parsons, 1884) Turbine was the prototype of the steam turbine that became the power source for most of the electric power generating capacity in the world up until the last decades of the twentieth century, and retains a commanding position in this market. As such, it enabled the expansion of the power industry and, by extension, the twentieth century's reliance on electricity. It was preceded by only two years by another turbine design credited to Gustav de Laval, but differed from the de Laval machine in such a way that the two could be distinguished by size potential: that of the de Laval Turbine was limited while that of the Parsons Turbine was, for practical purposes, not. At least part of the success of the

Figure 2.1
Otto Cycle Engine, Upright Version

Note: Courtesy Lindsay Publications, Inc.

Parsons Turbine was that the design took advantage of adiabatic expansion of steam by passing the steam through the blades of a series of rotors of increasing size, and this gave it material advantages in fuel economy (Figure 2.2).[23]

Both the Parsons and de Laval turbines faced the requirement for precise balance and extremely close coincidence of the axis of the rotor and the center of mass of the rotor. This requirement could not have been met much before the late nineteenth century due to inability to machine to very close tolerances and lack of materials with sufficiently high heat tolerance. In the first decade of the twentieth century, when expansion of power-producing capacity growth accelerated, generating units with capacity of 25,000 kilowatts operated at 400 degrees Fahrenheit. This called for steels that would tolerate these temperatures, including stainless steels with 18 percent chromium and 8 percent nickel. Among the driving forces for early work on steam turbines was the appearance of a number of application markets demanding higher rotational speeds than could be provided by reciprocating steam engines.[24] The early dynamos presented this problem, and attempts to drive these with reciprocating steam engines required up-gearing to be successful. Moreover, the Parsons design was not limited in size, and the capacity of the machine could be increased in accordance with the needs of the growing electric power industry with its central power stations.

The other major market that found the Parsons Turbine attractive was the marine power market. The Royal Navy specified the turbine as primary power for H.M.S. Dreadnought (launched 1906), and by 1910 was the preferred power plant for large vessels. Ocean liners with turbine power included the *Mauretania* and *Lusitania*. The steam turbine had a very high power density, with respect both to weight and volume, and an obvious attraction in the marine market

Figure 2.2
Westinghouse-Parsons Turbine, 1901, Longitudinal Section

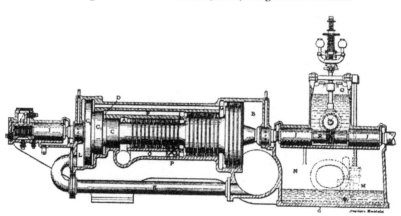

Note: Courtesy Lindsay Publications, Inc.

was the power that was available in a fraction of the space requirement of the reciprocating steam engine, the saved space being available for other needs, such as revenue cargo. Marine applications of the Parsons Turbine were enabled by a machine tool industry which gained the capability for making large and precisely dimensioned reduction gears. These were needed because turbine rotation speeds greatly exceeded desired propeller shaft speeds.

Diesel Engine

Rudolph Diesel, the inventor, took out his first patent in the year 1892 but his first successful working model appeared in 1897. Diesel benefited from what was arguably the best education in the science of thermodynamics that the world offered at the time, so that Diesel's engine, by proceeding from science, was an early example of a phenomenon that became familiar in the twentieth century. It is clear that the first commercial diesel engines were designed for the stationary engine markets by the fact that many of the earliest of Diesel's licensees were established builders of stationary steam engines. These included Societe Française des Moteurs R. Diesel (France), Maschinebau-Actiengesellschaft Nürnberg (Germany), Burmeister&Wain (Denmark), Mirlees (United Kingdom) and others. Diesel's idea had been that the diesel engine's high thermal efficiency and comparative compactness would make it attractive for stationary applications, and this did prove to be the first market. However, with the development of the electric motor, the diesel was often utilized to drive an electric generator rather than to drive factory machinery directly, the advantage of this arrangement being to eliminate the steam boiler from the power system.

A second major market opened up after about 1903 in the form of the marine engine, as diesels were first tried in a few river craft. There was much interest in the diesel as surface power for submarines, a ship type that was of much interest to various navies in the first decades of the twentieth century, and by the onset of World War I, it was established surface power for submarines in the major powers' navies. It was the surface cruising power for Germany's U-boats. Following World War II, the diesel became the standard power plant for the world's seagoing merchant fleet and for inland waterway power.

Hydraulic Press

In the mid-seventeenth century, the French scientist/philosopher Blaise Pascal stated what has become known as Pascal's Law, which states that pressure in a liquid confined in a vessel is transmitted equally in all directions. In the hydraulic press, a force is applied to a piston working on a liquid confined in a cylinder that has been fitted with a second piston with larger area than that to which the force is applied.[25] The force on the smaller piston is transmitted through the liquid to the larger piston, which moves with greater force than that applied to the smaller piston, but with less speed. Pascal's first working model of the press resembled a modern syringe, but the machining capabilities of his day were inadequate to achieve the tight tolerances to make the hydraulic press usefully industrially.

About a century later, the Swiss physicist, Daniel Bernoulli, formulated the law that states energy in a fluid is due to elevation, motion, and pressure, and if there are no losses due to friction and no work done, the sum of the energies remains constant. The two laws form the physical basis for all modern systems of hydraulic motors and controls. In Pascal's time, getting pressures much in excess of what nature provided was difficult, but in the nineteenth century, the development of pumps ultimately created a wide range of possibilities for hydraulic energy. In order to give Pascal's principle practical usefulness, it was necessary to have the pistons fit their cylinders with very tight tolerance. This was achieved with the improvement of machine tooling; in fact, the hydraulic press took its first practical form in the hands of Joseph Bramah, the English machine tool pioneer. One resulting system; built in London in 1882 for driving machinery, lifting bridges, and operating hoists; was still in use in the early 1980s.[26]

Early innovation based on Pascal's principle can be found in military applications. One of the earliest was in recoil mechanisms of artillery pieces, in which the force of recoil was made to force a liquid (at first, water) confined in a cylinder through a small orifice, thereby braking the recoil. Restoration of the barrel to its firing position was by a spring. These recoil systems appeared in the late nineteenth century. This system anticipated he automobile shock absorber by many years. An improvement was made in 1906 when oil was substituted for water as working fluid in a system installed to control the elevation and traverse of guns of the battleship *USS Virginia*. The main advantages of hydraulic systems are flexibility, the ability to multiply forces efficiently, and fast and accurate response to controls. Fluid power can provide a force either of ounces or thousands of tons. Unlike other innovation streams spawned by prototypes from before 1910, such as that from the Otto Engine, that of the hydraulic systems did not really begin to have substantial impact in the wider economy before the interwar period of the twentieth century. Moreover, the real acceleration in the innovation of this technology did not occur until after World War II.

Medical

Arguably, the germ theory of disease was the most important discovery of the nineteenth century, and with it came modern medicine. This was a true prototype discovery. Credit for the discovery of the relation between microorganisms and human disease goes to a number of individual researchers. One of the first people to observe microorganisms was Anton van Leeuwenhoek, a cloth merchant of Delft in the Netherlands in the 1660s. Van Leeuwenhoek by avocation was a maker and experimenter with lenses. He became highly skilled at grinding of lenses, and produced products of high quality, even by today's standards, and is credited with the invention of the simple optical microscope, with which he achieved magnification of up to 300X.

It was in using one his own instruments that he noticed what we now know as microorganisms, but in his time there was no body of theory or knowledge that would have enabled him to recognize what he was seeing. Van Leeuwenhoek's observations and those of others eventually came to be explained by the theory that the objects observed, while apparently alive, had evolved from inanimate matter. This is known as the theory of spontaneous generation. Early microscopes simply did not have the power to resolve *individual* microorganisms, and what *was* being observed was colonies of them, which seemingly were appearing from nowhere. This theory did not offer any suggestion as to how disease is transmitted, and as long as it was accepted widely, sparked little interest in investigation of how disease could be controlled.

Spontaneous generation had begun to be questioned in the nineteenth century, partly because of increased resolving power of microscopes. Ignaz Semelweis, a physician practicing in the 1840s, was able to show that washing the hands before delivering a baby greatly reduced the incidence of childbirth fever, a notorious killer of mothers and infants at the time. Louis Pasteur, working in the 1860s, conducted a series of experiments that showed that microorganisms existed in air, but were not created by air, a result that went directly counter to the spontaneous generation idea. Pasteur was instrumental in the development of vaccines, such as for rabies, and is famous for his method of preventing spoilage of stored liquid foodstuffs, such as milk, wine, and beer.

In the 1860s, Joseph Lister, working with the results from the work of Pasteur and Semmelweis, developed the technique of washing wounds with phenol (carbolic acid) prior to surgery, and thereby achieved chemical inhibition of infection. While these developments sound like swift progress when related years after, it should be borne in mind that antiseptic practices such as Lister's took some time to become general practice. The military hospitals of the American Civil War were notorious for losing patients to infections and gangrene after surgery. Statistically speaking, making it to a field hospital after being wounded may have improved chances for survival over simply lying on the ground where hit, but by how much is open to question.

In the 1870s, Robert Koch systematically worked with the notion that a specific diagnosis can be linked to a particular microbe. His systematic approach yielded discoveries of bacillus anthrasis, mycobacterium tuberculosis, and vibrio cholera. Koch's methodical approach resulted in Koch's Postulates, a sequence of steps of experimental logic for relating microbes with illnesses. Koch's postulates were as follows. (1) The suspected organism must have been established as present in every instance of the infectious disease; (2) the organism must be cultivable in the laboratory; (3) inoculating an animal with this culture would reproduce the disease; and (4) the organisms could be recovered from the diseased animal and grown again in a laboratory culture. Koch used this method to discover the baccilae that cause tuberculosis (1882) and cholera (1883). Koch's logical procedures were emulated by his contemporaries and resulted in the discoveries of the baccilae that cause typhoid, diphtheria, pneumonia, gonorrhea, undulant fever, meningitis, leprosy, tetanus, plague, syphilis, and a number of others. In the 1890s, Dmitri Iwanoski discovered the first virus, the cause of tobacco mosaic. By the turn of the century, the idea of chemical therapy was firmly established.

An advance that was at least of the magnitude of that of Koch and Pasteur was made by Paul Erlich, who was working toward correlating the chemical structure of a synthetic drug with its biological effects. His greatest discovery in 1910 was of the arsenical drug Salversan, which proved to be effective in the treatment of syphilis. The overriding importance of Erlich's work was twofold. First, it was one of the earliest drugs to follow the principle that the drug introduced into the body of a patient should destroy an invading organism without doing serious damage to the patient. Second, the idea that the biological effects of a drug could be predicted approximately from its chemical structure meant that as knowledge of molecular biology expanded, researchers had an increasingly powerful means of narrowing the field of the drugs that they had to test in the context of a particular illness. The body of knowledge growing out of this kind of correlation continues to grow in the early twenty-first century. Table 2.2 summarizes late nineteenth- and early twentieth-century medical discoveries.

Other Technology

Gyroscope

Jean B.L. Foucault, a French scientist, is responsible for the gyroscope—a rotor mounted in gimbal rings that permit the rotor to spin in a fixed plane

Table 2.2
Medical Discovery Highlights, 1880-1910: The Germ Theory

Inventor/ Discoverer	Discovery	Year
Louis Pasteur	Anthrax vaccine	1881
Louis Pasteur	Rabies vaccine	1882
Robert Koch	Rel. of specific microbe to specific disease	1870s
Emil von Behring	Antitoxins: tetanus and diphtheria vaccines	1890
Karl Landsteiner	Discovers existence of different human blood types	1901
Frederick Hawkins	Proposes vitamins and their relation to scurvy and ricketts	1906
Paul Ehrlich	Chemotherapeutic cures; chemical cure for syphilis	1907

independently of the orientation of the mount. Working in the 1850s, Foucault demonstrated that the rotor, spinning with high angular velocity, would hold its orientation in space regardless of the Earth's rotation. The gyroscope eventually became the basis for accurate navigation in marine, aviation, and space settings in the twentieth century. Needless to say, the ability to produce a balanced rotor of substantial mass depended, as did the rotor of the steam turbine, on the capability to produce a wheel with near-perfect balance on its axis—which the machine tool industry was able to provide only in the late nineteenth century. Because of this, it was 1908 when the first workable gyrocompass was developed by the German inventor H. Anschütz-Kaempfe for use in a submersible—thus providing a second element, along with the diesel, in the development of the modern submarine. In 1909, Elmer Sperry built the first automatic pilot using the direction-keeping properties of the gyrocompass to keep aircraft on course. The gyrocompass was developed into the basis for highly accurate marine navigation systems by the time of World War I.

Airplane

The entire history of powered flight occurred in the twentieth century, and the invention is noted here by way of introducing the origin of a particularly powerful stream of innovation. Even though the airplane was a twentieth century development, it benefited from a crucial enabler: the gasoline engine. The Wright brothers made use of the two-cycle engine, a variant from the Otto Cycle, which was characterized by minimal weight per horsepower. In the two-cycle engine, each down stroke is a power stroke. Its advantage is compactness and a high power-to-weight ratio, decidedly an asset in powered flight. As powered flight was improved, its advance was enabled by another nineteenth-century product, aluminum. A substantial amount of the impetus toward early development of the airplane came from the world's militaries. In the early stages of World War I, the airplane was regarded as a scouting and reconnaissance tool, but it became an effective combat weapon with the invention of the synchronized machine gun, which allowed guns to fire through the propeller. Indeed, the early attempts to arm airplanes for combat with other airplanes were motivated by the desire to prevent enemy aircraft from observing and photographing field fortifications of ground forces.

Mechanical Refrigeration

Modern mechanical refrigeration began with the work of Ferdinand Carré of France, who built an apparatus using ammonia as refrigerant in 1860. Prior to Carré, there had been many attempts to cool air in a room, many were based on ice and went back to ancient times. During the mid-1870s, Carl von Linde of Munich began developing improved ammonia refrigerators, and in 1877, he obtained a patent on a mechanical refrigerator. The Linde apparatus found a ready market, as one of the first examples was installed in an Italian brewery in 1878.[27] Von Linde's work was based on his understanding of the emerging science of thermodynamics, and he showed how the efficiency of mechanical refrigeration machines could be calculated and how efficiency could be increased over time.

While refrigeration machinery based on von Linde's patents found almost immediate markets in several industries, refrigeration spawned innovation in several directions distinct from industrial (and eventual) commercial and residential cooling. As early as 1877, oxygen was cooled to the temperature at which it became liquid (90 degrees Kelvin or -183 degrees Celsius), and in 1895, air was cooled to 40 degrees Kelvin at which point it is a liquid and separates into its principal components. Cryogenic air separation, which became the basis for the industrial gas industry, thus had a first-stage innovation period from 1878 to 1895 in which progress came from improving refrigeration machinery. Another important line of innovation started with the discovery of superconductivity in 1911. Improving ability to cool led to the discovery of the ability of some super-cooled metals to conduct electricity without resistance. By 1960, temperatures as low as 0.000001 degrees Kelvin could be obtained. Temperatures lower than 3 degrees Kelvin are produced almost exclusively for laboratory research.

Solar Cell

The first description of the photovoltaic effect was due to Antoine-César Bequerel in 1839. Charles Fritts, an Englishman, constructed the first true solar cells in 1886 by coating a sheet of the semiconductor selenium with gold hammered to such thinness that it could pass light.[28] Fritts' cells were highly inefficient by today's standards, as they converted less than one-tenth of 1 percent of solar energy into electric current. Efficiency has been greatly improved since that time, especially with the substitution of silicon for selenium as the semiconductor in the 1950s. Even though inefficient, these early cells fostered visions of abundant electric power among some people.

The solar cell attracted much interest early on, and the underlying principle was finally explained by Albert Einstein in a 1905 paper in which he envisioned light as composed of quanta (photons), which, striking atoms in certain materials, released electrons.[29] Einstein's paper led to the founding of the modern theory of quantum mechanics. However, the solar cell was not of much interest to engineers until 1953 when Calvin Fuller, a Bell Labs chemist, inadvertently stumbled on silicon as a basis in the solar cell that was far superior to selenium. This was the approximate beginning of the solar cell's stage-two development that proceeds today. Solar cell efficiency has been greatly improved through the use of better materials. Early silicon-based solar cells were capable of a 6 percent energy conversion rate. Energy conversion rate had climbed to around 20 percent by the late 1980s. Solar cells have found a number of *niche* markets, where there is a need for small quantities of power in locations remote from the power distribution grid, such as in space vehicles. The list of terrestrial applications is growing. It includes offshore navigation aids, crossing warning systems on remote parts of railway systems, pumping power, small power needs on offshore oil and gas platforms, and many other applications.[30] A number of small applications have developed on the basis of the high reliability of solar cell-battery systems in situations where reliability in excess of that offered by the power grid is considered decisive.

Nevertheless, in terms of the more optimistic visions for the solar cell's future, it has so far not attained the efficiency required to fill the need for base load power generation. The problem has been that as much improvement as has benefited the solar cell, the cost of its power has always been substantially higher than costs associated with alternative power technologies, because of the high costs of the materials used to make the most efficient solar cells (especially silicon) and because of the lack (so far) of technology for storing electric power. This topic is developed in Chapter 11.

Fuel Cells

A fuel cell is an electrochemical device that produces electricity from an external fuel supply. In light of the attention that fuel cells have received in relation to power supply problems of the early twenty-first century, it is of interest to note that the fuel cell is a nineteenth-century invention. Credited to Welsh scientist Sir William Grove, who built the first working example, the idea of the fuel cell goes back to a paper published by Swiss scientist Christian Friedrich Schönbein in the January 1839 issue of *Philosophical Magazine*. This article guided Grove in his work. Little happened with respect to this invention until 1932, when some successful fuel cells were developed by British engineer Francis Thomas Bacon. In 1958, a team led by Pat Grimes of the US built a fifteen kilowatt fuel cell as power for a farm tractor built by the Allis Chalmers Company and demonstrated across the country at state fairs. This cell used potassium hydroxide as the electrolyte and compressed hydrogen and oxygen as the reactants. Development proceeded, and it was used as a source of power and drinking water in the US. space program in the 1960s. A more complete discussion of the working of fuel cells is deferred to Chapter 12.

Telephone

The story of how Alexander Graham Bell happened onto the telephone during the course of his attempt to invent a hearing aid is well known.[31] The telephone is an example of a device for which demand growth depends on growth of a network—that is, it becomes useful only when more than one person uses it, and usefulness expands with expansion in the number of users. By 1900, telephone networks had been established in major cities. Long-distance telephony came into its own after about 1910 on the basis of amplification made possible by the invention of the triode. J.P. Morgan was said to have opined that the telephone did not have much of a business future, and it is not difficult to visualize what he might have been thinking. The industrial world already had a telecommunications network in the form of the telegraph. Because the sender of a telegraphic message paid by the word, the telegraph encouraged economy of words. It was regarded as expensive and was therefore used largely for business where succinctness of expression was held to be a virtue. Among individuals, a telegram came to be regarded as bad news, and indeed, telegrams to individuals often informed of a death.

What Mr. Morgan overlooked was that the wordiness allowed by the telephone had an *entertainment* value that assured its popularity in non-business settings. The telephone was a completely novel means of contact with a wider world for housewives and others normally at home during the typical day. The early growth of the industry was rapid: in 1893, there were 266,431 telephones in the US; but by 1907, there were over six million.[32] This growth occurred during the era in which most telephone traffic was local, and reflects the presence of telephones in many private residences. The mass-consumer market for telephones was extremely important, for it provided the telephone industry with a revenue base by which it could finance the investments called for by the advances in telephone technology. The telephone also became a highly important business tool, despite Mr. Morgan, but one has to ask the question of whether this could have happened had the development depended exclusively on business' adaptation to use of the telephone in its daily activities. As it was, the development of the telephone as a business tool was greatly aided by its cost reductions that resulted from the growth of the consumer mass market.

The telephone industry early on came to be dominated by the Bell System under the corporate umbrella of AT&T. Some of the early non-Bell telephone companies opened service to communities that Bell did not serve, but some duplicated services that Bell provided. It was these latter situations that created duplicated telephone transmission facilities and fed the notion that the telephone business was a "natural monopoly," a notion that was widely accepted without too much critical analysis. The result was regulation, a pattern that was not broken until the later decades of the twentieth century. The domination of the telephone industry's growth by AT&T had one very important consequence from a technological point of view. The company was in position to deploy its profits towards innovation invention activity that was necessary to maintain and increase the efficiency of the system. An alternative industry structure, such as a nationwide patchwork of independent local telephone companies would not, in all likelihood, have been able to do this. This point is developed in Chapter 14 below.

Notes

1. Prototype discoveries in germ theory of medicine are treated separately below.
2. There had been earlier attempts to industrialize small arms manufacture in France, and Jefferson's interest had been honed by a personal encounter with the French efforts.
3. What the American system achieved was to cut out enough of the labor-intensive work that the skills shortage was not an insurmountable restriction.
4. The private mass-market idea was not foreign in the nineteenth century. Eli Terry succeeded not only in establishing a works for producing clock movements (made of wood) before 1820, but he also, through pure salesmanship, created a market for clocks at a time when people had no apparent use for them. They needed little more than the span of daylight to regulate their daily activities. Hawke, David Freeman, *Nuts and Bolts of the Past*, (New York: Harper & Rowe, 1988), pp. 69-73.
5. Tool steel with a carbon content ranging from 1 to 1.2 percent was used by early machinists. Tools of this material tend to lose cutting ability at about 400 degrees Fahrenheit, a temperature easily reached due to the friction of machining. In 1900, the introduction of high-speed steel enabled the tripling of machining speed and a corresponding increase in the capacity of machine shops. A typical high-speed steel contains 18 percent tungsten, 4 percent chromium, 1 percent vanadium, and only 0.5 to 0.8 percent carbon.
6. Replacement reflected steel's greatly superior wear and strength characteristics relative to iron.
7. Even in the early twenty-first century, the traveler occasionally finds himself in neighborhoods where there are old factory buildings. In such places, it is easy to identify structures that were built before 1890 by the shape of their windows and doors: in the older buildings, walls above windows were supported by masonry arches. When structural steels became available, window supports became based on it, and therefore were square, not arched. The other crucial enabling invention in the appearance of high-rise buildings was the safety elevator.
8. Out of the early necessity for creating markets for the product, aluminum became the only one of the major nonferrous metal industries to be integrated through the final consumer.
9. The Goodyear Tire & Rubber Company is named for Charles Goodyear, whose name was prominent in the rubber business at the time the company became involved in it; but apart from that, there appears to have been no connection between the company and Charles Goodyear.
10. Synthesis gas can be made from any hydrocarbon feed, from methane to coke and coal.
11. Baekeland at this time was an established inventor and successful businessman. He had invented a dry-emulsion photographic printing paper around 1890, which he marketed under the name Velox. He enjoyed considerable personal wealth as a result of the sale of this venture to George Eastman.
12. "Photography," *Encyclopedia Britannica Library,* CD ROM, 1997.
13. Credited to Zenobe-Theophile Gramme.
14. Westinghouse was an established inventor, having been credited with the railway air brake system.
15. *Encyclopedia Britannica,* (Chicago, 1995, vol. 18), p. 194.

16. Strelingers was a Detroit-based hardware distributor with a huge range of product offerings from hand tools to boilers and steam engines. A reprint of Strelingers' 1895 catalog has over 500 pages, within which the "electric dynamos and motors" entry occupies less than four column inches and contains this interesting wording: "Taken as a while, the Electrical business is a very complex one. We don't know much about it . . . ." Charles A. Strelinger, *A Book of Tools, Machinery, and Supplies,* (Bradley, IL: Lindsay Publications reprint, 1991), p. 332.
17. *Scientific American Supplement,* (no. 761, Aug. 2, 1890), p.12160.
18. *Scientific American Supplement,* (no. 1293, Oct. 13, 1900), p. 20798.
19. For example, Hiram Maxim, who was later the inventor of the modern machine gun, also built a serviceable incandescent lamp about the time of the Edison lamp.
20. Earlier attempts to make an incandescent lamp, going back to the 1840s, failed for lack of ability to create a partial vacuum in a tube. The mercury pump (inv. 1865) solved this problem adequately.
21. The four-stroke cycle was first patented in 1862 by Alphonse Beau de Rochas, but inasmuch as Otto was the first to build a working engine utilizing the principle, Otto is remembered as the inventor. In 1886 Otto's patent was revoked when Rochas' prior patent was brought to light.
22. Cummins, Lyle, *Internal Fire,* (Wilsonville, OR: Carnot Press, 2000), pp. 163-182.
23. When steam does work, its temperature declines and its volume increases, hence, the increasing size of the rotors. This is but one example of compounding, or the use of what would be waste heat to do useful work. A recent example of this is the use of waste heat to power a supercharger for the purpose of improving the performance of a heat engine.
24. De Laval became interested in turbine design out of frustration at not being able to find an engine that could spin his centrifugal cream separator sufficiently fast. Abbott Payson Usher, *A History of Mechanical Inventions,* (Boston: Beacon Press, 1959), p. 393.
25. During the course of his experiments, Pascal invented the syringe. This device makes the difference between Pascal the scientist and Pascal the inventor.
26. *Encyclopedia Britannica,* (Fifteenth edition), vol. 6, p. 188.
27. The importance of the brewing industry as an early user of mechanical refrigeration shows in its rapid proliferation in American industry, per the 1880 Census of Manufactures. See Henry Hall, "The Ice Industry in the United States," in U.S. Department of the Interior, Census Office *Power and Machinery.* (Tenth Census, 1880: reprinted by Lindsay Publications, Inc., 1994), p. 21. A second major user of mechanical refrigeration was the cattle industry. Moving meat by rail was in refrigerator cars cooled with salted ice. These were stocked with ice from dedicated ice plants located at strategic points along the line. Mechanical refrigeration of the cars themselves and of highway trucks became feasible only in the 1950s.
28. One of the properties of gold that makes it valuable industrially is it ability to be hammered to extreme thinness. This was the basis for range-finding devices used in cameras and in the military.
29. The Nobel committee that awarded Einstein the Nobel Prize in 1921 cited his theory of photoelectric effect as his most important work. This was in spite of the theory of relativity (1915). The numerous experimental verifications of general relativity were just beginning to appear at that time, and the Nobel judges were then reluctant to confer the award on the basis of what they regarded as largely unverified theoretical work.
30. In many such applications, the economics are favorable to solar power in the form of avoided construction that would otherwise be necessary to connect the power

load to the grid, even if it were nearby. Railroads use a solar-powered system to power switches in rail yards, replacing heavy manually operated machinery. The economic attractiveness comes from a combination of reliability and avoided workers compensation claims, as the manual system required a worker to bend over to move heavy machinery, a source of back injuries, hernias, and other woes.

31. This story and others have given the impression that much of invention and discovery are largely accidental. This is true of Bell only in the narrowest possible sense, for his mind was so steeped in the work he was doing and he was such a careful observer that he immediately recognized the significance of what he had.

32. Sharkey, William W., *The Theory of Natural Monopoly,* (Cambridge: Cambridge University Press, 1982), p. 206.

3

Innovation and Prosperity: 1900-1930

A Rich Base of Technology: What was Done with It

The nineteenth century had yielded a variety of solutions to technological problems that must have seemed amazing to anyone who reflected on the matter and who had been aware of the situation of just twenty years earlier. If only power sources were considered, there were the internal-combustion engine in its Diesel and Otto forms, the steam turbine, and the electric motor. The technological change of the next several decades can be understood as the process of applying the new tools to possibilities that had not even been dreamed of through most of the previous history of civilization. Rapid technology flux is very familiar to people in the late twentieth and early twenty-first centuries, but it is easy to lose sight of what was perhaps equally rapid change in the early twentieth century. One can just begin to get a feel for how rapid this change was by perusing the Census Bureau's compendium of power sources in manufacturing circa 1880. This document is a snapshot of manufacturing technology in its time and is as interesting for what it does not include as for what it does. For example, it contains not one word about electric motors or electric power, and its chapter on refrigeration focuses mostly on ice-handling tools and only briefly mentions the primitive mechanical refrigeration technology of that time.[1]

Four industrial development streams based on new technology dominated overall economic growth in the first three decades of the twentieth century. Two of these were completely new: the automobile industry, including the industries that grew up to support it, and the electric power industry, including the delivery of electric power to masses of industrial, commercial, and residential consumers. The third was a physical restructuring of previously existing industry of revolutionary proportions based on electric power and mass production for mass markets. Prior to 1930, rapid growth in these three underpinned generally rapid overall economic growth in the United States and Western Europe—although the latter's growth was set back by the destruction of World War I. The fourth was the telephone, which enabled the growth of telecommunications well beyond what had been possible with the telegraph. The 1900-1930 period probably would have been a time of prosperity in the US if the new technology had done nothing more than enabled improvement in the cost efficiency of prior-existing industries. As it turned out, however, the new technology enabled completely new lines of industry whose economic impacts multiplied the economic impacts that resulted from reforming existing industry. For the most part, the determination of which technologies succeeded in which applications was due to market forces.[2]

Technology and Prosperity

Mass Markets and Mass Production

The important link between technology and growth in the period from 1896-1929 was the emergence of mass markets and mass production. These two factors assured that the scale of production would be much larger than in previous cyclical episodes, and created the possibility that the scale of any ensuing failure would be large. Mass markets grew around the automobile and its supporting industries, as well as the telephone and electric power. The significance of the mass market is that it justifies investment in tooling much of whose usefulness is peculiar to the production of large quantities of one or a limited number of products at low cost. This production strategy, known commonly as mass production, has a long history. Indeed, it is well described by Adam Smith in the eighteenth century in his famous description of the pin factory, which is embedded in a larger discussion of the productive potential arising from the division and specialization of labor.[3] The obvious example from the early twentieth century was the Ford Highland Park plant, in which there was a vast amount of machinery dedicated solely to moving parts and subassemblies to the points where and when they were needed to further the assemblies of finished Model Ts. The cost of such a manufacturing facility is high, but its productivity is so high that the average capital cost per unit produced is much lower than the cost level attainable with more direct production methods. The relationship between the various prototype inventions and low-cost mass production is circular; the new technology resulted in lower costs and expanded demand. Expanded demand justified the investment in facilities that permit full realization of the reduced production cost possibilities created by the new technology.

Figure 3.1 schematically illustrates the relation between production scale and unit costs. In it, Technology A represents a labor-intensive technology in which one worker or one team of workers takes responsibility for one or a few finished items, such as Model Ts or Cadillacs. In addition to being labor-intensive, Technology A is skill-intensive, for the members of each worker team have to be able to perform all the skilled tasks involved in assembling all the

Figure 3.1
Relation of Unit Costs to Production Scale

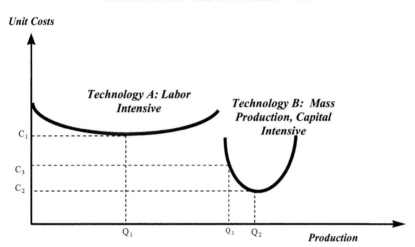

components of the completed consumer product. Technology B is the full mass production technology. B is far more capital intensive than A, and it is far more productive. Also, it is much less dependent on workers with high skill levels. Technology B has higher capital costs, but lower unit costs; its adoption represents substitution of capital for skilled labor. The "U" shape of the Technology B cost curve reflects that unit costs rise sharply when production falls below its optimum level at Q_2. With Q_1 as the lowest-cost output for Technology A, it is clear that A is preferable to B cost wise at production level Q_1, and the reverse is true at Q_2. If business falls off such that the market calls for production at Q_3, costs rise sharply with Technology B, for it has not generally been possible to adjust all that finely-coordinated machinery to perform optimally at a lower output level, especially in the pioneering days of the production line. The shapes of the unit cost curves of Figure 3.2 reflect this: that for Technology A is much flatter than that for Technology B. Cost flexibility is greater with Technology A because its costs (labor being a proportionately greater component than with Technology B) can be adjusted more closely in proportion with demand by laying off or hiring workers.

Inability to adjust costs easily to accommodate temporary variations in product demand has long been recognized as a problem with mass-production facilities, and was especially acute in the early days of this technology. Indeed, the inflexibility showed up in the changeover from Model T to Model A production in 1927 and 1928; production had to be shut down entirely for almost eighteen months. As techniques of mass production developed in time, large companies engaged in this strategy typically operated with more than one production facility. When demand fell off, it was sometimes possible to shutter a whole plant and concentrate the production into the remaining plants, but this strategy has never been what one would call finely tunable in response to demand variability. Adoption of a highly capital-intensive, mass-production strategy was a rational response to the growth of mass markets and a shortage of highly skilled workers, but it was achieved only with the loss of flexibility to adjust costs to high variability in product demand. Moreover, it was at first not possible to admit any significant variation in the product within the production run. With time, the production line was improved to tolerate some product variability, but the solution to the problem of accommodating the production line to major variations in product demand did not find solution until the modern era of computer-aided design and manufacturing.[4]

How do mass markets develop? Clearly, there is no reason for them to develop in the absence of the incentive to consumers of low price at retail. Yet, there is little obvious incentive for producers to invest in the tooling for mass production if they do not have some sense of a large potential market. In this chicken-and-egg situation, part of the answer has to be in the producer's gaining an idea of the sensitivity of his market size to price. If the producer deems the market to be highly sensitive, based on his experience with the results of his pricing strategies, then he has an incentive to get his production costs down to the point where the firm can be profitable at a lower price with a higher volume. A product price reduction can be viewed as, among other things, an experiment that will result in information as to what might happen in reaction to further price reduction. It is a feedback process.[5] The decision to build the Highland Park plant did not emerge, like Aphrodite, full-blown, but only with the experience and knowledge from many pricing experiments. Also, the plant that Ford opened in 1913, for all its sophistication, was less productive than the same plant that was shut down for a model changeover in 1927. A producer with a vision of high-volume, low-cost production will sometimes induce the future of the market by pricing a good at below production cost before economies of mass production have actually been realized. This sort of short-term sacrifice of near-term profitability in favor of higher future profitability at high volume is a recurring theme in recent times in Silicon Valley, among other settings. Another element in the formation of the twentieth-century mass market was rising income levels, especially among the middle classes. The fomenters of mass markets did explicitly attempt to promote this process by means of their wage policies. The story of Ford's five-dollar-a-day is well known. The primary purpose of this wage policy was to reduce worker turnover, but it was also justified as a way to fill the employee parking lot with the company's products. However, not even a large manufacturer could single-handedly account for the generally rising standard of living during the first three decades of the twentieth century.

One variation on the theme of creating a mass market is to give away some element of the process and thereby create a demand for another part of the technology. This is the origin of the story of the old Standard Oil's giving away lamps to Chinese people with the intent of selling them the illuminating oil necessary to use the lamps. This strategy is echoed in the practices of some internet service providers in their free distribution of software for the access to their service with the plan of making money by usage fees for the service and from advertising. Another recent class of examples occurs in the field of medical test equipment, where the test apparatus is sold at little or no profit and the producer makes money on a necessary complementary product, such as test strips for assessing blood sugar levels used by diabetics. What was historically unprecedented in the early twentieth-century experience with the growth of large markets was scale. In the time of Eli Whitney, no market, not even that of the War Department, was remotely as large as the markets for automobiles, telephones, and electric lighting that developed in the twentieth century. An important difference between the private mass market of the twentieth century and the government market that Eli Whitney approached was that Whitney did not have to create the market. Whitney's first problem was that of interesting specific individuals in the War Department in his interchangeable parts idea and gaining the contract. The second was that the materials and machining technology of the time did not fully support his ideas for parts interchangeability. However, the modern problem of proposing to supply the Defense Department, reduced to its essence, is little different from that faced by Whitney. The modern defense industry faces the problem of acquiring costly, specialized tooling for the purpose of creating sophisticated weapons systems for which there is only one customer in immediate prospect and a limit on the size of the production run.

If producers were able to encourage the development of mass markets by their pricing policies, they faced the problem of keeping their costs low enough that they could experiment with reduced prices. The progenitors of mass production knew they wanted to produce large numbers of products cheaply, but they learned how to do this only a step at a time. Henry Ford was extremely good at recognizing the value of the ideas of others and readily adopted them. While he did not literally invent the automotive production line, he seemed to have a sure instinct for what would contribute to such a line. Division of labor, for example, was an appreciated principle well before Adam Smith described it. Interchangeability of parts is an idea that went back to and before the time of Whitney and Colt, but it had been elegantly introduced to the nascent auto industry in Ford's time by Henry Leland.[6] Ford himself had observed that workers in his pre-Ford Motor auto ventures spent much time walking to and from storerooms to get parts and assemblies and worked with the idea of bringing the work to the worker rather than the other way around. He took the basic idea of the moving line from Gustavus Swift, the meat-packing pioneer. The final result at Highland Park in 1913 achieved the ideal that workers could do their work without having to move their feet.

In the early twentieth century, the introduction of mass-production techniques in American industry often was in lines of activity that had not existed in the previous century, and therefore resulted in the creation of jobs that had not previously existed. However, the increase in industrial employment was not as great as it would have been had it been attempted to organize the new industries with nineteenth century technology, but that option was probably out of the question for other reasons. The mass market was made possible by mass production at the same time that mass production encouraged the mass market with falling prices. Much has been written about the social impacts of mass production. While this production strategy involves substitution of capital for skilled labor, it does not imply increased unemployment under conditions of rapid market expansion. The technique was a superb response to a shortage of highly skilled mechanics and machinists, and its effect historically was an absolute increase in the number of factory jobs for people of limited skill. Complaints about mass production center around, but are not confined to, the nature of the work, which is reputed to be mechanical, repetitive, grueling, and boring. The grueling part came from having to stand in one place doing the same thing for an entire shift; workers would go home essentially worthless for any purpose other than sleep. The wages were good by the standards of the times, but the reason they were good was that Henry Ford realized that he had the problem of worker turnover. The response to this problem consisted of wage increases and bonuses for longevity. He became famous for his five-dollar-a-day minimum wage, which was widely attributed to caring for workers' welfare, but it came about only as a result of the aforementioned problems, and as a result of the idea Ford workers could also be Ford customers. The grueling nature of production line work definitely figured into the success of industrial unionism in the 1930s.

Economic Growth and Prosperity

The prosperity that more-or-less prevailed in the first three decades of the twentieth century has been described extensively. Prosperity was underpinned by the availability of new technology and novel combinations of technology of substantially greater productivity potential than any productive capital that existed prior to 1900. These decades were characterized by rapid growth in manufacturing output: the index of manufacturing production in the US expanded at the average rate of 4.49 percent per annum in the 1899 through 1929 period. By contrast, the average annual increase in the 1929 through 1939 decade was essentially zero.[7] The new automobile industry was at the heart of this prosperity. Not only did the assembly of increasing numbers of cars for an expanding demand underpin industrial growth, so too did expansion of a myriad of support industries that supplied parts and sub-assemblies to the auto assembly plants. The new owners of the cars were able to commute to work from residential neighborhoods that were not within walking distance from a street railway, thus making urban sprawl feasible. Also, car owners as a group constituted an increasingly powerful political interest in favor of public construction of paved roads and highways, a huge boost to the construction industry. All this and more can be attributed, in one way or another, to the availability of low-cost vehicles and other articles emanating from the new mass production. Whatever its social problems, mass-production technology underpinned an economy on an enhanced scale.

A Large Commodity Economy?

If the ingenious organizations of the new tools appeared to be novel, there was at least one long-established principle at work in the automobile and other new industries: when producing a commodity product, the only viable business strategy in the long run is to be the low-cost producer. In the early twentieth century, this was an important driver in the development of mass-production methods. A commodity product is one wherein one producer's output is essentially indistinguishable from that of any other producer. The classic example, and one that is trotted out in countless economics classrooms, is that of a major farm crop. Yet, the world of commodities is surprisingly full of products that incorporate much value-added processing. Is a car a commodity product? Henry Ford evidently thought of it that way. The first Ford production line was designed to produce large numbers of identical vehicles, which for many years had not even a variation in color. This production plan amply answered to Mr. Ford's desire to produce a very basic car that almost anyone could afford.

It will quickly occur to the reader that the commodity nature of the car did not last. The essential difference between a large-volume agricultural commodity and a value-added product such as an automobile is that demand for the latter is more sensitive to income level. This opens the possibility for achieving at least the perception of product differentiation by adding features that enhance comfort, ease of operation, and other traits. From the earliest days of mass production, at Ford and elsewhere, there was a powerful competition-driven impetus toward enhancing the flexibility of the production line with the object of differentiating the product from that of competitors, but without sacrificing mass-production economies. The hoped-for payoff to product differentiation was pricing power, something that would be totally lacking if all competitors' products were widely perceived to be the same. In spite of such later success at product differentiation in autos and other products as was achieved, the automobile of the 1910s and twenties had many commodity-like aspects. Technical improvements tended to appear first on luxury models and spread down the price levels, but these improvements tended to be small relative to the overall cost of the vehicle. Then, as now, the customer's decision was likely to rest on the nature of the deal offered more than dramatic technical differences as between vehicles. Moreover, the commodity nature of industrial products of this era was far from confined to automobiles. It is not necessary to appeal to the opening decades of the twentieth century or to the automobile industry for examples of commodities with high value-added content. Consider the personal computer. It has developed to the point that each producer's product delivers about the same utility as that of other producers.

The development of the modern automobile and other products having substantial value added was enabled by the late nineteenth-century technology legacy. Once the opportunities opened by this technology had been largely exploited, manufacturers found themselves using the no-longer novel technology to produce products that were alike in essential ways. This is the process of commoditization. The response to this process has given rise to a large industry whose main purpose has been to create the illusion of differentiated products among consumers: advertising. This industry has bulked large in overall economic development since the early twentieth century. The tendency for products once seen as novel to become commoditized is a familiar one to this day.

It is of interest to take this idea a step further. There have been many products that have through the years emerged as the result of stage-two innovation of some prototype invention. Innovation tends to result in declining costs of applications of the combinations of technologies involved. As costs decline, the field of applications of the technology widens. The process of bringing down the costs associated with a new technology tends to involve some sort of mass production, which at first results in production of many commodity-like products. Further development of the technology can be thought of as an ongoing tension between a tendency toward commoditization that is inherent in mass production and the effort to differentiate products in order to gain or maintain some sort of monopoly pricing power. The automobile is a classic example of such a product. There is no better illustration of this idea than the contrasting strategies of Ford Motor and General Motors in the 1910s and 1920s. Ford was committed to the "commodity car" or the Model T. The early General Motors, under Alfred P. Sloan, accumulated a portfolio of marquees under the general strategy of having a model suitable to the customer who was becoming more prosperous and wished to upgrade his or her car. Ford's negligence of the customer with rising income resulted in that company's loss of market share in the 1920s, and the somewhat belated abandonment of the Model T in 1927-1928.

Readers who have been trained in basic economics will recognize what is at the heart of this contrast: that quantity demanded responds not only to price but also to income. Ford's problem was a massive commitment to a low-price strategy to the exclusion of the income side of demand. Indeed, by making personal transportation more affordable, Ford increased the affordability of other items that might interest customers, including comfort and convenience features on their automobiles. In a time of generally rising incomes, this very much played to the demand for intermediate levels of luxury, such as Buicks and Oldsmobiles, if not Cadillacs. In the late twentieth century, the outstanding example of commitment to a low-price strategy is to be found with Wal Mart.

Low everyday prices produced high volumes and strong profitability. The enabling element was ingenious application of the possibilities of late-twentieth century information technology. Wal Mart's profitability growth appears to have slowed considerably as of this writing. The competition has discovered how to appeal to consumers, not exclusively by matching Wal Mart's pricing, but by offering greater variety of merchandise and at least the illusion of superior quality. Like Ford in 1920, Wal Mart appears to have neglected the income element in demand.

The Tools of Prosperity

The fundamental role of the nineteenth century's technology legacy in early twentieth-century prosperity is readily apparent in a review of new tools of industry that underpinned the prosperity. These combined to enhance the productivity of labor to the point of being able to serve the needs of an expanding population while at the same time supporting an improving life quality per capita.

The Line-Shaft

In this, the common factory power arrangement during the nineteenth century and before, power for machinery was obtained by belting individual machines to a central rotating shaft running the entire length of the factory building (see Figure 3.2).[8] At the time Eli Whitney was first making his proposals for small arms, the need for direct water power limited locations to the fall line of the rivers (where the rivers fell from the Piedmont Plateau to the coastal plain) and to navigable water. Therefore, it was not accidental that work on Whitney's contract and that of other small arms makers working for the Army occurred at places like Hartford, CT; Springfield, MA; and Harpers Ferry, VA; for these locations were all convenient to direct water power. By 1860, the combination of railroads and the stationary steam engine as line-shaft power had made industry independent of locations with waterpower potential and navigable waterways. By 1900, industry was geographically dispersed, but in the US, most was north of the Ohio River and east of the Mississippi River.

The line-shaft of 1900 had some clear disadvantages. First, it was powered by a steam engine, well under 10 percent of whose input energy was converted into useful work. Second, substantial additional energy losses occurred in the belting. Third, it was inefficient because it was seldom under full load, for it had to be "on" even when the work being done was confined to only a few of the connected machines. Also, when the line-shaft system was down for repair, the whole factory was necessarily down also. A fourth problem arose because a boiler was needed to provide steam for the shaft engine. A fifth problem arose because of the tendency for pipe joints to leak steam badly when restarted after having been completely cold. This necessitated a skeleton crew to tend the boiler even on occasions when it was desirable to shut down the plant generally.

Figure 3.2
A Line-Shaft Machine Shop, Circa 1895

Note: Courtesy Lindsay Publications, Inc.

Inventive activity in the last three decades of the nineteenth century had provided several potential alternatives to the steam engine as line-shaft power in 1900 in the form of the gasoline engine, the electric motor, and the diesel engine. The first of these had achieved significant penetration of the line-shaft market almost since its first availability after 1876. In its earliest days, it was an attractive option in situations demanding relatively small horsepower, and during the course of its improvement, it became available in larger-sized units. It could be shut down and restarted without penalty, unlike the steam engine, and it eliminated the need for the boiler. During this same time period, the electric motor was substantially improved, and while it was an ideal driver of factory machinery through a line-shaft, it required a source of electric current, placing it at a disadvantage in the choice of power units. The lack of a power distribution network began to be corrected, especially after 1900. Large industrial establishments could produce their power on site, using either steam or diesel power. In 1900, the diesel appears to have been too new to have appeared on most engineers' radar screens. However, by 1910, it had achieved a penetration into the factory power market, more often than not as the power unit for a dynamo rather than as a direct driver of a line-shaft.

It should be made clear at this point that plant designers faced with a choice among multiple power options encounter one of two basic situations: the power requirement is for a new establishment or it is for an existing one. In the former case, the design choice can be made solely based on the economic characteristics of the equipment options, there being no existing plant to constrain design. In the latter case, any proposed replacement power system must be sufficiently better than the existing system, while remaining compatible with the existing plant design in order to justify the change economically. In 1908, the gas (Otto Cycle) engine was selected for powering both blast furnace blowers and electric power generators for the Gary Steel Works, a completely new works—a "greenfield" plant in industry parlance. Electric motors were used to power rolling mills. The gas engines not only had capacities for this facility's needs, but they could use blast furnace gas as fuel, a resource previously wasted for lack of a market.[9] This example reflects plant designers' economic evaluation of the different power possibilities just prior to 1910.

As power technology developed in the opening years of the twentieth century, it was becoming increasingly evident to design engineers that the simple problem of replacing the line-shaft drive power was evolving into a much larger problem in which an increasingly feasible option was the complete abandonment of the line-shaft. Power system changeover in existing plants took place slowly and in several steps.[10] The line-shaft was first modified by cutting it into several or more segments and powering each segment independently with a gasoline or electric motor. This evolved to the modern system of having each machine powered by its own dedicated electric motor. The changeover was slow because any radical change in the design in the basic plant can be justified only if the increased profitability under the new design supports the cost of scrapping the existing system. By 1930, however, the trend toward replacing the line-shaft was obvious to all observers.[11] The steam engine builders attempted to ward off the changeover with incremental improvements such as superheated steam and uniflow cylinders, but they were working with a mature technology whose most fundamental problems remained problems.[12] Replacement of the line-shaft was an important element in the improvement of productivity, especially in the decade from 1911 to 1920. Some line-shaft plants managed to hang on into the depression of the 1930s. Many companies that managed to stay in business were doing little more than holding on, and to these, the economic attractiveness of fundamental plant redesign did not justify the expense. The labor-saving potential of redesign did not seem worth much against the background of wage levels that reflected 20 plus percent unemployment rates. Both labor and energy were cheap by the standards of the 1990s. The low cost of energy was caused by a number of very large oil discoveries, such as the East Texas Oil Field.

If the replacement of the line-shaft with a much more efficient and productive alternative seems to have been leisurely by present-day standards, it should be remembered that several factors affected the speed of the changeover. The first was the size and power characteristics of the electric motor. In 1900, which was early in the days when the use of dedicated electric motors was coming to be seen as a plant design strategy,[13] the problem was that the units were too large and bulky for the power requirements of individual machines. By 1915, there had been dramatic improvements in the power density of motors, and motors of low and fractional horsepower had become a reality. The other big holdup for the retirement of the line-shafts was the availability of cheap electric power. The power distribution grid was in its early development during this period. At first, industrial establishments requiring electric power were forced to create their own in on-site generating facilities. This meant that electrification of industry proceeded down the scale of establishment size as the availability of commercial power became increasingly widespread. The entire episode of the replacement of the line-shaft was a superb example of the adoption of a technology as quickly as several enabling technologies became available.

Electric Power Generation

Early power distribution engineering faced a choice between direct-current (DC) and alternating-current (AC) systems. The former was prominent in the earliest systems as power plants tended to be located in built-up areas near to the markets they served. Edison's pioneer systems were of this design in New York and other cities. In many cities, the power company had its origins as the power plant supplying the local street railway, from which power in excess of street railway needs was sold to the public. There were many examples of the interurban railway—cousin of the municipal street system—started by real estate developers to connect their developments with nearby town centers. In these situations, power production took place sufficiently near to the loads served that DC's severe line loss characteristic was not considered a serious problem.[14] The use of DC distribution of power did not disturb at least one prior-existing arrangement, namely, the location of certain industries in large cities. The apparel industry, for example, was centered in locations such as Manhattan, close not only to power stations, but also to low-cost immigrant labor during the years when immigration was essentially open. This did not begin to change until the immigration laws were tightened around 1920. The combination of the new immigration laws and depression conditions had the effect of reducing immigration to a trickle in the 1930s.

Alternating current systems of power distribution enjoyed the advantage of lower line loss than DC systems, and the higher the voltage, the less (proportionally) was the line loss. This advantage became decisive with the development of the Niagara Power Complex in 1895, whose power output was in excess of local needs, and which was to be transmitted several hundred miles to the New York City market. Other early opportunities to take advantage of falling water to produce power created the needs for transmission over long distances. In 1908, a project in California included the need to transmit power over a distance of 155 miles. Alternating current distribution of power enabled the development of the central power station. The transition to AC distribution enabled the dispersion of some industries away from the centers of large cities, such as New York.

In the pioneer municipal street railway systems, electric generators were driven by stationary steam engines, gas engines, and early diesel engines. None of these types were able to produce the power output and high rotational speeds necessary to produce AC at sufficient voltage, however. Once the AC system was recognized as the electric generating technology of the future, only the Parsons Turbine was ideal for the task. Moreover, the adoption of AC systems tended to force abandonment of the earlier small municipal systems because of lack of an efficient means of rectifying AC to DC current. The two systems were thus incompatible even locally.[15]

The power industry grew rapidly in the early twentieth century. In the 1900-1910 decade, a thermal generating unit with a 25,000 kilowatt capacity, with pressures up to 300 pounds per square inch at up to 500 degrees Fahrenheit was considered big. By 1930, the standard of big had grown to 208,000 kilowatts at 1200 pounds per square inch and 725 degrees Fahrenheit. By 1930, central power stations had essentially taken over the large city markets, and the efficiency of the central station was such as to bring a dramatic decline in the quantity of fuel consumption per kilowatt and the cost to the consumer. Small local power companies survived in a number of smaller cities, and typical power by 1930 was the diesel engine. By 1930, the urbanized part of the country, down to some very small places, had been electrified, but many rural areas still depended on kerosene lamps for lighting. In that year, 84.8 percent of urban and nonfarm rural residences had electric service, but only 10.4 percent of farm residences had electric service.[16]

Samuel Insull, British-born, was a pioneer in the electrification of industry and households in the United States. As a young man, Insull worked for and was a protégé of Thomas Edison. After a tour as Edison's secretary, Insull was appointed manager of a power plant in Schenectady, New York, a part of Edison's group of companies. When he lost his job in a merger in 1892, he went to Chicago, where he bought a power plant financed with a personal loan. This was one of approximately twenty plants in the Chicago area that served a total of five thousand customers in a metropolitan area with a population of over one million. At the time, electric power was an expensive luxury. In the ensuing three decades, Insull expanded this initial holding into a large utility combine serving customers in seven states. The expansion of the customer base and other measures that went along with it are credited with making electric power affordable to large numbers of small businesses and residences. A comparison illustrates this. The average cost of electric power for all urban and rural nonfarm residences was 16.2 cents per kilowatt-hour in 1902; by 1930, this had fallen to 6.03 cents.[17]

Insull achieved this by means of two basic measures. First, he exploited the potential for scale economies in power generation, driving down costs of production. Because this involved fewer and larger plants serving larger areas, he adopted alternating current, as transmission line losses could be held to reasonable levels with AC technology. The second major measure was that he recognized that different prices could be charged to different classes of customers and at different times of day. This allowed for charging higher prices during time blocks when demand tended to be high, and lower prices when demand was low. With this combination of measures, he was able to make electric service attractive to small businesses and residences.[18]

Clearly, the development of the large central generating station and relatively efficient power distribution enabled a huge variety of industrial development, and residential electrification enabled the growth of the household appliance industry.[19] Many very large industrial establishments could justify their own dedicated power stations; for example, the Gary works already mentioned. Indeed, when the Gary works was built, its designers really did not have a commercial power option. The steel industry was favorably situated to generate its own power, for an integrated mill had blast furnace gas for fuel, which otherwise most likely would have been flared as a waste product. Ford's Rouge Complex had its own power generation capacity. Commercial power continued to grow in capacity, and, in the post World War I period, as many of the earlier dedicated plant systems were old and needed replacement, commercial power became a genuine option.

The Automobile Industry

The automobile dates from about 1885 and its invention is commonly credited to Karl Benz and Gottfried Daimler; who both applied Otto cycle gasoline engines to carriages. By 1900, development of the horseless carriages had split into three branches based on three different types of power: steam, electric, and internal combustion. In that year, 1681 steam cars, 1575 electric cars, and 936 gasoline-powered vehicles were produced in the United States. The automobile then was little more than a plaything of the well-to-do. Electric cars depended on batteries and had driving range limited by battery storage capacity, but were popular because of quietness, cleanliness, and ease of starting.[20] Given that maintained roads seldom extended beyond the limits of cities, range was not viewed as a serious competitive disadvantage for electric cars. The steam car was based on small engines and flash boilers, and had the disadvantage of taking a long time to start cold. Given the "hobby" nature of the automobile, this was also not a serious drawback. Imagine the problem of one attempting to predict the future of the automobile in 1900!

By 1910, however, the automobile was beginning to be an object of mass consumption. Henry Ford sold his first Model T in 1908 for an asking price of $950. He introduced his production line in 1913, and this ranks as the major innovation in the history of the automobile. The idea of the production line goes back at least to Whitney's small arms venture of a century earlier, and in the late 1800s, the meatpacking industry had a production line of a kind as workers worked on carcasses moving on a conveyor belt. Ford took the idea to a much higher level, as his Highland Park, MI plant included multiple conveyor belts moving subassemblies to a main production line. The impact of the assembly line is evident in that by 1927, the end of the Model T production run, the cars required less than six hours to assemble, down from twelve hours before 1908, and were priced at $290.

Several very important innovation inventions supported the development of the early automobile into its present-day form and hastened the expansion of its market. These included the electric starter and the hydraulic brake system. The former appeared in 1909 and is credited to Charles F. Kettering and the Delco Corporation, which Kettering and others organized around their work on ignition systems and self-starters.[21] The starter was immediately applied to products of the Cadillac Motor Car Co., but it gradually became applied throughout the industry. This was an early example of how a new innovation is first offered to the public in a luxury model and then proliferates to less costly models. The expanded market for the feature enables larger-scale production and reduced cost to the consumer. The electric starter was an especially important innovation, for it opened the way for the automobile to become a machine of mass utility. Previously, it had been largely the toy of hobbyists, mostly men who had the physical strength and coordination to crank the engine. The impact of the electric starter is evident in the increase of auto registrations in the US from 64 thousand in 1908 to 356 thousand in 1912.[22] Hydraulic braking was developed during the 1910s, an application of the hydraulic press principle. The American Corporation was founded in 1924 by Vincent Bendix to produce automotive braking systems.

The efficiency of the internal combustion engine is related to but one parameter: compression ratio. This fact suggested that an Otto Cycle Engine should be designed to have as high a ratio as possible, but the process of achieving high compression ratio encountered a serious problem in the early days—knock. Knock occurred when the air-fuel mixture ignited prior to the piston's being at the top of its stroke and had a tendency to hasten the destruction of the engine. Much work was directed toward eliminating this problem in the 1910s, and the solution came in 1921 in the form of tetraethyl lead (TEL). This was a cheap additive to gasoline that eliminated knock even when applied in small amounts. TEL is credited to the work of Thomas Midgley, Jr. who was working in the General Motors research laboratories under Vice President of Research Charles F. Kettering.[23] TEL was the leading antiknock strategy until the early 1980s when it was phased out for environmental reasons.[24]

From the auto industry's US inception about 1885, the number of automobile producers grew rapidly to an estimated peak of 272 companies in 1909. While there were new entrants to the industry after that year, the net additions were negative, with a number of companies having been absorbed and others having failed. By 1920, the number of companies making automobiles was just over 140, but following that year there was a substantial shakeout that got well under way during the 1920s and was exacerbated by the depression; by 1941, the number of US automakers was down to 9. Concentration in the industry increased as the number of makers declined. In 1911, General Motors and Ford Motor shared 38 percent of the market. Chrysler appeared in the early 1920s as the result of a merger, and during the 1920s, Ford, GM, and Chrysler shared over 60 percent of the market in the US. This number rose to over 80 percent after 1930.[25]

The Petroleum Industry

No other growth enabler for the automobile industry was as important as the petroleum industry. The work Thomas Midgley and others was driven by a strongly perceived need to increase the efficiency of the gasoline engine. Naturally, the growth of automobile ownership in the 1910s massively increased the demand for gasoline and other petroleum-based fuels, and raised questions regarding future supply of crude oil.[26] It was such a period of fuel supply worry that spurred the antiknock research effort.

The really powerful enabler for automobile development came from the refining side of the oil industry. For most of the twentieth century, progress in refining in the US can be summarized succinctly as the getting of more gasoline out of a barrel of crude oil. The original problem was that crude oil, a natural mixture of many different hydrocarbon substances, only rarely contains the proportion of gasoline that matches the proportion of gasoline in consumption.[27] An early breakthrough on this problem occurred in the 1910s and is known as thermal cracking. In cracking, large molecules are "cracked" into smaller molecules, many of which are in the gasoline boiling range.[28] The feed to cracking units, known as gas oil (boiling in the 440 to 800 degrees Fahrenheit range), a direct product of atmospheric distillation of crude oil, was previously regarded almost as a waste product. Thermal cracking takes place at temperatures above 900 degrees, and while it achieved the job of increasing the gasoline yield,[29] it was itself a significant fuel consumer.

The economics of cracking were substantially improved in the late 1930s with the introduction of catalytic cracking ("Cat cracking") associated with the name of Eugene Houdry. Cat cracking, in which the cracking took place in the presence of a catalyst, allowed the process to proceed at considerably lower temperature than in thermal cracking. The Houdry process greatly improved the productivity of cracking and was partly a source of the supply of 100 octane aviation gasoline in World War II, which is credited with a tactically important aircraft performance edge for allied aircraft.[30] The cracking process has become even more efficient since Houdry's time with the development of modern fluid catalytic cracking.

Interestingly, thermal cracking has not disappeared, for the process survives in the process known as delayed coking. The economics of refining dictate that each processing stage use as feed only that material with which it can be most effective. This is often the residuum of the preceding process. Thus, the vacuum distillation unit takes only the residuum of the atmospheric still and achieves a further separation into various products, such as heavy gas oil, which goes to the catalytic cracking unit. The residuum from the vacuum still goes to the coker, which produces a light product that can upgraded into a gasoline-blending component and a residuum in the form of petroleum coke. The coking unit basically subjects its feed to severe thermal cracking. As will be seen in later chapters, the market for petroleum coke changed dramatically in the post-1970 era of the Clean Air Act. The problem is that a large share of the sulfur content of the original crude oil becomes concentrated in the coke and is extremely costly to remove. Today the most promising means for disposing of sulfur-laden petroleum coke is to convert it to synthesis gas in a gasifier, which is then burned in a gas turbine to produce electric power for the refinery, the sulfur having been removed.

The Electric Motor

Increasing power density and the development of the electric power distribution grid in the early 1900s went hand-in-hand to encourage the use of the electric motor. However, in 1900 the motor was still a bulky device whose markets were largely limited to industrial applications. The role of the electric motor in the replacement of the line-shaft has already been discussed, as well as the early applications to transportation in street railways in line-haul railways. The automotive and other mass-production lines would not have been possible without electric motors. Increasing power density opened the way for the beginnings of the home appliance industry, which grew in the 1920s. This growth was aided in no small part by the emergence of the consumer finance business, which itself drew impetus from the expansion in the use of the private automobile.

Innovation after 1930

Prototype Inventions, 1900-1940

The US economy of the first three decades of the twentieth century prospered on the foundation of some economically potent innovation streams that grew out of prototype inventions from before 1900. Yet, the growth of the "big four" innovation streams; automotive, electric power, telephony, and electrification of industry and homes; slowed or ceased by 1930. It was not as though there were no more prototype inventions after 1900 to generate new waves of innovation, for there were, but these collectively had far less impact than was needed to counteract the effects of maturation of the "big three," especially the automobile industry. Many of the important prototypes from this period appear in Table 3.1. The reader will note that of those listed five came after 1930. Of the five pre-1930 prototypes, two did not really have much commercial impact until the eve of World War II. These were synthetic rubber and penicillin. Television's large impact came after the war.

The reader will note that the technology-based boom that preceded the depression years got started only after a substantial delay following the emergence of the underlying technology—sometimes of multiple decades. During the 1910s and 1920s, resources of finance and talent were diverted to the exploitation of the late nineteenth-century, new technologies and away from innovating on those prototypes that appeared after 1910. While these were just beginning to achieve the economic force of their stage-two innovations, the economy experienced the depression. The post-1930 prototypes were too late to have much impact on the economy of the 1930s. Three—sulfa drugs, nylon, and radar—were catapulted into wide use only with the onset of World War II. Sulfa drugs proved literally to be a lifesaver in the war. Radar, developed independently in the US and the UK, was applied successfully to air defense in the Battle of Britain. It was also applied to naval gunfire control, and after the war was the basis for commercial and military air traffic control systems. Nylon, which found its first market as a consumer product—women's hosiery—was used for a variety of applications during the war, including parachute silk. Two inventions, the jet engine and helicopter, did not become heavily exploited during the war but had substantial impact in the postwar period.[31]

Synthetic Rubber

An elastomer is a material composed of long chain molecules having the property of returning to its original configuration after being distorted by stretching. The first elastomer in common use was natural rubber, derived from the latex of certain trees that were grown commercially in Brazil, and later in Malaya. For various reasons, there were a number of attempts to create a synthetic material that would emulate natural rubber before 1900, and while these yielded some interesting results, none succeeded in its basic object.

After 1900, the focus of these efforts fell on butadiene, and at about 1910, S. B. Lebedev of Russia succeeded in polymerizing it.[32] During World War I, a synthetic rubber industry that produced methyl rubber based on the Kondakov process, developed in Russia in 1901, was established in Germany. This was an inferior substitute for natural rubber, and once the wartime blockade was lifted, German industry returned to the use of natural rubber. In the Soviet Union, which under Stalin had embarked on a massive effort to achieve materials self-sufficiency during the 1930s, a synthetic rubber industry based on

Table 3.1
Prototype Inventions 1910-1940

Invention	Inventor	Date of Prototype	Date First Appl.	Physical Principle
Nylon	Wallace H. Carothers—Du Pont	1934	1938	Organic fiber science
Synthetic Rubber	Various	1910	1941	Organic chemistry understanding
Triode	Lee De Forrest	1907	*	Physics of electronics
Sonar	Paul Langévin	1916	1920s	Physics of ultrasound
Radar	Various, U.S. an U.K.	1930s	1938	Reflected radio waves
Television	Philo T. Farnsworth	1925	1939	Broadcast technology
Penicillin	Alexander Fleming	1928	1939	Biological
Sulfa Drugs	Various	1935	*	Biological
Jet Engine	Various	1935	1941	Thermodynamic laws
Helicopter	Igor Sikorsky	1939	*	Aerodynamic principles

Note: *Very soon after invention

the Lebedev process was set up and by 1940 was the world's largest synthetic rubber industry.

Continued research in Germany produced styrene-butadiene rubber (SBR), which proved to be an adequate substitute for natural rubber. Synthetic rubber had little commercial impact in the US in the interwar period, but it became intensely interesting with the onset of World War II, after the Japanese army occupied Malaya, previously the principal source of natural rubber. Huge amounts of SBR were produced during the war: 100,000 tons per year in Germany and the Soviet Union, and 800,000 tons per year in the US. In the postwar period, improvements in manufacturing process and materials led to synthetic rubber's becoming competitive with natural rubber from both a cost and performance standpoint. World consumption of synthetic rubber was approximately 9 million tons in 1993, with about 55 percent going into automobile tires. Other markets include mechanical parts such as engine mountings, belts, hoses, and gaskets, and consumer items such as shoe heels, furniture, and toys. Until immediately prior to the war, however, automobile tires were made from vulcanized natural rubber strengthened with carbon black.

Triode

A triode is an electronic vacuum tube with three electrodes: a cathode, an anode, and a control grid. When a small current is applied to the control grid, a larger current can pass between cathode and anode. Thus, the triode can be used to amplify a weak signal, or it can be used as an on/off switch controlled by the control grid. Its inventor, Lee de Forest, thought of it as a receiver of radio signals. The triode was subsequently adapted to radio and telephone signal amplification by Bell Labs scientists, thereby enabling long-distance telephony. The electronics industry, which grew from the vacuum tube, was largely concerned with the broadcast industry prior to the war. By 1930, radio, benefiting from effective amplifiers based on the triode, had found its way into a sufficient number of homes that its political potential was recognized.[33] It also enabled the birth of broadcast evangelism. Television had a stage-one development until 1939, when an RCA team led by Vladimir Zworykin demonstrated the first commercial-quality television system at the New York World's Fair. Radio broadcasting was a "growth" industry during the depression, a time in which there were not many such.

Sonar and Ultrasonics

A major prototype invention of the World War I years was sonar, an acronym for "sound navigation and ranging," in 1916. The inventor was, Paul Langévin, a French physicist. The stage-one innovation of sonar extended beyond the end of the war, but its strategic importance was fully appreciated by all the world's major navies. Sonar has been improved by innovation ever since its invention. It had been developed to an operational form by the eve of World War II. However, sonar's innovation stream was confined to the military and had little impact on the economy at large.

Interestingly, Sonar was among the first important applications of what has come to be known as ultrasonics, or the study and application of high-frequency sound waves, which later spawned a large number of important applications. The science underlying ultrasonics goes back into the previous century, to the work of Pierre Curie,[34] who around 1880 did extensive research on the piezoelectric characteristics of various materials. These are materials having the property of producing an electric current when vibrated, or, of vibrating when subjected to a current. Ultrasonic applications all rely on this type of material to generate sound waves, usually at frequencies in excess of 20,000 cycles per second, or well above human capability for hearing. Langévin had been a student of Curie at the Ecole de Physique et Chemie Industrielles during the time of Curie's interest in piezoelectric materials.[35]

The term "ultrasonics" brings to mind a well-known medical imaging technique, but there were a number of non-medical applications that preceded the medical. High-frequency sound waves have a low rate of attenuation in solids, especially metals, and were early adapted for the purpose of detecting fissures in steels that threaten to develop into failures. Thus, it is no surprise that the pioneers of medical ultrasound imaging should find equipment at boiler works and shipyards with which to begin their investigation; these establishments were using the technique to assess the integrity of welds. Railroads use ultrasonics to detect fissures in rails. In this application, special vehicles equipped with ultrasonic transducers move along the track, and the machinery is set up such that when a problem is detected, a spot of paint is placed on the rail to guide track maintainers to replace the faulty rail segment. Ultrasonic analysis is used to assess the integrity of pipeline welds.

One of the most common uses of ultrasonics is in cleaning metal parts of oil and oxides. This is done via the cavitation effect, in which the part to be cleaned is immersed in a liquid that is subjected to ultrasonic waves. These cause tiny bubbles that, when they break up, emit tiny shock waves to do the cleaning. The technique is used, among other things, for preparing the parts for soldering and welding. The technique is capable of cleaning oxide from aluminum such that it can be soldered. Aluminum, when exposed to air, forms a micro thin oxide that protects the metal from further oxidation; indeed, this is the characteristic that gives aluminum its normally desirable weathering traits. However, the oxide coating does not adhere to solder.

The Modern Chemical Industry

As noted in the preceding chapter, the modern chemical industry has its origins in the nineteenth century, but it became far larger and more complex in the opening decades of the twentieth century. Several examples will illustrate the nature of this expansion. The development of chemical fertilizers underwrote a major increase in agricultural production, and created a large demand for such inorganic chemicals as sulfuric acid, which today is the largest-volume chemical produced in the world. The growth of the petroleum industry created a demand for chemicals used in oil production, but it also introduced a host of large-volume raw materials derived from oil and gas processing that could be developed into various products by the chemical industry. Examples include ethylene and propylene, the bases of the petrochemical industry and the largest subdivision of the overall chemical industry. Other chemical feedstocks emanating from petroleum refining included the "BTX" (benzene, toluene, xylene) chemicals. The BTX group of chemicals can be derived from the destructive distillation of coal tar, and their availability as refining by-products underpinned a fundamental shift in the basis of the US chemical industry from coal to petroleum the early twentieth century. The European chemical industry continued its basis in coal for years after the US industry shifted to petroleum. To a great extent, this difference reflected the difference in natural resource endowments in the two regions. However, the shift to petroleum-based feeds was far from complete in the US. Modern plants that produce coke from coal for steel production are equipped with facilities for recovering usable products from by-product gases from coking, and these include BTX. Still, today, most BTX in the US comes from petroleum refineries. Chemical industry expansion was enabled materially by the development of commercial electric power.

The polymers of ethylene and propylene emerged as highly useful products in the 1930s.[36] Industrial production of polyethylene originally was based on work at ICI Chemicals starting about 1933, and low-density polyethylene (LDPE) became a commercial product in 1939. Subsequent improvements in LDPE have centered on lowering its cost by developing increasingly efficient catalysts. This work, at a number of companies, has resulted in high-density polyethylene (HDPE), a very strong and light material used in packaging such as grocery sacks.

The discovery and subsequent development of nylon in the DuPont Laboratories is a suburb example of successful invention and innovation in a corporate research and development division. Nylon, credited to a team led by Wallace H. Carothers, was first marketed as a fiber in the form of women's hosiery. The development of this market was interrupted by World War II and the activities of the War Production Board, but was expanded after the war. Nylon found a number of wartime uses such as parachute silks. GIs found that a pair of nylon hose could be traded for any of a number of items in the European Theater of Operations. Nylon also found significant markets as a molded plastic.

Medical Discoveries

The emergence of the modern pharmaceutical industry was an outgrowth of the development of the chemical industry. The science of pharmacy is very old and has its origins in the empiricism of the herbalist, but by the turn of the century, there had been solid achievement in the analysis of existing drugs and in the preparation of new ones. Table 3.2 shows some of the medical highlights of the interwar period.

Two important pharmaceutical materials appeared in the early 1930s: sulfa drugs and penicillin. The former is a popular name given to sulfonamides, which turned out to be the active ingredient in Prontosil, the first antibacterial drug marketed. Prontosil was the discovery of a team led by German chemist Gerhard Domagk—who reported the results of his work in a 1935 paper after his employer, I.G. Farben, had obtained a patent on the substance. Prontosil was a red azo (nitrogen-containing) dye that had proved to be effective against streptococcal infections in mice. With war in prospect, the allies were interested in breaking I.G. Farben's patent, and it was the effort to do so that identified sulfonamide as the active ingredient. Sulfa drugs were mass-produced in time for the war, and each individual allied soldier was issued a packet of sulfa powder with the instruction to sprinkle the substance on any open wounds. Over 5,000 sulfa drugs have been compounded, but most are no longer used because of the development of resistant strains of bacteria. The sulfa drugs have been replaced by antibiotics, of which penicillin is the earliest important example.

Alexander Fleming is credited with the discovery of penicillin. Fleming worked with his discovery but found the culture difficult to grow. The techniques for penicillin's mass production were developed by Howard Florey and Ernst Boris Chain, and it was through their work that penicillin was ready in mass quantities prior to the war. Fleming, Florey, and Chain shared in the Nobel Prize in medicine in 1945. Penicillin proved effective in the treatment of a wide variety of infections, including streptococcal infections, syphilis, and Lyme Disease.

Table 3.2
Medical Discovery before 1940

Inventor/Discoverer	Invention/Discovery	Year
Edward Mellanby	Vitamin D and its relation to rickets	1921
	First vaccine for diphtheria	1923
	First vaccine for pertussis	1926
	First vaccine for tuberculosis	1927
	First vaccine for tetanus	1927
Alexander Fleming	Penicillin	1928
	First vaccine for yellow fever	1935

Growing Use of the Diesel Engine

One innovation stream from the nineteenth-century legacy that was beginning to have wide impact during the 1920s and 1930s was the diesel engine. The diesel had realized initial success in the stationary engine market for power for electric generators and as a marine engine. By the onset of World War I, the diesel had been developed as a marine power plant, in which use its fuel economy and use of low-volatile fuel made it especially desirable. The diesel proved especially advantageous as power in submarines for surface cruising and battery recharging power.[37] After 1920, it was made into a preferred power plant for highway trucks, partly through the efforts of Clessie Cummins at Cummins Engine and Charles Kettering at General Motors. This development could progress only as rapidly as the country's net of improved highways developed. As indicated above, the diesel was adapted for railroad freight power, a market which it captured completely after the war.

Diesel power for the railways was enabled by some critical technology that was developed around 1920. When designing a straight electrification system, the designer customarily assumed an infinite supply of power in the trolley wire. When the available power was confined to the output of an on-board engine-generator set, as with a diesel-electric, the designer had to recognize that power was limited. The engineer would have to control the speed of the prime mover, that of the traction motors, and a number of other functions, such as braking, to the point that the task appeared beyond the capabilities of an engineer with only two hands. The solution to the problem came in the form of an automated system where the speed of the prime mover would respond to the engineer's setting of the traction motor speed.[38]

Notes

1. U.S. Dept. of Commerce, Bureau of the Census, *Power and Machinery Employed in Manufacturing,* (Washington, DC: U.S.G.P.O.; Reprinted by Lindsay Publishing Co., 1885), various chapters.
2. The universality of market forces in this selection has come under challenge by recent scholarship. See Chapter 14 below.
3. Smith, Adam, *An Inquiry into The Nature and Causes of the Wealth of Nations,* (New York: Random House, Modern Library ed., 1937), p. 3f.
4. For example, in the early 1950s, the General Motors assembly plant at Arlington, TX, produced Buicks, Oldsmobiles, and Pontiacs on a common line. A number of options involving such features as engines, body color, body styles, and other things were possible within each marquee.
5. This is an example of the kind of experiment that working business organizations have to make in the course of making their pricing decisions, often in the face of rapidly changing demand conditions.
6. Leland, like Ford, had grown up on a farm and was a self-educated mechanical engineer. He founded the company that became the Cadillac Division of General Motors, and later headed the Lincoln Motor Co., which built liberty engines for

aircraft during World War I. When the war was over, he took his company into the auto business and thus became the founder of both of the main American luxury brands of the post- World War II period.

7. U.S. Dept. of Commerce, *Historical Statistics of the United States,* (U.S.G.P.O., 1970), p. 668.

8. The universality of the line-shaft in industry before 1910 is illustrated by the existence of a variation of it in the oil patch. In this, pumps at the individual wells were connected by steel wire rope to a centrally located steam-powered draw works. One outstanding example of this arrangement was the Electra field, which took its name from the nearby town of Electra, Texas. The cable operation lasted until about 1950.

9. United States Steel Corp., *The Making, Shaping, and Treating of Steel,* (Eighth ed., Harold E. McGannon, ed., 1964), p. 590.

10. In 1923, the author's father was a teaching assistant in the mathematics dept. of a small liberal arts college in North Texas. A group of students had approached the faculty expressing interest in a course in engineering thermodynamics. This seemed to confound the faculty because the school had nothing resembling laboratory facilities needed to support such a course. My father was offered this challenge and his solution, which seems to have worked well, was to contact local industries and arrange a program of field trips. This was a farming community, with a number of farm-related industries, such as cotton gins, compresses, grain elevators, etc., and all of these had power requirements that they generally met with stationary steam engines. There turned out to be more than enough examples to make a plausible course. Some of these engines were still there when the author knew that neighborhood in the late 1940s.

11. Line-shaft shops were notoriously dangerous places to work because of all the belting. Complete replacement of line-shafts was spurred after 1911 by a desire to reduce workman compensation claims under state laws passed starting in that year and that became almost universal by 1921.

12. Because of physical limitations on the length of a line-shaft imposed by characteristics of the steels used in the shaft itself, factories could be only so long, and shaft length beyond the physical limit had to be on a different floor, with both shafts belted to the a common drive engine. In such a system there was much lifting of work in process from one floor to another. This source of inefficiency could be eliminated with the abandonment of the line-shaft power system, and its demise enabled a revolution in factory architecture, away from multi-story buildings to single-story designs.

13. An early technical note illustrates the view of dedicated electric motors in 1900. The note contrasted the US practice of using motors of five horsepower minimum for driving machine tools with the German practice of using smaller motors even though the smaller motors were considered less efficient. *Scientific American Supplement,* No. 1293, (Oct. 13, 1900), p. 20728.

14. The close relationship between the early power companies and street railways is illustrated in Martin, Jean, "From Mule to MARTA," *The Atlanta Historical Bulletin,* Vol. XX, No. 4., (Winter 1976), 1-13.

15. Direct current was able to hold onto street railway, interurban railway, and some steam railroad applications; some of these systems survived the depression and lasted into the 1950s and 1960s.

16. U.S. Department of Commerce, *Historical Statistics of the United States,* (Washington, DC: U.S.G.P.O., 1973), p. 827.

17. U.S. Dept. of Commerce, *Historical Statistics of the United States*, p. 827.

18. Insull's companies were controlled by means of the holding company arrangement, which allowed him to build up a large group of controlled companies quickly. However, this expansion was financed heavily by debt, and much of the equity had been pledged as collateral. When the depression came, the collapse of stock prices caused Insull to lose control of his companies, and he himself ended up with a huge negative net worth. Equity holders were wiped out. The political reaction was that Insull was indicted for fraud, but he was acquitted at trial. He died a broken man several years afterward.

19. The demand for household electrical appliances appears to have begun in earnest about 1912, enabled by a general reduction in electric power rates in that year and the development of small, fractional horsepower motors.

20. Ferdinand Porsche, a name not normally associated with electric automobiles, was an early pioneer of this power form. One of his electric designs took the *Grand Prix* at the Paris Exposition of 1900. Porsche was fully aware of the electric's range problem, and to solve this, he designed a car that used an internal-combustion engine to power a generator that charged a battery pack whose power was used in the hub motors of his earlier all-electric designs. He, thus, anticipated today's hybrids by almost a full century.

21. Leslie, Stuart W., *Boss Kettering*, (New York: Columbia University Press, 1983), p. 38f.

22. U.S. Dept. of Commerce, Bureau of the Census, *Historical Statistics of the United States.* (Washington, DC: U.S. Government Printing Office, 1975), p 1718.

23. Kettering's Delco Corp. was acquired by General Motors on the last day of 1918.

24. TEL was found to kill the catalyst in catalytic converters whose function was to eliminate some of the more toxic products of gasoline combustion. It was also a direct polluter because of the lead compounds that went out the tailpipe into the atmosphere, but this was a relatively minor pollutant when compared with the other products of gasoline combustion.

25. The numbers cited in this paragraph are from Smith, Phillip H., *Wheels Within Wheels,* (New York: Funk and Wagnalls, 1968), as quoted in Klepper, Steven, *The Evolution of the U.S. Automobile Industry and Detroit as its Capital,* Pittsburgh: Carnegie Mellon University, (mimeo), pp. 3-5. The Smith estimates have eliminated all but firms that actually produced cars, having excluded firms that were incorporated but never actually produced cars.

26. This questioning has been a recurring theme in American history. In 1930, it appeared that the country had ten years' supply at current rates of consumption, but massive new discoveries in the next several years put these worries to rest. There was a notable such period of running-out-of-oil worry in the 1970s, and we are experiencing one now.

27. Gasoline is itself a mixture of many substances. The definition is in terms of a range of boiling points: gasolines boil at between 90 degrees and 210 degrees Fahrenheit.

28. "Pyrolysis" is a more general term than "cracking," which seems peculiar to the petroleum refining industry. Both terms mean the subjecting of large molecules to high heat in the absence of oxygen such that the molecules break down into smaller molecules.

29. It raised the gasoline percentage to 25 percent within a few years after its application in a Standard of Indiana refinery in 1913.

30. The antiknock qualities of gasoline are proportional to the octane rating. On this scale, isooctane, which has very desirable antiknock characteristics, is assigned a

value of one hundred, and normal heptane, which has very undesirable antiknock characteristics, has a rating of zero. The octane characteristics of any other fuel can be established with laboratory equipment on the octane scale.

31. The jet engine was enabled, among other things, by the availability of austenitic stainless steels from which the engine's turbine blades could be made.

32. Polymerization is a process that combines small molecules into larger molecules that consist of many of the smaller molecules connected by chemical bonds.

33. President Franklin D. Roosevelt made famous use of radio broadcasting with his "fireside chat" broadcasts during the depression. Before him, his opponent for the 1932 Democratic presidential nomination, Governor Al Smith of New York had campaigned on the radio.

34. Pierre Curie is best remembered as the co-discoverer of radium with his wife, Marie Curie.

35. Paul Langévin had a particularly distinguished career. He was a contemporary of Albert Einstein and became an early acceptor of Einstein's equivalence principle. Einstein later wrote of Langévin that he had the mental wherewithal to understand the general relativity theory, and that had he himself not proposed it when he did, Langévin would have.

36. Ethylene and propylene are examples of a family of organic chemicals known as olefins. An olefin is created when two hydrogen atoms are stripped from a saturated hydrocarbon and replaced with a double bond between two of the carbon atoms.

37. The U-boats were diesel powered. The operational success of these influenced design in all navies of the industrialized world.

38. The circumstances under which the development of the technology for controlling a limited power source was the interest of a number of railroad companies in a railcar, using a reduced-size crew, that could replace a conventional steam-drawn train on lightly-trafficked branch lines, especially in the Midwest. The earliest of the railcars was powered by a small steam engine, but the control problem was solved in the course of adapting the gasoline engine and the diesel to railcar service. See Hamley, David H. and Corley, Raymond F., "How to Control an Engine of Limited Power," *Trains*, Vol. 34, No. 3, (January 1974), 26-28.

4

Technology and Depression

As the economies of Europe and the United States developed in the nineteenth century, it became obvious to students of economics that economic progress was anything but uniform; it was interrupted by setbacks sufficiently often that they could not be brushed off as infrequent anomalies. Study of the matter by statistical pioneers suggested a long cycle, consisting of economic boom, crisis, downswing, and depression of about fifty years. Moreover, the long cycles were divided up into two or more shorter cycles. The long cycles also tended to be triggered by major introductions of new technology. For example, in the US, one major cycle coincided with the early building of railroads following the recession of 1837 and ended in the depression that followed the year 1873. The long cycle of interest in these pages started following the severe recession of 1896 and the upswing was massively sustained by investment activity growing from the nineteenth century technological legacy. It collapsed into the Great Depression of the 1930s. When an invention emerges from its stage one innovation there can be a substantial return to investments that apply it, especially if the invention has prototype potential. The technology that it embodies has a lifecycle in that returns to investments in its application tend to be high when it is novel and to diminish as the underlying technology becomes widely adopted and commonplace. Consequently, investment activity itself will eventually diminish as the possibilities of the driving technology become fully exploited.

Investment Failure and the Depression

The economics profession largely failed to anticipate the Great Depression of the 1930s and argued over its causes for several decades following the event. There were exceptions to the failure to anticipate the Depression publicly. In the 1920s, Ludwig von Mises and F. A. Hayek anticipated trouble, basing their arguments on an anticipated investment failure, but they did not venture a timetable. Their apprehensions were based on what has become known as the Austrian School capital theory, and their pessimistic prognostications came from their perception of an unsustainable imbalance between investment and consumption in the second half of the 1920s. As a result of this call, the view that the Depression was the result of an investment failure was widely accepted for a number of years. This view has been largely superseded from approximately the time of the publication of the *Monetary History of the United States* by Milton Friedman and Anna Schwartz in 1963. This work defended the hypothesis that the Depression was the result of a series of misguided monetary policy actions in the early 1930s. It convincingly demonstrated that monetary policy of the early 1930's was indeed ill inspired, and the monetary explanation of the Depression in combination with an investment failure constitutes a highly plausible total explanation for the disaster. For various reasons, however, the Monetarist School has seemingly rejected the notion that an investment failure was even involved. In looking over the period, this seems strange.

As of the early 1960s, the accepted causes of the Depression had taken the form of a short list of preconditions that individually might have produced nothing more than a recession, and a number of fiscal and monetary policy steps that succeeded in turning recession into disaster. The preconditions included deteriorating incomes in the agricultural sector, brought on by a downward trend in agricultural commodity prices during the 1920s, international financial instability, and the end of the boom in construction activity, both residential and nonresidential, which had begun in 1921.[1] The failure of the nonresidential construction boom is of special interest because of what it implies for business investment in general.

How the construction boom ended is highly important. To quote Professor J.R. Hicks, "There are thus at least two quite different ways in which a general boom can be brought to an end: it may be killed by credit restriction or it may die by working itself out."[2] Hicks went on to explain that in the former case, the supply of investment opportunities upon which the boom had been feeding might not be exhausted, and there may remain investment opportunities that can help a recovery once another cause of the setback, such as a credit crisis, is resolved. However, if the boom has ended because of exhaustion of the fund of investment opportunities from what this book terms stage-two innovation streams, then it can end in a severe recession, for recovery has to await the development of a fresh stream of innovation. While there were prototype inventions between 1910 and 1930, they did not result in innovation streams and resultant investment sufficient to offset the maturing of the new industries enabled by the nineteenth-century legacy and that grew after 1900.

Much of the extraordinary investment activity of the first two decades of the twentieth century was induced by second-stage innovation streams based on inventions that completed first-stage innovation after 1880. This investment activity induced a large and growing demand for labor to work the resulting capital plant and construction to house the new industries. Three of the major innovation streams—the automobile industry, the electric power industry, and the electrification of industry and households—underwent high growth through the 1910s and into the 1920s, but by 1930 all had reached a plateau of maturity. Thus, the end of the construction boom is consistent with exhaustion of the fund of investment opportunities that grew out of the new technology available at the turn of the century.

The history shown in Figure 4.1 reveals what will surprise no one: that automobile factory sales responded to depression conditions in the 1930s. The more interesting point is what the history prior to 1930 as this chart depicts says to the question of whether or not the pattern of development of the automobile industry was an explicit contributor to the Depression. The striking thing about Figure 4.1 is the near cessation of growth in automobile factory sales after 1923. Average annual growth in sales from 1900 to 1923 was approximately 34 percent; but after that year to the end of the decade, growth was essentially zero if one ignores the spike of 1929.[3] Table 4.1 summarizes the growth of the electric power generating industry from 1912 to 1940. The outstanding aspect of this record is the rapid expansion of capacity up to 1930, followed by a dramatic reduction in growth rate. The industrial component of total generating capacity, which reflects the largest of industrial establishments, expanded at a much lower rate than did the utility capacity. In the very early part of the century, the power distribution grid was just beginning to be developed, and use of commercial power was not an option in many locations. The largest establishments were forced to produce their own power. As commercial power increasingly became available, however, it became available to industrial establishments of increasingly larger size. Thus, the comparatively low growth in the industrial power capacity reflects that the industrial sector's growth was increasingly supported by the expanding utility power grid.

Consider what the maturing of the automobile industry meant for the industries that supply automaking, such as hydraulic systems (brakes), tires, steel, glass, and the capital goods industries of industrial machinery and industrial construction. Prior to 1923, these had grown their capacities to accommodate the growth of the auto industry, but after that year their orders would not have reflected what had been the growth component of the automobile business. Figure 4.2 shows that industrial construction did not level out after 1923 as did auto sales, but grew through 1929. This suggests the growth of an imbalance between consumer demand for autos and investment in the facilities for making them. This realization that rapid growth in auto sales was over sufficiently to halt investment in construction and capital equipment related to auto production would not have been immediate. Expansion of automobile sales from the turn of the century to 1923 had not been uniform, as there had been reversals in growth around the time of World War I and in the sharp recession of 1921. These setbacks had proved transitory, and therefore the pause in automobile sales growth after 1923 would not immediately have set off alarms affecting capital expenditure decisions. There is a well-documented tendency among business managers to assume that because things have been going well recently,

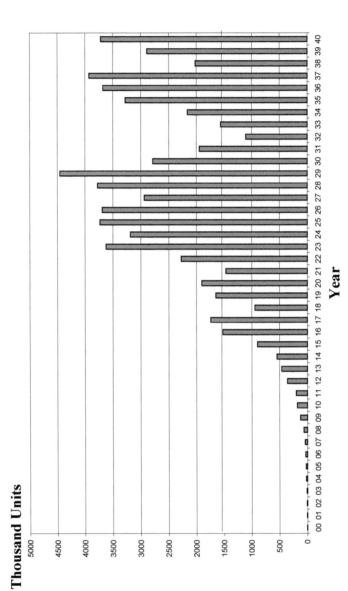

Figure 4.1
Automobile Factory Sales, 1900-1940

Thousand Units

Year

Source: Historical Statistics of the United States.

Table 4.1
Installed Electric Generating Capacity

Average Annual Percent Change

	Year		
	1912-1921	**1921-1930**	**1930-1940**
Total Capacity	7.24	7.99	2.16
Electric Utilities	11.28	10.19	2.12
Industrial Establishments	2.22	2.39	2.33

Source: US Dept of Commerce, Historical Statistics of the United States.

they will shortly resume doing so. This shows up in the form of a "business as usual" attitude and would partly explain why there was an industrial construction revival in the late 1920s. Consider that by 1923, senior business managements in automotive-related industries had had close to twenty-five years' experience whose lesson was that their business was growing and required an expansive capital investment policy. Only the persistence of the major downturn that developed after 1929 would have been sufficient to convince these people that the prosperity was gone and unlikely to return for a while. Thus, the business optimism that was characteristic of the mid-1920s may have kept the investment growing well after growth in the downstream consumer business had subsided, creating an imbalance between investment and consumption in the late 1920s.

When did the investment activity induced by the stage-two innovation streams that grew out of the nineteenth-century legacy of invention have their maximum impact on the economy in terms of the rate of productivity increase? The record of changes in overall industrial productivity hints at the answer. Of course, investment activity and the resultant incremental upgrading of the stock of productive capital in most times has a tendency to improve general productivity. However, the industrialization of the early twentieth century was built on the possibilities created by the rich technology endowment of the late nineteenth century. What is of interest here is when (in broad terms) did the results of exploiting the rich possibilities of this endowment coalesce to the point of raising *overall* productivity growth above the background level.

Figure 4.3 presents the productivity record for the period 1911 through 1932. The pattern that is most apparent here is that the largest annual productivity increases, on the average, occurred in the 1910s, rather than the 1920s. The 1911 through 1921 average compounded annual productivity growth rate was 2.32 percent in spite of setbacks in 1913, 1916, and 1919; as compared with 1.77 percent for 1921 through 1929. The reader will remember that the 1910s saw

Figure 4.2
Industrial Construction Put in Place, 1915-1940

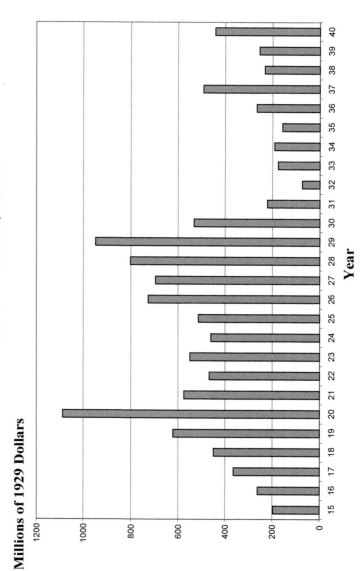

Source: Historical Statistics of the United States

Figure 4.3
Productivity Growth: US Real GNP per Man Hour, 1911-1932

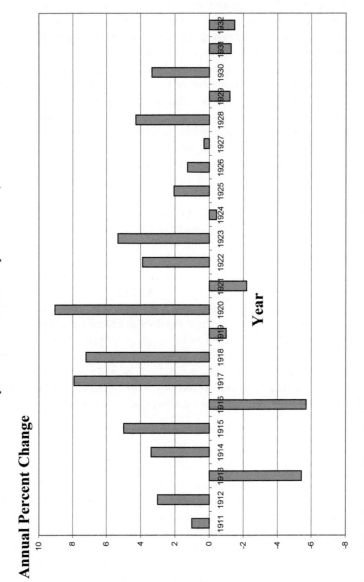

Source: Historical Statistics of the United States

such developments as the establishment of Ford's Model T production line, a technique of production organization that was emulated in other industries.

What accounts for the large year-to-year increases in productivity that took place in the 1910s? An obvious suspect is the automobile industry itself on the grounds that it rapidly grew into the country's largest industry. While productivity statistics for the auto industry itself during the 1910s is lacking, one can gain a hint of its influence with a comparison. In 1921 through 1929, when motor vehicles industry data were available, output per worker-hour expanded at an average annual rate of 7.3 percent in the motor vehicles industry; this compares with 4.4 percent for all of manufacturing in the same period.[4] This comparison hints that the capital investment in the manufacturing sector had its principal direct impact on labor productivity in the auto industry. The electrification of industry had the initial effect of increasing the efficiency of factories with respect to energy consumption and in the handling of goods in process. This was the time when the traditional line-shaft factory was being replaced by electrified plants, and the factories built for the new industries were designed from ground up around the possibilities of electrification. Ford's Highland Park plant would not have been possible without electric motors to drive its highly coordinated production lines. At the same time, the rapid expansion of the automobile industry called for expansion of industries supplying automobile production, and this expansion permitted a number of industries besides auto manufacturing itself to achieve large-scale production economies. As Henry Ford came to realize, the automobile is an *assembled*, rather than a manufactured product, and the parts and subassemblies that were brought together in the auto assembly plant were often manufactured off-site. These included electrical systems, hydraulic braking systems, glass, paint, and many other items and assemblies, all of which achieved increased labor productivity with expansion. The general productivity growth slowdown following 1930 reflects the slowdown of the central industry of the period, automobiles, and the induced slowdown of the supplying industries.

Investment and the Depression

Indeed, it was in the 1910s that large-scale production for a mass market really made its appearance. By the 1920s, improvements to production tooling were still occurring, but as increments on top of the fundamental productivity gains that had been achieved in the 1910s. Why is this interesting? It is because all schools of economics at the time and since, Austrian, Chicago, and Keynesian, appear to have concentrated on events in the 1920s in their explanations of the Depression. The investment failure of the 1930s, which was noted widely among the various schools of thought, provoked much discussion as to its origin. The investment levels that characterized the post-1925 years do not appear to have been justified by the consumer demand growth that actually took place, and this imbalance between consumption and investment has drawn much attention from those attempting to explain the Depression. To the Austrian

School, this condition suggested some form of the overinvestment explanation of the bust following a period of high prosperity. By contrast, the Monetarist School concentrated on a purely monetary explanation for the debacle to the point of denying that any imbalance between investment and consumption had even occurred.

It was the Austrians who explained the onset of the Depression as a classic failure of capital investment following an artificially induced investment boom.[5] In their view, the imbalance between investment and consumption that seems to have grown up in the late 1920s was the product of attempts by the Federal Reserve to reduce interest rates. This effort succeeded in reducing rates below their "natural" level. The concept of the natural rate of interest comes from Swedish economist Knut Wicksell, and is defined as that rate of interest at which "the demand for loan capital just equals the supply of savings."[6] In the Austrian model, interest rates below the "natural" level would induce a boom in capital investment, and therefore explain the imbalance between investment and consumption.

Wicksell argued that if the market rate is below the natural rate, general prices will rise, and if it is above the natural rate, prices will tend to fall. [7] The Monetarists have taken the Austrians to task over this, for the Austrian model of the time appeared to call for the imbalance between investment and consumption to have been induced by measures that would also have brought on a period of general inflation. The Chicago School has always emphasized the testing of all theory against the available data, and rejection of all hypotheses that do not pass muster on the empirical test. These scholars have made a solid case that the 1925 through 1929 period was *not* characterized by any general inflation; indeed, if anything, the tendency of general price levels was *deflationary* during this period. The absence of evidence of inflation in the second half of the 1920s[8] has been a conspicuous element in the Chicago-Austrian argument ever since.

Could the apparent discrepancy between consumption and investment have developed in the late 1920s *without* the appearance of general price inflation? After all, the Monetarists' focus on the absence of inflation in making their case rested on the statistical record of wholesale prices from the period, and that record only measures the *net* impact of all the influences bearing on prices at the time. There were, in fact, several powerful influences whose impacts were to reduce prices, especially prices of goods. First, the heavy capitalization that characterized the automobile and other mass production industries resulted in falling costs of production for the consumer goods produced, and these would have been reflected in declining prices.[9] Second, the general prices of agricultural commodities had begun a decline after the First World War,[10] and there was no uptrend in the 1920s. Together, these two trends could have offset any tendencies for broadly defined indexes of prices to rise due to monetary measures in the 1920s. One counter to this argument is that while the prices of *consumer* goods may have tended to fall, those of the factors of production would have tended to rise under pressure of demand for labor to work in the new factories. Any tendency in this direction would have been offset, however, by a large supply of new workers from immigration. Up until about 1920, immigration was essentially open. While many of the new immigrants were unskilled, it was the characteristic of mass-production methods to use unskilled workers who could be trained in the minimal skills required in a short period of time. Indeed, one of the bases commonly cited in explaining the success of industrial unionization in the 1930s was the tightening of the immigration laws in the early 1920s. The Austrians point to the Federal Reserve's attempts, especially in 1924 and 1927, to lower US interest rates in an attempt to assist Great Britain in her effort to resume and maintain the gold standard. In their view, moves such as this would have contributed to the development of an investment-consumption imbalance in the mid- to late 1920s.

In their explanation of the Great Depression, members of the Chicago School placed heavy emphasis on the FRB's pursuit of monetary policies that had deleterious consequences for the US banking system in the late 1920s and early 1930s. Part of the reason for such policies was the FRB's bemusement with policy goals in the international sphere where the policy prescription was in conflict with the needs of the domestic economy. At any rate, a low interest rate relative to the natural rate would be a theoretical argument explaining investment performance of the late 1920s. It was the FRB's bemusement with international financial instability that contributed to its policies that resulted the approximately 33 percent decline in the US money supply between 1930 and 1933, including the failure of a large percentage of the banking system.[11] It was this decline in the money stock on which the Monetarist School has concentrated for its understanding of what caused the Depression instead of a mere recession. The Monetarists' points are well taken, for the adverse monetary policy of the early 1930s certainly would have had a severely deleterious effect on the economy even without a failure of investment. It seems strange that the double punch of unfortunate monetary policy and investment failure is not more universally accepted as the explanation for the Depression. Yet, the Monetarists have almost to a scholar rejected the notion that there was anything remiss about the relationship between capital investment and consumer demand in the late 1920s. The unanswerable question is what sort of setback would the economy have experienced had monetary policy *not* been a source of damage? It appears as though there could still have been at least a very severe recession.

The investment failure itself deserves more attention based on the rapid growth of the auto-related industries prior to 1923 and the productivity record of the 1910s. The scholarly focus has been heavily on the decade of the 1920s, but the cyclical dynamic of the post-1929 failure was forming in the 1910s and before. Growth rates achieved by the automobile industry and its supporting industries prior to 1923 were unsustainable and had to come to an end at some time. Set against the background of the extremely high rates of growth attained by some industries in the first two decades of the twentieth century, including automobiles and supporting industries, the general investment failure following 1929 suggests the accelerator principle. According to the this principle, a change in the rate of growth in the production of a consumer good will induce a proportionately much larger change in the growth of production of capital goods used in its production. If demand for a durable consumer good, say the automobile, is increasing, then the producing industries have to be able to produce not only to offset the wear and tear on the existing fleet of autos, but also to be able to support the demand growth. When demand for the durable consumer good ceases to grow, demand for the supporting capital plant tends to fall absolutely sooner or later. The accelerator principle is not itself an independent explanation of the business cycle, but it fits comfortably with the overinvestment theory advocated by the Austrian School.[12]

How does the acceleration principle work at the actual decision level? Producers of a consumer good that recently have been facing an increase in demand encounter two problems. The first is foreseeing how big the future increase in demand is. The second is deciding how much capital expansion is needed to accommodate the anticipated increase in demand. It would be very surprising if the typical answer to the first problem in the 1910s and 1920s were not tinged with optimism. As already noted, this kind of decision is not made in temporal isolation, but is made very much in the context of preceding experience, which in the early 1920s consisted of years of predominant annual increases, interrupted infrequently by downturns that proved to be transitory. In this atmosphere, it is not surprising that the post-1923 maturation of the automobile market would not have immediately been recognized as such. Hindsight reveals that continued investment by auto and related industries served to create a capital plant that was increasingly in excess of anything justifiable by consumer demand. The makers of industrial investment decisions, of course, were not beneficiaries of hindsight. Their awakening to the realities of market demand had to wait for several years of disappointed market growth expectations.[13] The essential point revolves around the length of time it takes for this kind of recognition to permeate the thinking of those making the investment decisions. The longer this takes, the greater will be the accumulation of capital plant that is surplus to need in the ensuing downturn. To the extent that growth optimism survived into the late 1920s, the spike in auto factory sales (see Figure 4.1) must have been seen widely as vindication of optimism, wrongly, as it turned out.

The usual simple illustration of the accelerator principle, both in past literature and in recent textbooks, does not recognize the possibility of a delay between the onset of unfavorable market fundamentals and the perception thereof. What is being suggested here is such a delay, which implies that at some point there developed a disconnect between the common valuations of assets and fundamental market conditions that might justify such valuations.

This suggests a bubble. Economic theory in the twentieth century emphasized the rationality of economic actors, and as a result, shed little or no light on the topic of bubbles. The definition of a bubble as a disconnect between perceived asset valuations and market fundamentals describes the bubble phenomenon, but offers little help towards the problem of recognizing the onset of bubble conditions in markets.[14] Indeed, one of the possible preconditions for the formation of bubble conditions is the combination of a slowdown in the rate of expansion of a consumer market (an accelerator condition) and a delay in perceiving that the slowdown is permanent. The academic study of bubbles has attracted much recent interest following the experience with valuations of technology stocks in the late 1990s and with housing values in 2006 and 2007.

That the acceleration principle was not entirely neglected in explaining the Depression is made clear in the following quotation from Professor Hicks.

> I do not see that there is any adequate reason to suppose that the *real* boom of 1927-9 was at all an exceptional boom; if the accelerator mechanism, and nothing else, had been at work, it should not have been followed by an exceptional slump. But the slump impinged upon a monetary situation which was quite exceptionally unstable. The main cause of this instability was not the purely superficial speculative over-expansion which had occurred in New York in 1928-9; its roots went much farther back. The monetary system of the world had never adjusted itself at all fully to the change in the level of money incomes which took place during and after the war of 1914-18; it was trying to manage with a gold supply, which was in terms of wage-units extremely inadequate. Difficulties in the post-war adjustments of exchange rates (combined with the vast changes which the war had produced in the creditor-debtor position of important countries) had caused the consequential weakness to be particularly concentrated in certain places; particular central banks, as for instance the Bank of England and the Reichsbank, were therefore particularly incapable of performing their usual function as "lenders of last resort" . . . [15]

It is suggested here that Professor Hicks' failure to find the origin of a major acceleration dynamic after 1927 can be explained by the absence of such events in this time period. The origins of the really serious acceleration effects that worked after 1929 are to be found prior to 1923, especially in the 1910s.

If an investment failure had been a central factor in bringing on a severe recession in 1930, why was investment revival so delayed as to prolong the recession into the debacle that actually occurred? The above-cited Hicksian distinction between a setback due to exhaustion of the investment opportunities on which the previous boom had been feeding and one wherein unexploited investment opportunities remained points to the answer to this question. In the normal course of events, a firm's investment is tied positively to its decision regarding desired productive capacity. Presumably, any investment increment results in at least a small operating-cost improvement in comparison with the capital that it replaces. However, under depression conditions, the capacity decision in a large segment of industry is not likely to be for expansion of any kind; it is more likely to be for the opposite goal, namely, a *decrease* in productive capacity. The only way an investment could have produced a positive business result in 1930-1933 would have been in response to an investment opportunity of sufficient cost-reduction potential as to offset the effect of decline in market demand for products of firms. If, as suggested above, the lines of investment opportunity that had fueled industrial expansion of the preceding decades were exhausted, then no investment strategy could have made much sense in the early 1930s. As pointed out in the previous chapter, inventive activity produced some results that eventually turned out to have prototype magnitude, but the innovation of these was not sufficiently developed to be able to underpin investment-induced growth in the 1930s.[16]

The Nature of Capital Formation

It is of interest to note that two kinds of capital formation were taking place in the first three decades of the twentieth century. The first was the type that went directly into the new mass-production facilities. This kind of machinery was not generally applicable in situations other than the highly-coordinated production lines and did not have much use outside of that context. When the failure of consumer demand came, it was this specialized capital that became underemployed. Inasmuch as a substantial amount of the capital formed in the 1910s and the 1920s was specialized to the mass-production techniques that were adopted, as in the automobile industry, the economy was especially vulnerable to the maturation of the consumer markets to which the mass-production industries sold, especially automobiles. The second kind of capital was that exemplified by the capital involved in electrifying industry, commerce, and households. This was based on such technologies as the electric motor, electric lighting, hydraulic control systems, and others, and these were not industrially specialized. It was the capital involved in the replacement of line-shafts, among other things. It was an apparently non-specialized form of capital in that it could be retrofitted in a wide variety of preexisting factories with the effect of increasing productivity.

The Austrian capital theory, as elucidated by F.A. Hayek and Ludwig von Mises, held that once capital was committed to specific production processes, it became inherently specialized. This view appears to fit the preceding observations regarding capital committed to automotive production lines. It attained a great deal of acceptance in the early 1930s, having gained force from Hayek's and Mises' having anticipated the Depression (as to its nature, not timing). Moreover, much of the apparently non-specialized capital that was formed was in industries whose prosperity was tied in one way or another to the automobile industry; and it was, therefore, as subject to the dynamics of maturing consumer demand for autos as the highly specialized capital that composed the assembly plants themselves. As the decade of the 1930s wore on, however, there developed a curious academic controversy between Hayek and Frank Knight. Knight, for various reasons, rejected the Austrian view of capital and instead treated it as a homogeneous stock. The implication of this view is that capital can be applied across a wide variety of industry with positive impact on productivity regardless of where or how applied. The living example of that period was the electric motor and commercial power. Knight's view was especially influential among the Chicago economists, including Milton Friedman.[17] In looking back at the actual patterns of capital investment of the time, it seems as though neither side of this controversy was looking at the totality of the actual experience.

Technology and Industry Structure

One of the implications of the foregoing discussion is that the overall US economic growth was not balanced in the sense that it disproportionately rested on rapid growth in the manufacturing sector, and within manufacturing, on the growth of the automobile market. Indeed, the extremely rapid growth of that industry, especially before 1923, is the point conveyed by Figure 4.1. It was this unbalanced quality of growth that made the economy vulnerable to a failure of investment in what had been its most rapidly growing component. The lack of industrial balance in US growth is likely to have influenced the configuration that the growth industries assumed, namely, a vertically integrated structure.

Table 4.2
National Income

Shares Originating by Major Industry Group

Industry Group	Year	1929*	1900**
		Percent	
Increased Share			
Manufacturing		22.2	18.6
Transport, Communications, and Public Utilities		11.2	10.3
Finance, Insurance, Real Estate		14.3	12.7
Government		8.7	6.0
Unchanged Share			
Services		10.4	10.3
Mining		3	2.9
Contract Construction		4.2	4.3
Decreased Share			
Agriculture		11.5	18.2
Trade		14.5	16.6

Notes: *Estimated 1926-1929 average
        **Estimated 1899-1903 average
Source: Historical Statistics of the United States, p.240

In a balanced growth situation, one would expect for national income shares originating in a wide spectrum of industries to remain about the same over the period studied. Table 4.2 makes this comparison for major industry groups for the years 1900 and 1929, and it is clear that there was a lack of balance. Of the four groups that increased share, three (manufacturing, public utilities, and finance buoyed by consumer finance) are known growth groups from earlier discussion. Agriculture and trade both lost share. Within the manufacturing group, production workers in the "motor vehicles and motor vehicle equipment" subgroup accounted for 5.3 percent of the total of manufacturing production workers, as compared with .9 percent in 1910.[18] As rapid as this expansion was, it fails to impart the full impact of automobile-related activities on the total economy. The auto industry expansion induced rapid growth in basic steel, raw materials transportation, automobile marketing, highway construction, and many other activities.

One characteristic that appears to have been shared by many, if not all, of the large industries spawned in the early twentieth century on the basis of the late-nineteenth-century technology is vertical integration of some sort. The details vary. For example, both basic steel and automobiles tended to integrate upstream: they sought to take supply functions in-house. Aluminum, by contrast, integrated both upstream and downstream. Nonferrous metals other than aluminum tended to be integrated from mine through metal ready for fabrication, but not forward into markets. Telecommunication became integrated into the development and manufacturing of switchgear. The petroleum industry became highly integrated, but if one probes into the pattern of integration, what is found are good examples of both downstream and upstream integration. Sears & Roebuck was once integrated into manufacturing, sometimes in the form of an equity stake, and in other examples, full ownership.

This question has interested a number of economics writers. Some of the resulting literature has emphasized the idea that the taking of supply firms in-house reflects a tradeoff between the costs of contracting with multiple outside suppliers and the administrative and other costs of operating these functions within the corporate fold.[19] If the former is deemed to be higher than the latter, then integration becomes an attractive option. This theory has never had to stand the test of empirical verification because of lack of a suitable database. However, it seems broadly consistent with a number of observed circumstances. Ford and other automotive pioneers of mass production created enterprises that were very large in comparison with firms with which they had to deal for component assemblies. Large numbers of such suppliers would have exacerbated the costs of contracting, which would be more sensitive to numbers of contracts than to the average size of the contractor enterprise. From the auto assemblers' standpoint, it was desirable to reduce this cost, and one option for doing this would have been to acquire makers of subassemblies, such as hydraulic braking systems. It is interesting to note that in the opening years of the twenty-first century, both Ford and General Motors have fallen into difficulties that are well documented and need not be elaborated here. Both companies are undergoing processes known as "restructuring," which involves solving a number of serious problems. The restructuring programs for both companies as described to the security analyst community include the goal of reducing the number of component suppliers, as well as reducing the number of dealer outlets. Broadly speaking, these goals seem consistent with the transactions costs hypothesis for explaining integration.

If the transaction cost hypothesis appears consistent with experience at least in a broad circumstantial sense, this does not rule out the possibility that there were other motivations that also pointed early-twentieth century managements toward integration. If one delves into the expressed reasons for early-twentieth century integration, one finds several themes, among which the transaction costs motive almost never appears explicitly. One common theme is that upstream integration (toward supply sources) reflected managerial desires to assure a reliable supply of inputs to some asset whose profitability depended on operation at or near full capacity. This would apply to the automakers and to steel companies. The former would have focused on their new mass-production facilities, whose costs were sensitive to deviations from full-capacity operation, and the latter would have been concerned with keeping their basic steelmaking facilities full. Alcoa, based on a new and largely unfamiliar material, aluminum, not only integrated upstream into mining and the various stages of bauxite processing, but also integrated forward into consumer markets.[20] AT&T integrated into development and manufacturing of switchgear. The automakers integrated downstream into consumer finance to support auto sales.

The petroleum industry provides an insight into the complexity of the early-century integration process. In the nineteenth century, the Standard Oil monopoly was formed around the strategy of owning the refining facilities and controlling the supply of crude oil inputs to the refineries, not through direct ownership of crude producing facilities, but through ownership of the pipelines that moved the oil away from the wells (the gathering lines). This strategy served the purpose of keeping the refineries full and profitable. The effective breaking of the monopoly resulted from the discovery of very large oil deposits in the Texas and Louisiana Gulf coasts in 1900 and in the mid-continent region where Standard Oil then had no presence.[21] Even after the effective diminution of the Standard monopoly, however, the strategy of control via control of gathering lines persisted. The Humble Oil & Refining Co. was founded in 1917 by three individuals in the crude oil production business who thought they were being squeezed by owners of gathering lines and other downstream assets. Their purpose was to integrate *forward* into downstream assets, including refining and marketing.[22] In the industry's subsequent history, the pattern has been of a mixture of forward and backward integration. For example, Amerada Petroleum, a crude producer, was acquired by Hess, then an independent refiner and marketer.[23] Both parties to this merger regarded it as beneficial integration.

The Sears & Roebuck case is especially interesting. While this case fits with the overall pattern of assuring a reliable supply, other conditions are likely to have entered into the integration decision. Returning to the extant business conditions of the 1920s and 1930s, the industrially unbalanced nature of the overall economic growth of the era resulted in the rapid growth of some enterprises to very large size in relation to many companies that were their suppliers. There were cases in which a retailer the size of Sears was capable of absorbing the entire output of a supplier. The original Sears business plan consisted of a mail-order business serving farmers, a group the founder had identified as having limited access to quality products at low prices. Farmers were becoming less distinctive as a market in the early twentieth century due to rising cash incomes and increased mobility (enabled by motor vehicles). Sears adapted to the change by a decision to become a general retailer appealing to an urban as well as a rural customer base. One of the problems the company encountered early in its venture into stores was lack of suppliers of goods of the desired quality in sufficient quantities. Part of Sears' response to this problem was to finance the creation of suppliers. As a result of this strategy, by the 1960s, Sears owned or controlled the manufacture of over 60 percent of the goods it sold in its stores.

Consider the vulnerabilities that dependence on a single customer/supplier creates. A supplier with one customer lives with the risk that the customer will find a more advantageous supplier. This condition would have a number of adverse effects on the supplier company, not the least of which would be an unfavorable credit rating that would adversely affect its access to working capital. In this position, the supplier wants more from Sears than a handshake; it wants assurance of a long-term relationship, especially if the provision of Sears' requirements calls for the commitment of additional capital on the part of the supplier. On the Sears side, the interest is, as suggested, on supply reliability, and there may be weak confidence in the ability to replace existing supply arrangements. Partial or full integration is a way of dealing with this situation that may have been especially appealing under conditions of the early twentieth century when there were major size disparities between suppliers and supplied.[24]

Even though the detailed patterns of integration seem highly variegated, there are at least two common themes. First, integration was in most cases a strategic reaction to the problem of protecting the profitability of some capital-intensive process in the chain of production, processing, and/or distribution steps. For the automakers, it was the assembly plants; for the petroleum business, it was either the market for crude oil or the reliable supply of crude to the refineries. This was done by means of some combination of downstream and upstream integration. Second, full integration, while not the only strategy option, was an attractive one under the generally unbalanced growth conditions of the early twentieth century. It frequently happened that both sides of an acquisition deal were intensely interested in assurance of a future business relationship. *An integration strategy was an insurance strategy.*

One way to judge the strength of the insurance hypothesis for vertical integration is to study the pattern of what eventually happened to some of the integrated structures that originated in the early twentieth century. What one finds is a number of examples of divestiture, wherein activities previously conducted within the corporate organization have in one way or another become stand-alone enterprises. The proximate reasons are disparate, but there is a common thread of consistency with broadly changed business conditions. The petroleum industry affords a prime example. Between the late 1960s and 1980, a number of the largest of the integrated oil companies lost their internal self-sufficiency in crude oil—that is, they no longer owned sufficient crude production worldwide to supply their refineries. The loss of owned crude was partially due to nationalizations of producing facilities in host countries. This induced these managements to revise long-standing attitudes in which refining and marketing assets had been regarded as mere adjuncts supporting the crude production business, and during the 1970s, there was a movement to reorganize these functions as profit centers within the corporation. In some instances, such reorganizations led to more efficient operation of the functions involved; and, in other cases, there were divestitures. Often the proximate reasons for such divestitures as became public did not cite poor performance of the divested function. For example, some of the Exxon and Mobil refineries were sold when those two companies merged in order to forestall an unfavorable intervention in the merger process by the US Department of Justice. There has been much divestiture by the automakers. The recent divestiture of General Motors Acceptance Corp. by GM was clearly an effort to raise cash to finance ongoing restructuring in the face of a down rating by credit evaluating companies. The breakup of the old AT&T was proximately in response to an antitrust suit. However, there is a confusion of reasons here. The company had successfully defended against this suit for years. It has been said that AT&T "agreed" to be broken up. Behind this speculation lies the realization that broken up, the parts of the old AT&T were worth more on the stock market than the integrated whole had been. This idea was indeed vindicated for some years after the breakup, during which time the total market value of the AT&T pieces was well in excess of the pre-breakup market value of the integrated company.[25] If there is a common thread in this collection of divestitures, it is that managerial confidence in the market's ability to supply company needs reliably was much greater in the post-1980 era than in the 1920s and earlier—such that earlier justifications for integration could more easily be trumped by recent and current considerations.

One of the best examples comes from Sears & Roebuck. Whirlpool was founded by Sears in the 1920s as a supplier of white goods appliances; and, as long as all appliances sold by Sears were made by Whirlpool, Sears owned

Whirlpool. During the postwar era a decision was made that Whirlpool, which had grown to a large size as supplier to Sears, should sell appliances directly to the public under its own brand. As a result of this decision, Sears sold part of Whirlpool in a public offering and presently holds no equity interest in Whirlpool. The divestiture of Whirlpool reflected two realities. First, Sears was no longer dependent on Whirlpool for an adequate supply of appliances since a number of suppliers had appeared in the market. Second, Whirlpool evolved from a captive supplier to Sears into a competitor, which meant that the two companies' interests diverged. The conditions behind the original founding of Whirlpool had changed radically, and the original vertical integration no longer made sense.

Why did vertical integration last so long? Once an integrated company is created, the reasons for its integration will often not be challenged until there is some external pressure on the integrated structure. There is also an important matter of corporate culture that works for acceptance of whatever structure current management grew up with, especially in a company used to drawing its top management internally. In regards to automakers, the integrated structure may have been an advantage during the Depression. The 1950s saw the heyday of the "big three." In the 1980s, there were some attempts to expand cross-ownership networks, a strategy borrowed from the Japanese. These arrangements were supposed to resolve the interests of suppliers and supplied by aligning the interests of both; it can be regarded as clinging to the idea of integration. However, these arrangements had been mostly abandoned in the 1990s in favor of reliance on the market to support mass-production facilities. It is much easier to squeeze your supplier on price when you do not have a serious equity position in that supplier.

Another hint as to the change in basic business strategy can be had from examples of acquisition patterns of some of the large companies that have emerged in the last two decades of the twentieth century, such as Microsoft, Intel, Cisco Systems, and Oracle. These companies have acquired many smaller enterprises, but the reasoning behind the acquisitions activity seems to have been the desire of the acquiring company to possess itself of technologies that many of the acquired companies own. The classic idea that integration is a way to assure input supply seems remote. The recent acquisition pattern seems more reflective of conditions of high technological flux affecting final products rather than the means of production of standardized products, such as Model Ts.

Returning to the early twentieth century, the insurance hypothesis suggests that a rapid but narrowly -based overall growth promoted integration strategies whose results persisted well into the post-World War II era. This was an indirect shaping of the configuration of industry by technological development that enabled the rapid growth. The transactions cost basis for integration has probably been present all along. However, it does not seem to have been the overriding motivation behind the industry structure that developed in the early twentieth century. In the late twentieth century, companies have resorted to a variety of ways of dealing with the problem of reliability of supply/reliability of markets other than acquisition.

Innovation and the Economy after 1930

While there were a number of prototype inventions that became commercially usable between 1910 and 1940, as detailed in the previous chapter, the collective investment impact of these was clearly insufficient to have compensated for the downturn of 1929 through 1930. The likely reason is that the resources that would have been needed for the stage-two innovation of these prototypes during the 1920s had been claimed by the requirements of stage-two innovation that led to growth of autos and related, electric power, and others. This assumes that the post-1910 slate of prototypes even had the potential to underpin industrial growth of the required magnitude even if their stage-two potential had been pursued. There were attempts to induce economic activity by using government expenditures *in lieu* of absent private investment, but the overall failure of these measures raises the question of whether investment, private or government, can be an effective tool of stimulation in the absence of new technology that offers substantially improved productivity.

The Private Sector

In the mid-nineteenth century, it was relatively easy for an individual inventor to form a company for the purpose of exploiting an invention if he could provide for his livelihood during stage-one innovation. This could be the path to great wealth for an individual inventor if he had business acumen as well as inventive skills. As the innovation process moved increasingly into the hands of large organizations, however, the individual inventor became increasingly rare. A formal business organization organized around something like the telephone faced substantial risk by waiting for individual inventors to come up with the innovation inventions needed to exploit their prototypes. The telephone, for example, required a series of follow-on inventions to be commercially successful, such as switchgear, amplifiers, transformers, and transmission mechanisms.

The industrial response to these emerging conditions was the research laboratory, or an invention mill. Each worker in one of these organizations could gain recognition for an invention, but worked under a contract obligating him or her to sign over patent rights to the employer. Thus, credit for twentieth-century inventions can frequently be assigned unambiguously, more likely than not, to an employee of a research arm of a corporation. Thomas Edison was a transitional figure between the individual inventor and the emergence of industrialized research, and some of the companies he founded evolved into today's General Electric Corp. (GE). General Electric research labs was formally started in 1900 under the direction of German immigrant Charles P. Steinmetz, and this organization became famous for innovations in X-ray technology, radio, lighting, and a number of other fields. GE research was carefully studied by other companies interested in how to organize research, including General Motors, DuPont, and Eastman Kodak.

What sort of inventions came from these famous corporate research shops? From the earliest attempts at organized inventive activity, there has been a tension between a very basic kind of research guided by science and a much more product-oriented kind of research.[26] In the conditions of the early twentieth century, there was much to do in the way of innovation invention to support the rapid development of major industries. Also, who were the inventors but engineering graduates not in a financial position to venture on solo pursuits who hired out to the corporate research shops? It is tempting to suppose that, faced with myriads of money-making opportunities, industrial companies sought innovations in established lines of development. With this strategy, private companies avoided the high-risk activity of stage-one innovating in favor of the much less-risk prone activity of advancing a technology that was already commercially established.

This is not to say that there were no prototypical breakthroughs. Four of the most important inventions of the postwar era, the transistor, the integrated circuit, optical fiber, and the laser, came from major corporate research shops. However, Bell Labs parent, AT&T, declined to take full commercial advantage of either the transistor or the laser, which proved to be unusually powerful generators of innovation streams. Corporate parents of even successful research arms had a tendency to see the future of inventions only in terms of existing products and lines of business. On the whole, the record of their handling of truly novel inventions emanating from their own research shops has not been inspiring from the standpoint of their shareholders. This theme is developed in Chapter 13, below.

Government and Innovation

The US government played an important role in the first half of the nineteenth century by means of army contracts for small arms. These contracts supported entrepreneurs like Eli Whitney in the development of laborsaving production procedures in manufacturing, and thereby indirectly enabled some of the late-century machine tool developments that enabled important parts of the nineteenth-century technological legacy. As far as can be determined, however, the intent of the army was no more than to procure quality arms for its soldiers at as advantageous a cost as could be attained. There was nothing resembling a deliberate and comprehensive policy to promote general economic growth by means of these contracts.

Generally speaking, US participation in World War I was over too short a period to have fostered significant inventive activity. The army was built up quickly from the small establishment left over from the Spanish-American War, which had been very short and largely a naval conflict. Also, in many cases, especially in procurement of artillery pieces, weapons were based on the designs of allied militaries. As in any major war, government expenditures for the consumables of war, such as food and ammunition, became very high, but this kind of spending produced no economic development side effects analogous to the effect of War Department purchases on the development of the machine tool industry that took place in the previous century. Machine tooling development between the Civil War and World War I progressed because of the needs of private industry serving mostly civilian demand. In the 1910s, the government was no longer the only market big enough to justify specialized and mechanized tooling that it had been in the nineteenth century.[27]

Even though the World War I effort did not generate the direct fostering of technology by the government anywhere near at the level that was to be the case in World War II, it is not correct to say that the period saw no significant innovation, for several prototype inventions from before the war were developed by the military, especially the US Navy. The work on sonar has already been mentioned. There was also the gyrocompass, based ultimately on the gyroscope of Jean Foucault and enabled by the development of the electric motor, among other things. The gyrocompass, attributed to Elmer Sperry, was successfully employed on naval ships in place of the magnetic compass to improve accuracy of navigation. Sperry was a pioneer in the development of modern hydraulic control systems, and his systems were employed by the Navy in gunnery control. Another innovative thread was wireless communication, with which the Navy had been experimenting since the turn of the century. This field was extensively developed by Germany and the United Kingdom, and at first, the US Navy attempted to purchase components from these countries. The war cut off part of this supply, but there was a healthy US wireless innovation effort on the part of Lee DeForrest's company and others.

Prior to the war, the US had depended on the natural deposits in Chile for its supply of nitrates, the bases for all modern explosives.[28] It was feared that enemy submarines would cut off this supply of imported nitrates, and the alternative source was the Haber process for nitrogen fixation from the atmosphere. This resulted in the construction of the Wilson Dam at Muscle Shoals, Alabama, as a power project—for the Haber process was a heavy consumer of electric power. The Muscle Shoals project was completed in 1919, after the war's end, and it was decided that the complex would be adapted to the manufacture of chemical fertilizers. The complex eventually evolved into the National Fertilizer Development Center (NFDC), and during the 1930s, it was folded into the Tennessee Valley Authority project.[29] NFDC was an early support for the agricultural sector, to which the general prosperity of the 1920s did not extend fully.

At the outset of the Depression, the private electric power industry had covered as much of the US with service as it profitably could. However, there were substantial parts of the country that had been left out, especially many rural areas. In 1930, an estimated 84 percent of urban and nonfarm rural residences had electrical service; but only 10.4 percent of farm dwellings.[30] Much of the expansion of this industry in the 1930s was due to government investment under such programs as the Tennessee Valley Authority (TVA), the Rural Electrification Administration, and the Bonneville Power Authority. The first sought to harness the power potential of the Tennessee River Valley and the last for the Columbia River Valley. These projects were successful at promoting economic development in the areas of the country affected through making relatively low-cost electric power available for industry. One industry that was drawn to these conditions was aluminum, because the smelting step is a large consumer of electric power.

The War Department was not as active in the interwar period as it was in the 1800s; the army had again been reduced to a small establishment following the war. At the outset of the Depression, the navy's size was restricted by the terms of the Washington Treaty of 1921.[31] Navy procurement was not entirely dead, for the US had never built its cruiser strength up to the level allowed by the treaty. At one point, the Roosevelt Administration, in order to generate employment in shipyards, used WPA funds to build four cruisers for the Navy. Needless to say, these "treaty cruisers" proved to be highly useful when war broke out in the 1940s.[32] Despite the War Department's low involvement in war-related research and development, especially before 1930, there was dramatic progress in aviation technology. Part of this came from growing interest in the commercial possibilities of aviation. The Post Office had growing interest in the development of airmail. On the surface of these initiatives, public interest was maintained by pilot showmen, or "barnstormers," and by airplane racing. This latter activity encouraged the development of increasingly improved aero engines, such that by the time the government's interest in war preparations revived in the mid-1930s, there had been much technical progress in the fields of aircraft design and power.

Generally speaking, however, the government's impact on innovation during the Depression years may have overall been negative. The plethora of alphabet agencies that were formed for the purpose of stimulating employment often had the effect of stifling competition and thereby discouraging interest in innovation. The best known of these was the National Recovery Act (NRA), an outright attempt to cartelize American industry. While NRA was struck down as unconstitutional by court action, other agencies had a similar effect on competition. The problem was that once regulation was established, the regulators tended to form comfortable relationships with the managements of affected industries, thus, creating a situation in which there was little incentive to aggressive development of business through innovation or by any other approach. All parties became committed to a comfortable *status quo*.

Notes

1. Walton, Gary M. and Rockoff, Hugh, *History of the American Economy,* Sixth Ed., (New York: Harcourt Brace Jovanovich, 1990), p. 481.
2. J.R. Hicks, *Value and Capital,* (Oxford: The Clarendon Press, 1965), p. 297.
3. Based on factory sales data as published in US Dept. of Commerce, Bureau of the Census, *Historical Statistics of the United States,* (Washington: U.S.G.P.O.), p. 716.
4. *Historical Statistics of the United States,* pp. 950-951.
5. For a summary discussion of this type of theory of the economic cycle, see Haberler, Gottfried, *Prosperity and Depression,* (United Nations: Lake Success, NY, 1946), pp. 33-72. One group of writers in this vein has been called the "Neo-Wicksellian" school and includes prominent Austrians Hayeck, Mises, and Machlup.
6. Wicksell, Knut, *Vorlesungen über Nationalökonomie,* Vol. II, p. 220, as cited in Haberler, Gottfried, *Prosperity and Depression,* (United Nations: Lake Success, NY, 1946), p. 34.
7. The modern Austrian capital theory, briefly described here, was developed during the years immediately preceding and following the onset of the Depression. Thinking along these lines, the principal Austrian of the times, Ludwig von Mises, was the one of the leading economist of the period to predict troubles of the kind that materialized in the 1930s, but he declined to put a timetable on his prognostication.
8. The U.S. Wholesale Price Index (all commodities, 1926=100) was 97.6 in 1921 compared with 95.3 in 1929.
9. The Wholesale Price Index for all commodities other than farm products and foods (1926=100) was 104.9 in 1921 compared with 91.6 in 1929 for an average rate of decline of 1.19 percent.
10. The Wholesale Price Index for farm products (1926=100) was 88.4 in 1921 compared with 88.3 in 1929. This index had been as high as 157.6 in 1919, reflecting healthy export markets for agricultural products.
11. For example, Britain's abandonment of the gold standard in September of 1931 led to several other countries' moves to convert their dollar assets in the US to gold in the anticipation that the US would also abandon the gold standard. In general, this triggered a net gold outflow from the US. To counter this, the FRB moved to tighter money by means of drastic increases in the rediscount rate, a move that severely aggravated US domestic financial difficulties by reducing the reserves of US banks. The FRB could have offset the deleterious domestic effects by massive purchases of US securities in the open market but did not do so. One result was that between August of 1931 and January of 1932, 1860 banks with deposits of $1,449 million suspended operations. Friedman, Milton, and Schwartz, Anna J., *A Monetary History of the United States, 1867-1960,* (Princeton, NJ: Princeton University Press, 1963), pp. 315-318.
12. Haberler, 85-105.
13. There have always been some who saw the Great Depression as a confluence of a number of negative events. In addition to the monetary and investment aspects, a list would have to include dust bowl conditions in the west that were the result of years of aggressive grazing and cultivation practices, harmful fiscal policy as governments sought to offset falling tax revenues with higher taxation, and attempts by many countries to protect domestic employment by high protective tariffs.
14. Probably the greatest problem in applying economic theory to actual conditions is that of recognizing the timing of the onset of a condition. There is no better example

than the problem of recognizing the onset of a recession. The National Bureau of Economic Research has defined a recession as declines in real gross domestic product for two or more successive calendar quarters. This is of no help in short-run economic forecasting, for by the time a recession becomes official according the NBER definition, its onset is already three or four quarters in the past due to delays in assembling the GDP data. Broadly speaking, this is the problem that one encounters with recognizing a prototype invention, but in this case, the recognition delay is much longer than that of recognizing the onset of a recession.

15. Hicks, J.R. *A Contribution to the Theory of the Trade Cycle.* (Oxford: Oxford University Press, 1950), p. 163n.
16. At least some historians of the Depression have focused on business uncertainty induced by massive government experimentation during the first Roosevelt Administration as a cause of the dearth of investment. This chapter offers an alternative view. See Schlaes, Amity, *The Forgotten Man,* (New York: Harper Collins, 2007), pp. 147-172.
17. Skousen, Mark, *Viena vs. Chicago: Friends or Foes,* (Washington, D.C.: Capital Press, 2005), pp. 162-166.
18. *Historical Statistics of the United States,* p. 143.
19. See, for example, Coase, Ronald H., "The Nature of the Firm," *Economica,* 1937. As reprinted in Kenneth Boulding and George J. Stigler, *AEA Readings in Price Theory,* (Chicago: Richard D. Irwin, Inc.), pp. 331-351. See also Williamson, Oliver E., *Markets and Hierarchies: Analysis and Antitrust Implications,* (New York: The Free Press, 1975).
20. One of Alcoa's early consumer products was a line of kitchen ware—saucepans and skillets. This author's mother's first kitchen was so equipped.
21. The years-long effort to break up the Standard monopoly by antitrust action is well documented, and the record often reads like an act in futility. The real demise of the monopoly was the discovery of huge oil reserves over which the monopoly had no control.
22. This particular strategy has to be regarded as highly successful. The company evolved into the US arm of the Exxon Mobil Corporation.
23. In the petroleum business, the idea that the downstream assets (transportation, refining, and marketing) were primarily services in support of the main business, crude production, was very much alive as recently as the late 1960s. This was a prevalent attitude that this author found when working for a major integrated oil company starting in 1968.
24. This idea is very clearly expressed in Drucker, Peter, *Management: Tasks, Responsibilities, Practices,* (New York: Harper & Row, 1973, 1974), pp. 686-687.
25. The "breakup value" of a firm is very much an important consideration in latter-twentieth-century corporate strategic thinking. The stock market has always had a problem in valuing disparate assets that are collected under a single corporate umbrella. The tendency of the market has been to pigeonhole a company according to its perceived main line of business and fail to value assets that do not fit directly with the main line of business. There are a number of examples of corporate divestitures wherein the post-breakup market valuation of the assets is greater than the pre-breakup value, as in the AT &T case.
26. In a recent announcement of renewed emphasis on fundamental research, Jeffrey Inmelt, CEO of General Electric Co., pointed out that it is preferable to spend $10 million on developing an idea than to have to spend $100 million to acquire some other company who has brought the same idea to fruition. See "GE Finds its Inner Edison," *Technology Review,* vol. 106/ No. 8, (October 2003), pp. 46-50.

27. As noted above, the work that led to the application of ultrasonics to submarine detection commenced during the war, but this development went well beyond the end of the war and played no part in its prosecution.
28. Nitrates are water-soluble. Therefore, the only natural conditions under which they can be found in commercial-sized deposits are extremely dry places, otherwise rainfall and leaching would long since have dissolved and washed away the material. Sufficiently dry conditions are found in Northern Chile's Atacama Desert region.
29. This led to the curious arrangement by which TVA, not the Department of Agriculture, was the Federal source for fertilizer data and statistics. This arrangement was still in place in the late 1970s.
30. U.S. Department of Commerce, *Historical Statistics of the United States,* p. 627.
31. The Washington Treaty was an arms limitation agreement among the principal naval powers of the world. The treaty imposed ceilings on the tonnage of major ship types, and is remembered for the 5-5-3 agreement, which represented, respectively, the allowed relative battleship strengths of the British fleet, the US fleet, and the Japanese fleet.
32. This is an example of superb political timing. A real scandal might have erupted over this had the war not made the move look inspired and prescient.

5

Materials: Enablers in the Twentieth Century

The enabling technology behind many of the significant prototype inventions that have substantially furthered technological progress has often been in the form of newly available materials. This was evident in the developments accompanying the industrial revolution and goes back much farther. The importance of materials is especially great in technological progress since 1910. The materials here treated qualify as prototype inventions by all three standards of that term. Materials such as stainless steels and the superalloys originated from growing knowledge of the science of metallurgy. The plastics resulted from growing knowledge of hydrocarbons at the molecular level. In all cases, they moved rapidly from simple replacement of existing materials to wider fields of use. Some of the materials described in this chapter, such as stainless steels and aluminum, were available in the interwar period, but most have had their impact since 1940.

One interesting point about many of the modern materials, especially plastics and synthetic rubber, was that the stage-one innovations with attendant risks were generally borne by the private sector. In the cases of nylon and Kevlar, both developments of the DuPont Laboratories, the first-stage risks were borne by the DuPont Corporation. In the case of nylon, the War Department did become a large consumer during World War II, but this was simply the act of procurement of a needed war material and was not, as in some cases, the financial foundation for bringing the product to market. Nylon's initial market during the 1930s was women's hosiery that was less expensive and more durable than silk. Many of the modern metals, including the superalloys, were also private sector developments, although in the case of superalloys, the market was profoundly affected by demand from gas turbine engine manufacturers, which were responding to the needs of military before there was a large demand from civil aviation.

Polymers: Plastics and Fibers

The plastics industry's pioneering start came from several discoveries dating from before 1910: cellophane, rayon, and bakelite. However, the real emergence of this type of material dates from the interwar period. The discoveries that have led to the modern plastics industries mostly required research into basic structure of materials at the molecular level. This required a lot of costly laboratory equipment, and for this reason the development of plastics, rubber, and related materials, especially after 1930, was a natural capability of the corporate research laboratory.

One of the most important developments in the genesis of the modern plastics industry was the emergence of the petrochemical industry in the interwar period. The basic materials of the petrochemical industry are called olefins. Natural gas and crude oil are mixtures of many organic substances, but these include largely what are known as alkanes, or substances characterized by molecules that are linear chains of carbon atoms, each of which is chemically bonded to two or three hydrogen atoms. The rule is that each carbon atom wants to combine with four things, and in an alkane, those things are either hydrogen or other carbon atoms. An olefin is an alkane with two hydrogen atoms stripped away, and in which there is a double bond between two of the carbons. This double bond gives the olefins the characteristic of being highly reactive chemically, and that is why they are commercially important as basic building blocks in the production of many plastics and related products. The commercially most important of the olefins is ethylene, which has two carbons and four hydrogens. In the early days of the petrochemical industry, the most economical feedstock in the manufacture of ethylene was ethane, an alkane with two carbons and six hydrogens, which was a part of the natural gas mixture or a refinery by-product. Before the appearance of the petrochemicals market, ethane was either left in the natural gas mixture and burned as fuel, or flared as a waste product where no local fuel market was available.

There is an impression, which is dying a slow death, that plastics are a cheap substitute for metals and other more traditional materials. This could have originated in the early days of plastics with the appearance of countless small household items. What should be understood is that plastics are not cheap compared with metals, generally speaking. For example, plastics are used extensively on automobiles for such applications as bumpers and front-end body parts, traditionally made of steel. As for cost, the engineering plastics used cost $3.00 to $6.00 per pound compared with less than a dollar per pound for steels typically used. The plastics possess the strength characteristics to perform in normal operation, and are far lighter than steel, an essential consideration in producing vehicles that are reasonably economical with fuel. Plastics are made to meet all kinds of performance specifications. Where the objects are not under serious stress, the cost of plastics is commonly reduced by the use of filler materials, such as clay, but when plastics substitute for metals, it is generally true that the substitution results in superior performance.

Polyolefins

Polyolefins are olefins that have been combined into polymers, or substances that are composed of large molecules that are individual olefin molecules chemically bonded to one another.[1] Actually, bakelite and celluloid were polymers, but polymer materials did not really begin rapid growth as commercial products until polymer chemistry began to be understood in the 1930s and 1940s. Table 5.1 presents a brief history of the introduction of selected polymers. The most common example is polyethylene, a material that is manufactured in huge quantity all over the world and is the material of such common products as plastic bags and other packaging. Other common uses include wire coatings or insulation and squeeze bottles. Easily the second most important of the olefins is propylene, which has three carbon atoms, and its polymer is polypropylene. About 25 percent of propylene is used to make polypropylene, from which a number of other products are made. Major uses of propylene are the manufacture of propylene oxide (a chemical intermediate), isopropyl alcohol (a common household product), and acrylonitrile. Familiar consumer products from propylene derivatives include carpet fibers, rope, clothing, plastics in automobiles and appliances, toys, and epoxy glue.[2]

Another important group of polymers is the polycarbonates, so named because all the monomers in the polymers are joined together by carbonate groups (CO_3) in a long molecular chain. Polycarbonates are easily worked, molded, and thermoformed, and the polymer is a clear plastic. These materials have light transmission characteristics superior to most glasses but are more costly than glass. This limits their use as glass substitutes to applications where polycarbonates' great strength can justify the cost differential, such as aircraft windows, windshields on motor bicycles, and others. If not completely bulletproof, they are bullet resistant in that they will slow the bullet to the point that hopefully no damage accrues to persons behind the shield. A common household use is in CDs and DVDs. Polycarbonates are sold under the trade name Lexan (from General Electric) and Merlon (from Mobay Chemical Co.).

Nylon

Nylon was developed in the 1930s by a team headed by Wallace H. Carothers at E.I. du Pont de Nemours and Company. It was the first of what became a succession of materials developed by chemical synthesis from air, water, coal, or petroleum. It was first produced as a fiber, but it has desirable properties that made it useful for making objects by injection molding. First marketed in the form of hosiery, nylon was used in a number of products during World War II, such as parachute silks.

Kevlar

Kevlar is a stiff and fire-resistant fiber that is used in tirecord, structural parts (such as reinforced boat hulls), and in lightweight body armor. It was invented in the 1950s by Stephanie Kwolek, a chemist working for DuPont Laboratories. Kevlar has tensile strength five times that of steel and is light and flexible. It was originally marketed as a replacement for steel belting in tires. It has been found useful in a number of other applications, but the best known of these has been lightweight body armor widely used by police and the army.

Prior to the 1970s, soldiers had no better personal protection than heavy, bulky nylon flak jackets that were effective against shrapnel but not against bullets. Police forces found these minimally useful. The project that led to Kevlar's use as armor was directed by Lester Shubin and Nicholas Montanarelli who were working for the US government in the 1970s. The formal project that resulted in the armor vest was carried out by the National Institute of Justice (NIJ), an agency of the US Justice Department. After experimentation on goats, the vests were field-tested by police departments in areas that were especially crime-prone, and the final conclusion was that Kevlar made a highly effective armor. It was introduced to military ground forces in the early 1980s. Kevlar stands as a good example of the prototype invention: from replacing steel in an established use, it went on to a use—armor—that was probably not contemplated by its inventors and early developers.

Teflon

Teflon is the trade name of polytetrafluoroethylene, a strong, waxy, nonflammable resin that is almost completely resistant to attack by any chemical and has a slippery surface that does not stick to anything. It was discovered in 1938 by Roy Plunckett, a DuPont Chemist. During the war, it was used as an anticorrosion coating for metals, especially metal equipment used in the production of radioactive material. DuPont introduced its well-known line of teflon-coated cookware in 1960. Teflon's slippery surface made it useful in bearings in applications where high temperatures are not a problem.

Polyethylene Terephthalate

Terephthalic acid is made from para-xylene and its derivative, dimethyl terephthalate (DMT), which, when copolymerized with ethylene glycol, yields polyethylene terephthalate (PET). This material, from DuPont, when spun into fibers, is known commonly as polyester and on some clothing labels as Dacron. This material appeared in the 1950s. While PET was successful as a fiber, its really large market was the clear plastic carbonated beverage bottles that are so ubiquitous in the present time. This application came about due to the efforts

Table 5.1
Selected Polymers and Commercialization Dates

Polymer	Commercialization Year
Nitrocellulose (cellophane)	1869
Bakelite	1908
Polyvinyl acetate	1919
The polyacrylates	1931
Polyvinyl chloride	1639
Nylon and polystyrene	1938
Polyethylene and polyesters	1942
Epoxies	1947
Polycarbonates	1950s
Polyurethanes	1953
Polypropylene	1957
Polyamides	1964
Polybutylenes	1973
Linear Low Density Polyethylene	1977

of Nathaniel Wyeth, an engineer working for DuPont Laboratories.[3] Wyeth had wondered for some time why plastic was not used for soda bottles and was reluctant to accept colleagues' explanation that plastic was not strong enough to withstand the pressures within carbonated liquids. He knew that stretching nylon strengthens it because it forces the molecules to align themselves; however, for a soda bottle, he needed molecular alignment in two dimensions rather than just one, as in stretched nylon. He attained two-dimensional alignment by means of the design of a mold, and, after a search that tested many materials, discovered that PET was adequate to the task. He patented the PET soda bottle in 1973.

Metals

Stainless Steel

While efforts to improve the corrosion-resistant properties of carbon steels had been made in the latter decades of the nineteenth century, stainless became a usable material only about 1913, the result of work in Germany, the United Kingdom, and the United States. The defining admixture of stainless steel is the element chromium, and the minimum chromium percentage to qualify as stainless is 4 percent, according to the American Iron & Steel Institute (AISI). The actual chromium loading necessary to attain corrosion resistance depends on the corrosion agent to be encountered in service.

In 1912, Harry Brearly of the UK discovered stainless steel while in the course of research into the problem of eliminating fouling in gun barrels. He noted that a 12.8 percent chromium admixture in the steel greatly reduced the undesirable fouling. In the US, F.M. Becket, while investigating the effect of chromium alloying on corrosion resistance at 2,000 degrees Fahrenheit discovered that a 20 percent chromium admixture achieved the desired corrosion resistance. The iron-chromium-nickel alloys that have become so important in applications requiring high-heat resistance were developed in Germany starting in 1909 under the direction of Benno Strauss and Edward Maurer. These included the 18-8 specification (18 percent chromium and 8 percent nickel) that not only has been very successful in high-temperature applications but has become a quality standard for table ware.[4]

Stainless steels fall into three major groups: austenitic (16 percent-26 percent chromium and up to 35 percent nickel); ferritic (10.5 percent chromium and no nickel); and martensitic (11.5 percent to 18 percent chromium with nickel sometimes added). Day-to-day uses for austenitic stainless include kitchen equipment and utensils, dairy installations, transportation equipment, and oil, chemical, and food-processing machinery. Austenitic stainless steels were essential enablers for turbine blades in early jet engines and turbosuperchargers, of which the latter were manufactured in huge quantities during World War II for application to piston-driven aircraft engines. Ferritic grades have high temperature resistance and are used in furnace parts. Their high resistance to nitric and other acids makes them suitable for chemical processing equipment. Common uses for martensitic grades included cutlery, including surgical instruments.

Stainless steel contributed to one of the dominant technical trends of the twentieth century: the reduction in the weight of passenger vehicles. The massive lightening of automobiles that took place after 1970 under the impetus of the Clean Air Act was anticipated in the 1930s with the use of aluminum and stainless steel in the construction of railway passenger cars. Stainless steels were far stronger than the strongest available alternatives and permitted a much lighter structure with no loss in overall structural integrity. The Budd Company of Philadelphia had invented a process for welding stainless, known as shot welding, and by 1939 Budd was delivering full-sized passenger cars to railroad customers that weighed just over 50 tons each. This compared with the approximately 86 tons of conventional designs made of Cor Ten steel.[5]

Superalloys

The increased thrust of jet engines of the postwar era was bought at the cost of higher operating temperatures. Development of the superalloys has taken place largely to keep pace with the requirements of gas turbine engine makers and makers of turbosuperchargers. Some superalloys were based on iron, consisting of the addition of substantial amounts of the element cobalt to austenitic stainless steels. Another family of superalloys was based on nickel, of which an example is Hastelloy B (30 percent molybdenum, 5 percent iron, .12 percent carbon, balance nickel) and was used in early turbosuperchargers. This material was difficult to form, and was replaced with cobalt-based alloys such as vitallium, which could be cast accurately and was found to have good performance at temperatures above 1500 degrees Fahrenheit. Generally, the superalloys can be considered enablers for the postwar development of gas turbine engines.[6]

Aluminum

Aluminum the element had been known for many decades when, through the inventiveness of Hall and Heroult, it became widely available at reasonable cost in the mid-1880s. As a new material, the producer, which evolved into ALCOA, was required to promote uses for it. One of the earliest lines of product was kitchenware (skillets, saucepans, etc.). During the interwar period, aluminum's desirable strength-to-weight characteristic made it an essential enabler for the development of military and commercial aviation and, by 1940, aluminum had also achieved some penetration of the automobile parts market. During World War II, the massive increase in military aviation demand was met partially by new production capacity built by the government and operated by ALCOA.[7] Following the war, aluminum played an increasingly important role in automobile construction. This trend was pioneered in Europe, as cast aluminum alloys were used for such parts as cylinder heads, intake and exhaust manifolds, and others. The advantage of aluminum's light weight was clearly recognized in the US following the Clean Air Act of 1970, and it displaced traditional materials, such as cast iron in engine blocks.

As a good conductor, aluminum has displaced copper in the electric transmission grid. Today, the typical electric transmission line consists of a bundle of aluminum wires around a central steel core (for strength). Aluminum has almost completely taken over this market from copper, the exception being in areas near coasts where there can be salt in the atmosphere. Aluminum failed to displace copper in house and building wiring because it required special techniques at junctions: if aluminum wires are joined in the simple way that is appropriate for copper, the result is a serious fire hazard. The problem is that aluminum exposed to air forms a thin but dense surface oxide. This is the reason for aluminum's desirable weathering characteristics in most applications, but unless special techniques in joining aluminum wires are followed in junction boxes, the oxide is a point of heat buildup. Too many electricians proved to be ignorant of this characteristic of aluminum, and aluminum-wired houses had a tendency to burn down.

Titanium (Ti)

Titanium was first identified as an element by the German chemist M. H. Klaproth about 1793. Titanium metal was first produced in 1906 in the course of efforts to discover a superior material for incandescent light bulb filaments. This effort was guided by the belief that Ti metal had a melting point of 6000 degrees Celsius. The reality was discovered to be much lower (1800 degrees Celsius or 3300 degrees Fahrenheit), and its role in the search for filament materials was dropped. However, research with the isolated metal suggested very high strength-to-weight characteristics. The modern process for producing Ti metal from its ores was developed in the 1930s by William J. Kroll of Luxembourg, who combined titanium dioxide (TiO_2) with calcium. Kroll fled Europe at the start of World War II and came to the US where he continued his work as an employee of the Union Carbide Corporation.

Titanium's high strength and corrosion resistance qualities attracted the attention of the US Army Air Corps, whose interest in the 1940s was based on the need for strong but light metals whose performance could exceed that of steel or aluminum. The US titanium industry was started with the help of production incentives from the Department of Defense. More recently, gas turbines account for almost half of titanium production, with Ti alloys being used in inlet cases, compressor blades, disks, and hubs. The large high-bypass turbofan engines used on modern wide-body commercial aircraft are about 25 percent titanium by weight. Titanium also is used in landing gear supports, fasteners, springs, fail-safe straps, and in wing structural supports. Titanium metal has begun to penetrate the consumer market, where it has been used as golf club heads and shafts, and as outer cladding for some high-end miniature cameras. Titanium dioxide is high-end pigment, competing with kaolin in the markets for white pigments in such markets as house wire insulation and other markets for white pigments, especially high-end products.

Carbon Fiber Reinforced Plastics (CFRP)

The carbon fiber part of CFRP refers to carbon filament thread, or to woven cloth made from such threads. One of the more common methods of making carbon filaments is the oxidation and thermal pyrolysis of polyacrilonitrile (PAN), a widely used polymer in the creation of a number of materials. These fibers are extremely strong, and when encased in a matrix, the result is a material with exceptionally good strength-to-weight characteristics. CFRP materials have desirable mechanical properties such as high tensile strength and high specific elastic modulus. For a given weight, they are considerably stronger than either carbon steel or titanium. CFRP materials are also very costly, and their use, therefore, tends to be confined to applications wherein their high-strength and lightweight traits are valuable. CFRP materials have been generally available since the 1970s; they are truly space-age materials.

CFRP materials have been employed in airframe design where high strength and light weight are needed in applications not subject to extremes of high temperatures. As of this writing, the weight of carbon composite materials in military aircraft ranges from 30 percent to 35 percent. In the early 1970s, the F-15 had composite composition of 2 to 5 percent. This contrasts with the F-35 currently under development, which has 50 percent of its weight in the form of carbon composites. Carbon composite materials are important also in military pilotless aircraft as well as commercial aircraft currently under development, such as the Boeing 7E7. In addition to their obvious advantages in terms of strength/weight ratio, these materials also have the advantage of being formable into far more complex shapes than any metals. Fabrication of parts for aerospace applications proceeds by placing cloth woven from carbon fibers in a mold. All the air is then withdrawn from the mold, and epoxy is injected, and then cured by heat or air. Parts resulting from this process are extremely strong for their weight and highly resistant to weathering.

When CFRP materials were first introduced, they were used in a number of high-end consumer applications. Examples include golf club shafts and bicycle frames. The latter application sold well among very serious bikers and bicycle racers, where conditions called for riding for hours at a time. Another application is the helmet part of the HAN, a head-restraint system designed by James Downing for automobile race drivers. This system incorporates an efficient escape mechanism that can let a driver escape from fire. The HAN system became instantly popular with drivers after the death of Dale Ernhart when it was widely opined that he would have survived had he been wearing it.

Ceramics

Ceramic materials are usually understood to be inorganic, nonmetallic solids. These materials literally go back to the dawn of human civilization in the form of pottery, and are based on natural materials such as silica and alumina. Ceramic products from ancient and medieval times include window glass and porcelain, the latter a Chinese development from about 600 A.D. Refractory materials were important in the Industrial Revolution having been used for building bricks and for high-temperature applications such as linings of steel furnaces.

In modern times, starting in the 1950s, ceramic materials have been adapted to a number of industrial, medical, and other uses. Where traditional ceramic products included clay products, silicate glass, and cement, advanced ceramics consist of carbides (e.g., SiC), pure oxides (Al_2O_3), nitrides (e.g., Si_3N_4 BN), nonsilicate glasses, and others. These advanced ceramics are characterized by careful control of crystalline structure in manufacture in order to obtain desired properties. Modern ceramics generally are characterized by a number of properties, including a mostly crystalline structure, mechanical strength, brittleness, chemical inertness—especially with respect to the deteriorating effects of oxygen, acids, water, bases, and organic solvents—and ability to take a decorative finish.

One of the interesting examples of modern ceramics is Boron Nitride (BN). This synthetic material combines properties of near-diamond hardness, superior dielectric strength, machinability to very fine tolerances (in its solid form), and high thermal conductivity. After machining, it is ready for use without further heat treatment. It can withstand temperatures in excess of 2000 degrees Celsius (upwards of 3600 degrees Fahrenheit) in reducing or inert atmospheres. It can thus be used as a container for most molten metals such as aluminum, cryolite, sodium, iron, steel, silicon, boron, tin, germanium, and copper. BN has a variety of uses in high-temperature environments. Some of these are welding tips, plasma arcs, microcircuit packaging, microwave oven windows, glass-forming tools, and others.

Heat Engine Applications

The stability of ceramic materials in high-temperature environments has led to their application as wear-resistant surfaces in working parts of internal combustion engines, both reciprocating and turbine. These coatings permit higher operating temperatures, an advantage that pays in the form of more complete fuel combustion with consequent reduced fuel consumption and diminished production of waste products. These coatings are sufficiently thin as not to affect the basic design of the engine. Ceramics have also been applied in the form of heat-barrier coatings on piston heads, intake and exhaust valve heads, and cylinder heads. Ceramic materials have been formed into the complex shapes of rotor and stator blades in gas turbine engines, where their superior high-temperature performance and relatively light weight offer advantages in the form of lower fuel consumption.

During the 1970s, several of the major automobile companies investigated the idea of a ceramic engine block. The interest in this idea grew from some ceramic materials' stability at very high temperatures. An engine built this way could operate at over 6000 degrees Fahrenheit, more than hot enough to melt any metal material commonly used for this purpose. The advantage to be gained was high efficiency at a high operating temperature, lightness of weight, and reduced complexity, as the ceramic engine would not need a conventional cooling system. Some successful engines were built in laboratories, but in spite of the attractive characteristics of a ceramic-block engine, none are in production today. Mass production of ceramic parts is difficult. Even a minor imperfection in the ceramic material can develop into cracks. This remains an idea for the indeterminate future.

Automotive Applications

Possibly the best known of the automotive applications are the substrates for the catalysts in catalytic converters. Initially, there were two basic substrate designs: pellets and honeycomb monoliths. The latter design was perfected by the Corning Glass Works in time for application to autos in the 1975 model year, and eventually, this design of substrate became universal in the automobile industry. There was a period, however, in which General Motors used the pellet design. The pellets were approximately three millimeters in diameter and had ten square meters of surface each (exterior plus pores). A liter of these had up to 500,000 square meters of catalyst support surface. The honeycomb monolith structure was built of a ceramic known as cordierite, a magnesium aluminosilicate known for having a coefficient of expansion compatible with the metal used in the converter housing. Honeycomb monoliths have one to two thousand longitudinal pores that are coated with the platinum catalyst. The collective effect of these pores is a very large surface for the catalyst. The modern automobile has a variety of sensors that feed information to the computer that regulates engine operation, the catalytic converter, and other functions. Many of these sensors use ceramics. For example, dynamic pressure sensors, which monitor combustion chamber pressures, utilize piezoelectric ceramic materials, in which a current can flow when the material is subject to pressure. These ceramics are also used for acceleration-deceleration sensors.

Medical

High density, high purity, fine-grained polycrystalline alumina (Al_2O_3) has a number of properties that make it suitable for repair or replacement of musculoskeletal hard connective tissues. This material has excellent corrosion resistance, biocompatibility, high-wear resistance, and high strength. Other applications include knee prostheses, bone screws, segmental bone replacements, and maxillofacial reconstruction. Other bioceramic implants can serve as porous media to support the growth of new bone structure, as materials that bioreact with tissue.

Dental Ceramics

Dental ceramic applications include resin-composite restorative materials, cementation agents, and fixed prostheses. Resin composites, due to their attractive surfaces and due to concern over the mercury in dental amalgams, have found a growing market in the repair and restoration of teeth. Dental cements are used in the attachments of crowns and bridges and as bases under other restorative materials.

Consumer

The military encountered problems of dealing with frictional heat generated as high-speed missiles moved through the atmosphere well before this type of problem became publicized in the space program, and in the 1950s used a ceramic product of the Corning Glass Works as a heat-resistant nose cone for the Snark, a primitive cruise missile. The Snark has long since become obsolescent, but the nose cone material can readily be found under the brand name Corning Ware. Other products familiar to consumers include glassware, windows, pottery, magnets, dinnerware ceramic tiles, lenses, home electronics, and microwave transducers (ovens).

Ferroelectric and Piezoelectric Ceramics

These materials have specialized properties that have been utilized in a number of industrial applications. A ferroelectric material is one that can spontaneously generate a polarization, even in the absence of an electric field. A piezoelectric material has the property of producing an electric current when vibrated, or, oppositely, of vibrating when subjected to a current. These materials are used in such applications as microphones and ultrasound generators, and are the bases for generating sonar waves for detecting the presence of submarines. A pyroelectric material generates an electric field when heated.

Semiconductors

Solid-state materials fall into three categories: insulators (conductivity less than 10^{-8} siemens/centimeter), conductors (conductivity greater than 10^3 siemens/centimeter), and semiconductors (in between insulators and conductors). The conductivity of semiconductors can be sensitive to temperature, light, magnetic fields, and trace amounts of impurity atoms. In some cases, the addition of a tiny amount of an appropriate impurity can increase the conductivity of a semiconductor material by as much as four orders of magnitude.

Semiconductor properties of various materials have been recognized since well before the turn of the twentieth century. One of the first materials to be so recognized was selenium, the basis of the photoelectric cell of 1883. Since then a number of semiconductor materials have been examined. Single-atom substances that are semiconductors include silicon, germanium, and gray tin (column IV of the periodic table) and selenium and tellurium (in column VI). Several binary compounds have also been recognized as semiconductors. These include silicon carbide, gallium arsenide, and gallium phosphide. Both are combinations of gallium (column III) and arsenic or phosphorus (column V). A number of ternary compounds have been found to have semiconductor properties: for example, mercury indium telluride. Prior to the invention of the transistor (1947), semiconductors were used mostly as rectifiers and photodiodes.

Silicon

If any material has had a defining role in the second half of the twentieth century in a manner analogous with the role of steel in the second half of the nineteenth, it is silicon—for it has been the material basis of the transistor, the integrated circuit, and all their applications. In its silica form, silicon is at the heart of the fiber optic revolution in telecommunications. Silicon is the second most common element in the earth's crust, but it does not naturally occur in pure form. Common occurrences include sand, which is silica, an oxide of silicon, and quartz, also an oxide of silicon. Silicon has long been used in building materials. In its carbide form, which is a manufactured product, silicon is a common abrasive material and an emerging semiconductor material.[8] Sodium silicates are used in glass, soaps, and in the treatment of wood to prevent decay.

Inasmuch as silicon is not chemically active at ordinary temperatures, it has to be prepared at high temperatures. It is produced commercially by reducing its naturally-occurring oxides in electric furnaces. The pure silicon is then formed into ingots, the size of which has been growing in accordance with the integrated circuit producers and the technology of production. From the ingot form, the wafers on which circuits can be etched are created by machining. As of the present writing, the silicon industry has attained the ability to produce wafers of 12 inches in diameter.

Silicon has become the dominant industrial semiconductor material. In the early days of transistors one other semiconductor, germanium, was tried, but its relatively low tolerance for heat confined this material to uses in which it would not be subjected to high temperatures. Silicon overcame this difficulty with its melting point of 2570 degrees Fahrenheit. Production of electronics-grade silicon is costly, but performance has justified the cost in the host of electronic applications based on it. However, its costliness has been a barrier to the expanded use of at least one of its applications: solar cells. As discussed elsewhere, the high cost of materials in solar cells has confined their commercial application to limited *niche* markets. As of the end of 2005, there were reports that the rapid growth of demand for silicon for use in solar collector panels was outstripping silicon producers' collective capacity. This has produced upward pressure of silicon's price, presumably temporary. There appears to be a downward trend in the cost of silicon, and its continuation may hold the promise of bringing solar power costs to levels competitive with conventional power. At present, this competitiveness has not been achieved, and the growth in solar power production has been fueled by government subsidies.

Silicon Carbide (SiC)

Silicon carbide is an interesting material. Its traditional large markets have been as the abrasive in fine sandpaper and as a gem material. About 1950, it was discovered to be a semiconductor, but until fairly recently, this market for SiC has not developed appreciably. Interest in SiC as a semiconductor arises from its high-heat tolerance. Its melting temperature is above 2900 degrees Fahrenheit, making it desirable in circuits employing high temperature and voltage. It is reported to be able to withstand ten times the voltage that pure silicon can tolerate. Electrical resistance can be held to 1/80th of that of standard silicon integrated circuits, and the lower resistance means that devices can be physically smaller and require less heat dissipation capacity.

The reason silicon carbide has been slow to come into its potential as a semi-conductor is its extreme hardness, which makes manufacturing a problem (it has a hardness of 9.5 on the Moh Scale, as compared with the diamond at 10). Methods that have been successful in working with pure silicon do not work because SiC has no liquid state, as does pure silicon. Recent progress has been made due to the development of the modified Lely Method, in which the SiC is vaporized and crystals of semiconductor grade material are grown from the vapor. So far, however, the process has yielded crystals of 3 inches in diameter, in contrast with the 12 inches that are possible with silicon. Recent applications include use in light-emitting diodes (LEDs), especially the recently developed blue LEDs. Also, the modified Lely method still has some quality problems, including what are called "micropipes," which can affect the performance of integrated circuits made with this material. Because of the remaining manufac-turing problems, SiC can be regarded as a material still largely "in the wings," but recent progress suggests that the future is plausible. SiC's penetration of markets now held by pure silicon may not be rapid, for in order to take full advantage of SiC's heat and current tolerances, it will be necessary to redesign the circuits to which it is applied.

Electrically Conducting Plastics (ECP)

These interesting materials were co-invented by Alan Heeger who, as a result, shared in the 2000 Nobel Prize in chemistry. The first application of electrically conducting plastics was in flat-panel displays, based on their property of emit-ting light in response to an applied electric current. What makes these materials all the more interesting is the possibility of reversing this process: exposing them to light to produce electric current. Inasmuch as the chemistry of ECPs is not highly complex, manufacturing them should not be hugely expensive, and therefore the suggestion is that they might constitute the basis for a dramatic reduction in the cost of solar cells.[9]

Nanomaterials

The science of nanotechnology is very new and it has become recognized as a "hot" topic for research. Generally, the term refers to working with objects with dimensions measured in nanometers (billionths of a meter). The research has taken two broad directions: active and passive. The former refers to the construction of tiny mechanical devices, and the construction of these has proved difficult in the face of dimensions so small that the laws of quantum mechanics affect the effort. The passive branch of the research has yielded a number of possibilities, some of which may be close to commercial usefulness. These possibilities include coatings that greatly increase wear resistance and nanotubes, from which flow some promising developments for power transmission, computing, flat-screen display technology, and solar energy. Although the organization of startup firms aiming to commercialize nanoproducts goes back to the eighties, progress since 2001 has been encouraged by quickening interest by venture capitalists.

It seems inevitable that the results of nanotechnology research will affect the way people live in a variety of ways, but a number of researchers in the field are presently convinced that the greatest impact will derive from nanomaterials' electronic possibilities. One area of present development aims at the creation of very low-cost solar cells. The materials being developed rely on nanorods, semiconductor structures several nanometers in cross section and hundreds of nanometers long. These devices are mixed into a very thin film of electronically conducting plastic, which is sandwiched between two thin-sheet electrodes. When exposed to light, electric flows in the nanorods and the plastic matrix and can be used to charge a battery. The battery's output current an be used to power household appliances. The assembled film is sufficiently thin that it can be attached to ordinary roofing tiles.

The result is a solar cell that promises to be cheaper to produce than conventional cells based on silicon by a factor of four or five—for its materials are relatively cheap to produce and the production process does not involve high temperatures. When perfected, this cost property is of clear interest to solar power developers; to date the cost of materials going into solar cells is one of the factors that has kept this technology confined to a limited niche market.

The 1996 Nobel Prize in chemistry was shared by Richard E. Smalley, Robert F. Curl (of Rice University), and Sir Harold W. Kroto for their joint discovery of $Carbon_{60}$ (C_{60}, or Buckminster fullerene, or Buckyball), the third allotropic form of carbon (the other two being diamond and graphite). The naming came from the resemblance of the molecule to a tiny soccer ball or a geodesic dome. Following this discovery in 1986, it was theoretically predicted that there existed tubular versions of this same structure, and that such tubes would have some interesting properties. This prediction was verified in 1993 by Sumio Iijima and his team at NEC, and independently by a team at IBM laboratories. The IBM version of the nanotube had walls only one atom thick.[10] As will be discussed, the nanotube promises to become a basis for flat-panel displays and future computers far more powerful than today's computers whose silicon-based circuitry is anticipated to become increasingly difficult and costly to make smaller and more powerful.

The carbon nanotube has a tensile strength measured between sixty and one hundred times that of steel. It is highly conductive, with some variants demonstrating properties that are close to superconducting. This has generated speculation that nanotubes may someday be applicable to power transmission.

Nanotubes also conduct pulses of light, thus making them of interest regarding liquid crystal displays. As of September of 2005, Houston-based manufacturer Carbon Nanotechnologies, Inc. reported that carbon nanotubes cost between $375 and $2000 per gram. This suggests that most of the intriguing prospects for these materials are strictly in the future, but there is evidence that the costs of nanotubes are falling. In November of 2005, Bayer Material Science, A.G. announced that it planned to sell a line of nanotubes, trade named *Baytubes,* for $100 per kilogram.[11] Other nanotube applications under consideration include LCD TV screens, as a replacement for silicon in semiconductor chips, and as a replacement for silicon in solar cells. Clearly, considerable cost reduction will be necessary for these applications to be competitive.

Expanding Possibilities

Table 5.2 presents a history of the timing of the emergence of some of the principal materials in current use. By 1850, the possibilities from the established materials may have been approaching the technological limit. This was strongly

Table 5.2
Availability of Selected Nonfuel Materials from 1850

	Pre-1850	1850	1880	1900	1920	1940	1960	1980	2000	2005
Stone	X	X	X	X	X	X	X	X	X	X
Wood	X	X	X	X	X	X	X	X	X	X
Gold	X	X	X	X	X	X	X	X	X	X
Silver	X	X	X	X	X	X	X	X	X	X
Copper	X	X	X	X	X	X	X	X	X	X
Tin	X	X	X	X	X	X	X	X	X	X
Bronze	X	X	X	X	X	X	X	X	X	X
Wrought Iron	X	X	X	X	X	X	X	X	X	X
Traditional Ceramics	X	X	X	X	X	X	X	X	X	X
Carbon Steel			X	X	X	X	X	X	X	X
Tool Steels			X	X	X	X	X	X	X	X
Aluminum				X	X	X	X	X	X	X
Rayon				X	X	X	X	X	X	X
Bakelite					X	X	X	X	X	X
Synthetic rubber					X	X	X	X	X	X
Stainless Steels					X	X	X	X	X	X
Polyolefins					X	X	X	X	X	X
Nylon						X	X	X	X	X
Superalloys						X	X	X	X	X
Semiconductor Silicon						X	X	X	X	X
Kevlar								X	X	X
Modern Ceramics								X	X	X
Titanium								X	X	X
Carbon Fibers (CFRP)									X	X
Nanocoatings										X

hinted by frustration of the early efforts to achieve true parts interchangeability in the manufacture of metal objects. The following half-century saw the introduction of a number of new materials, such as cheap and abundant carbon steel and tool steels, which enabled more precise machining of materials better fitted to hold shape under the stress of machining. One thing can be said to have led to others, as the twentieth century has seen the introduction of a host of materials that have enabled a huge variety of novel and useful devices. It could be said that what was possible technically in 1850 could have been done with the same materials in 1750; but the same could absolutely not be said in 2000 in relation to 1950.

Notes

1. A polymer is a large molecule formed by joining small molecules, or monomers, with chemical bonds such that the weight of the polymer is an integral multiple of the weight of the monomer.
2. Burdick, Donald L. and William L. Leffler, *Petrochemicals in Nontechnical Language,* (Tulsa, OK: Pennwell Books, 1990), pp. 71-72.
3. Nathaniel Wyeth was a member of what is arguably the premier American artistic family; he was the son of N.C. Wyeth, the illustrator, and older brother of Andrew Wyeth.
4. United States Steel Corp., *The Making, Shaping, and Treating of Steel,* Harold E. McGannon Ed., (Pittsburgh, PA, 1964), p. 1111f.
5. Cor Ten steel is carbon steel alloyed with small amounts of copper and phosphorous and is stronger and more corrosion resistant than ordinary carbon steel. See John H. White, Jr., *The American Railroad Passenger Car,* (Baltimore: The Johns Hopkins University Press, 1978), p. 167f.
6. Battelle Memorial Institute, *Cobalt Monograph,* (Columbus, OH, 1960), p. 255.
7. Following the war, these government-owned plants became the basis for the entry of two other players into the aluminum business, Reynolds Metals and Kaiser Aluminum. This result came about as the outcome of a famous antitrust suit.
8. Silicon carbide is prepared by reacting silicon with carbon in an electric arc furnace.
9. Fairley, Peter, "Solar on the Cheap," *Technology Review,* vol. 105, No. 1. (January/February 2002), pp. 48-52.
10. Rotman, David, "The Nanotube Computer," *Technology Review,* vol. 105, no. 2, (March 2002), p. 38.
11. "Nanotubes: Everywhere Soon?" *Chemical Market Reporter,* (January 16-22, 2006), p. 13.

6

Government Fostering of Invention/
Innovation after 1940

An appreciation of the US government's roles with regard to technology during World War II and its aftermath is necessary to a full understanding of the progress of technology in the postwar period. A tracing of the history of many of the technologies that support the civilian economy in the early twenty-first century reveals the activity of the government. In addition to its role as a direct consumer of technology, government interaction with technology took several major forms: direct fostering, regulation, and tax policy. This chapter deals with the first, and the following chapter deals with regulation and taxation.

The Second World War brought the US Federal Government back into the business of fostering technology. This activity went beyond that of the nineteenth century in that not only was the government awarding contracts for small arms and the consumables of war (food, ammunition, and fuel), it was fostering research and development for the specific purpose of keeping up with or exceeding the weapons capability of the enemy belligerents. This effort produced a series of inventions and innovations on previous inventions, of which many eventually spawned important innovation streams in the private sector. What the government's fostering of inventive activity had in common with the nineteenth-century experience was that it sought no more than to provide for common defense; and there was generally no formal planning that took into account possible spillovers into the civilian economy from weapons-related research. Indeed, foreseeing the full range of commercial possibilities that subsequently flowed from these defense-related research efforts would have been far beyond the imaginations of individuals involved with the projects.

What the US did during and after World War II could not really be termed industrial policy. A formal industrial policy involves organized government effort to influence the allocation of capital to this or that industry. It places the government squarely in the role of making decisions regarding where profit is to be made in the private sector, and exposes industrial decision making to constraints that do not necessarily have anything to do with profit making.[1] In order to be successful as a fomenter of economic growth, a formal industrial policy demands from government planners a level of foresight not normally attainable by human minds. To appeal to one example, the Japanese experience suggests that bureaucratic industrial planners are backward looking.

At the war's end, the government sought to keep the spirit of invention alive by various formal means. These measures were spurred by the onset of the Cold War against the Soviet Union and its allies. The Soviets proved highly capable as appliers of technology to weaponry, and inspired competitive effort on the part of the US. The result was, as in the war itself, a series of inventions of which many were ultimately adapted and improved as commercial products by the private sector. Thus one of the most interesting aspects of postwar technology was the blurring of what once had been a clean division between military and civilian spending, i.e., the "guns vs. butter" choice.[2] Government spending for research and development, especially that related to defense,[3] eventually resulted in private sector activity, sometimes intensive, that enriched the general quality of life. In the course of postwar technology development, the government effectively shouldered the cost and risk of stage-one innovation in a number of lines of technology development. This included research activity that the private sector would probably have deemed to be unacceptably speculative. This was historically unprecedented because in all its previous postwar experiences the US had effectively disarmed. It was only in the post World War II period that the government responded to a need to maintain a large military establishment and to enhance war-making capability with the latest of technological possibilities.

Wartime Invention and Innovation

Wartime Regulation

During the war, research and development was focused narrowly on the immediate problem of prosecuting and winning the war. To this end, important sectors of industry that had sold to the private sector before the war were converted to war-related production, and important lines of civilian goods were shut down for the duration. For practical purposes, no civilian automobiles were produced during the war. Moreover, innovation in what were regarded as purely civilian lines of production was discouraged through the activities of the War Production Board (WPB). This was a wartime agency that had the power to review and approve or disapprove any design changes. In general, the policy of the board was to favor proven reliability in the form of existing items well tested in practice, and to reject any proposal that seemed to have any experimental element. The WPB also had the responsibility of allocating materials as between civilian and military uses. As a result of its efforts, for example, certain steel alloys all but disappeared from civilian markets.[4]

Radar

During the 1930s, scientists in Great Britain attempted to answer the question of whether an airplane could be shot down with radio waves. The answer proved to be "no," but it was established that radio waves of known speed reflected from an object such as an airplane could be used to ascertain the position of that object. Radar (for radio detection and ranging) was independently developed in the US and the UK and became an effective part of air defense in the Battle of Britain, and it also played a key role in the Battle of Midway. Radar has been subject to continuous improvements during and since the war; it has become the technological basis for both civilian and military air traffic control systems and a central tool of weather forecasting.

The wartime radar research program is of special interest because of the inventions that resulted from it, either directly or indirectly. Because microwave and radio radiation are adjacent parts of the total spectrum, a number of scientists who had been employed in radar research during the war shifted into microwave research following the war's end. One resulting invention was the microwave oven. The principle was discovered by Percy Spencer, who while working for Raytheon on magnetrons for radar sets, was working, so the story goes, near an active radar that at one point melted a candy bar that was in his pocket. He recognized what had happened and, by 1990, it was estimated that microwave ovens could be found in 95 percent of American homes. The microwave connection also played a role in one of the most important prototype inventions of the postwar period, for microwave research in the immediate postwar era led to the maser and the laser (see Chapter 9).

Computer

One of the employments for mathematicians and physicists during the war was the calculation of range tables for artillery and aerial bombardment. These took into account a variety of parameters affecting trajectories, including barrel angle, muzzle velocity, wind, and others, and involved very arduous calculations. In order to speed up the computation process, the government contracted with a group at the University of Pennsylvania for a machine to mechanize the calculations, and the result was ENIAC (Electronic Numerical Integrator and Computer) in 1946. ENIAC weighed 30 tons and could perform 500 arithmetic operations per second. It was not necessarily the first machine embodying most of the attributes that are commonly associated with the modern computer, for there were also computer projects in Britain and Germany at about the same time. However, ENIAC is commonly remembered as the start of the US computer industry of the postwar period. John von Neumann, who emigrated from his native Hungary to the United States, played a crucial role in the development of ENIAC and the British computer by having created a blueprint, or an "architecture," describing the logic of such machines. Inasmuch as this blueprint was conceptual, it allowed for considerable difference in the engineering detail of the machines that followed it.

Jet Engine

The jet engine, invented in the late 1930s, did not affect the progress of the war and appeared in the form of fighter aircraft only at the war's very end. Germany was just beginning to deploy the jet-powered ME-262 when allied ground forces overran the bases for these aircraft. This was fortunate for the allied air forces, for in its few combat appearances, the ME-262 demonstrated considerable superiority over the allies' best machines. Many ME-262s were captured on the ground after having been deliberately sabotaged by the Germans. The ME-262 was powered by an engine designed by Austrian Anselm Franz and developed in parallel with a British design. After the war, its engine was intensely studied and influenced early US and Soviet jet fighters. The British were able to license jet engine technology based on their own development after the war, and American designs did not come into their own until the 1960s.

Following the war, the military aviation of the principal military powers converted to jet propulsion within a period of several years. During the 1950s, jet power moved into the commercial aviation arena. Early pioneers of commercial jets included the Caravelle (Sud Aviation, France), the Comet (de Havilland Aviation, UK), and the Boeing 707 in the US. The 707 was a direct adaptation from a military project, as it was originally designed and built to satisfy the US Air Force's need for a refueling tanker that could keep pace with its jet bombers in aerial refuelings.[5] Conversion of military aviation to jet power started in the mid-1940s. The earliest jet applications were with small fighter aircraft, and subsequent development included, among other things, application to increasingly larger aircraft as the power of engines increased. Development of the KC-135 tanker and its adaptation as the 707 had become feasible by the mid-1950s.

Mention of the 707 leads to one of its more obvious impacts on civilian life: the spread of major league sports to the entire continental US. Historically, the range of major league professional sports was confined to the distance a team could travel overnight. When teams depended on rail transport, the extreme geographic sports range was from Boston to Kansas City, which a team could manage by scheduling an intermediate stop for a game and overnight stay. Piston-powered aviation speeded up planning, but jet power created the possibility for expansion of the leagues to the west coast. This was an example of how government-originating technology eventually created expanded commercial opportunities. In addition to this one example, the jet greatly speeded up general business communications.

Manhattan Project

The best known of the wartime prototype inventions was atomic energy, which grew out of the Manhattan Project. Apart from the direct weapons implication of atomic energy, there were three important branches: nuclear power, nuclear marine propulsion, and an assortment of civilian applications. Power was directly promoted by the Atomic Energy Commission (AEC), and in the two decades following the war, became an important source of electric power in the US and other countries.[6] By 1980, however, disposal of nuclear power generation's waste products was becoming a serious problem. Moreover, the economics of nuclear power had been severely compromised by an avalanche of safety requirements imposed by the AEC and successor organizations that had to be incorporated into current nuclear plant construction, and whose effect was to lengthen construction time and costs. This problem clearly reflected what can happen when a new and potentially dangerous technology is pushed into commercial application before there is a body of working experience with the technology. Safety concerns following the Three-mile Island mishap were powerful enough to turn public sentiment solidly against further development of nuclear power plants, and this, combined with the diminished economic appeal of the technology, effectively eliminated new nuclear power plant projects after 1980. Interest in this power technology was just beginning to recover against the background of high energy costs and air quality concerns of the early 2000s.

The safety problems that dramatically surfaced at Three Mile Island were found to be partly due to design flaws in the light-water reactor technology promoted by the AEC.[7] This design proved successful as a steam generator in marine propulsion systems used by the US and other navies. In the days of sailing vessels, the only limitation on a ship's cruising range was the supply of provisions for the crew that could be carried. When steam power replaced sail, the on-board fuel supply became a constraint on cruising range.[8] The introduction of nuclear steam generation meant that the fuel supply constraint was effectively lifted. This was especially important to the operations of submarines, as a boat could remain on station for months at a time, and such cruises became standard practice during and after the Cold War. Nuclear propulsion was never widely adopted with merchant shipping. There were a number of civilian applications traceable to the Manhattan Project other than nuclear power. These include imaging technologies used to detect defects in welds and metal parts. Several important medical applications relating to X-ray technology, MRI scanning, and CT scans have emerged (see Chapter 10).

Government Postwar Arrangements

RAND Corporation

Under the leadership of Gen. H.H. (Hap) Arnold, the Army Air Corps—soon to become the US Air Force—set up a "Think Tank"[9] named the Research and Development Corp. (RAND). It was established in 1946 as a freestanding division of the Douglas Aircraft Corporation of Santa Monica, CA. RAND was deliberately designed to have a flexible innovative intellectual environment with heavy emphasis on interdisciplinary contact. Indeed, the physical arrangement of offices was designed to promote casual as well as formal disciplinary interplay. RAND's charter included the right to initiate its own investigations and to refuse requests from the Air Force that management deemed to be not appropriate for the capabilities of the research staff.

RAND's early work largely revolved around problems posed by the Cold War, whose outlines were already becoming visible at the time of RAND's founding. These problems consisted of working out basing strategies for the Strategic Air Command and strategies for responding to a surprise nuclear attack by the enemy. RAND is credited with developing the science of systems analysis, which became a centerpiece of a variety of research efforts in the postwar world. By the late 1960s, RAND's exclusive relationship with the Air Force had become diluted, and the organization found itself working for other clients, including some from the business world. This trend has continued to the present time.

In 1960, Paul Baran of RAND was working on the problem of survivability of the military command and control system in the aftermath of a nuclear first strike against the United States. In basic concept, this system was built with a great deal of redundancy, the idea being that destruction of part of the system could be overcome by routing messages by a different path in the network. One problem was that even after redundancy was built in, how would a sender know which paths were open? Baran developed the concept of "packet switching," which enabled the operation of what has since become known as the Internet.[10] In packet switching, messages put into a network are cut into standardized blocks of data, each of which contains information about its origin, its destination, how long it has been in the network, and where it fits into the overall message. Whenever a packet encountered a defective node—e.g., one that had been destroyed in the attack—it would return to the sending node and leave instructions there not to send anything more by that route. When a packet found a functioning node, it would send a confirming message back to the previous node, and this process would take place until all packets were assembled into the full message at the intended destination.[11]

National Science Foundation (NSF)

The NSF, although usually not thought of as defense related, was inspired by achievements in science and technology that occurred during World War II. Founded by the National Science Foundation Act of 1950, it supports research in a wide variety of fields, including biology, physical sciences, geoscience, social science and economics, computer and information technology sciences, engineering, and arctic and Antarctic research. In the 1990s, the NSF provided approximately 20,000 grants in support of basic research in the United States.

National Aeronautical and Space Administration (NASA)

NASA was established in 1958 as an independent federal agency for the purpose of exploration of space both within and outside of the Earth's atmosphere. It oversaw a collection of assets formerly a part of the Army Missile Command, and its early senior technical management was composed heavily of individuals who had been involved in the German rocket programs of World War II, notably Werner von Braun. NASA's ties with national defense have been both overt and covert. For just one example, the launch vehicles for NASA's early space explorations were based on the same intercontinental ballistic missiles used by the Air Force's Strategic Air Command. Although the Space Program's impact on the civilian economy is not always in the forefront of the news media, for whom graphic results of deep space probes are of more interest, they are significant. The Space Program provided a powerful impetus toward miniaturization of various devices that is so much of everyday life in the early twenty-first century. There were some contributions other than these. The author once attended a NASA public presentation in which the speaker took questions from the audience. One questioner, who sounded hostile to spending money on space exploration, asked what NASA had contributed to the general economy. The speaker said "draft beer in cans." This riposte was not as facetious as it might seem. NASA wanted to provide this astronaut morale builder, but found that known pasteurization processes completely destroyed the flavor traits that beer drinkers prize in draft beer. Moreover, the FDA insisted on pasteurization of any fluid packaged in cans. The answer turned out to be an irradiation process on the filled can that preserved the draft quality characteristics.

The proximate impetus toward NASA's founding was the Soviet launch of Sputnik, the first artificial Earth-orbiting satellite in 1957. Its most famous achievement was the manned landings on the moon as a result of a project started in the Kennedy Administration. NASA has launched unmanned space probes for the extensive exploration of the solar system. Earth-orbiting satellites serve such functions as weather prediction, collection of data on terrestrial resources, and the global positioning system, which has proved a highly valuable tool of the military in the Afghanistan and Iraq campaigns and is becoming a useful tool in a number of civilian activities.

The Soviet achievement of being first into Earth's orbit, represented by Sputnik, came as a shock to many Americans with a strong sense of competitiveness in sports. However, this disappointment had some hidden benefits for the US that became extremely useful. Most importantly, the event provided the political cover for a civilian space program that continued what had previously been progressing as an arm of military spending. President Kennedy's moon-landing goal found considerable popular support. The effects for general economic growth are mentioned above. Moreover, the separation between the civilian space program and national defense was never as clean as was the general public impression. The military was at least as much a gainer from the space effort as was the civilian economy.

Defense Advanced Research Projects Agency (DARPA)

DARPA is the basic and applied research and development facility of the Department of Defense. Where RAND has been mostly in the concepts business, DARPA has been associated with considerable military hardware development, including the development of some ideas that involve such high risk as not to be attractive to the private sector. Much of this work has been done under contract, typically in university research facilities. DARPA has worked on a number of projects that have proved to have commercial development potential. DARPA was originally organized as ARPA (Advanced Research Projects Agency) in response to the Soviet launch of its Sputnik satellite in 1957.

The first primitive form of the Internet was established in 1969 by DARPA's predecessor, ARPA, to facilitate sharing of files among government research contractors, mainly universities. This system, known as ARPANET, made use of the new technology of packet switching, and included simple mail transfer protocol (SMTP, or e-mail) for short messages and file transfer protocol (FTP) for longer files. ARPANET quickly became an essential tool among university and other organizations' researchers. Not all DARPA-sponsored research has commercial applicability. Some recent projects include exoskeletons for human performance augmentation, human identification at a distance, global positioning experiments, friction drag reduction (for ships' hulls), and micro air vehicles (a small air vehicle a soldier can carry and launch to gather information about terrain ahead and enemy positions). An early DARPA-sponsored development was a ceramic heat-resistant material for the nose cone of the Snark—a primitive cruise missile—which is today commonly known by its commercial name, Corning Ware.

Internet

DARPA played a pioneering role in creating a small-scale version of the Internet among its research contractors. The National Science Foundation (NSF) worked with DARPA to expand the network to all the scientific and academic community. In 1985-1986, the NSF funded five supercomputing centers at five research universities and also the development of NSFNET to connect these centers. The present-day Internet was the result of the interconnection of NSF-NET, ARPANET, and other networks in the US and Europe, including various private sector networks. From its origins in the 1970s, control of the Internet has shifted from the government to the private sector, mostly with the government's blessing and encouragement. This shift was hastened as a result of a study by the NSF, which revealed that the private sector was providing a facility adequate to researchers' needs; this allowed NSF to withdraw financial support from NSFNET. Internet technical standards are maintained by an international body, the nonprofit Internet Society, with headquarters in Reston, Virginia.

Commercialization of the Internet has made rapid progress since the mid-1980s. The first big push came from the joining of e-mail service with the Internet in 1988. In 1993, the World Wide Web appeared. Early browsers were based on Mosaic, a program developed at the University of Illinois. Mosaic incorporated a number of access protocols and display standards originally developed by Tim Berners-Lee at the European Organization for Nuclear Research for the World Wide Web. In 1994, Netscape Corporation was organized to develop Mosaic, and later, the Microsoft Corp. became interested in developing its own browser, which, at least initially, was based on Mosaic. For the private sector, the problem of the Internet was how to make money with it. Internet startups were prominent in the stock market bubble of the late 1990s and also in the collapse of the bubble. Several firms managed to survive the downturn, and now appear to be commercializing the Internet successfully.

The Internet has two of the characteristics of the prototype invention in that it grew out of an arrangement for making military communications more secure in a war crisis, and it has come to be used for purposes far beyond the concepts under which it was originally organized. In its initial form, it embodied the assumption of legitimacy in use, and therefore has no defense against many of the problems which have become familiar to users. These include spam, crime of various sorts, and destructive mischief, such as virus-laden e-mails. These and other problems have elicited predictions of the Internet's collapse during the last ten years. Not only has this not happened so far, but the system seems to work with remarkable smoothness. However, the problems are coming into focus, and a number of thinkers are conceptualizing how the system may be fixed. One of the fundamental questions deals with whether the problems cited are really net problems, or do they simply reflect inadequate design of Internet application software.

Global Positioning System (GPS)

The GPS is a worldwide navigation system that relies on radio signals broadcast from a number of Earth-orbiting satellites. The signals can be interpreted by computer-directed ground instruments to determine terrestrial position with very high accuracy. The system was the result of an Air Force program begun in 1970, and it was developed under the direction of project director Col. Bradford W. Parkinson. It became operational in 1995 and is the basis for a number of weapons systems capable of delivering munitions to a small target with high accuracy and at greatly reduced risk to the lives of pilots of aircraft delivering the munitions. GPS is also enabling a number of technologies in the private sector.

The GPS relies on twenty-four satellites at an altitude of greater than 11,000 miles, of which eight have line-of-sight communication with any point on Earth at any time of day or night. Each signal contains information on the position of the sending satellite, and the broadcast time. A GPS receiver can use signals from a minimum of four satellites to determine latitude, longitude, and altitude. Accuracy is to within ten meters. Other broadcast signals, at first interpretable only by the military, increased the accuracy to within one centimeter. In 1996, however, an executive order made the higher accuracy version usable by all GPS users, civil as well as military.

GPS had a number of enabling threads of development. Satellite launching technology was the most obvious. Another was the atomic clock, having an accuracy of one second per 100,000 years. The atomic clock came from the late 1930s, and it was developed with a seemingly unrelated purpose: that of testing the predictions of Einstein's general theory of relativity. Each satellite is equipped with four atomic clocks. A third thread was the discovery, following the Soviets' launch of its Sputnik orbiter in 1957, that the precise orbit of a satellite could be calculated from the Doppler effect of the radio signals emitted by the satellite. The GPS reverses this computation in that it starts with a known orbit and calculates terrestrial position. There is a growing list of civilian applications for GPS. It includes enabling emergency vehicles to pinpoint destinations and routes, keeping track of truck fleets to speed deliveries, keeping track of the positions of ocean shipping, surveying, aerial navigation, and automotive navigation systems. Unlike most government research and development projects, many of the civilian uses of GPS were explicitly anticipated by the system's developers.

Energy Independence

In 1973, Egypt attempted to regain control of the Sinai Peninsula, which it had lost to Israel in the Six-Day War of 1967. A massive US airlift of war material to Israel so angered the Arab oil producing countries that they introduced the "oil weapon," an attempt to deny oil supplies to the US by means of an export embargo. The effect of this was a temporary shortage of motor gasoline for several months; this is remembered as a time of long queues at gasoline stations. In 1979, deteriorating relations between the US and Iran as a result of the Iranian Revolution led to another brief period of energy shortage in the US. It was from these shortage scares that the political impetus toward regaining US independence of foreign sources of petroleum arose. Part of the US Government's response to these conditions was the encouragement of a variety of alternative energy research projects in an attempt to find economical substitutes for petroleum products. These focused on oil shale, coal, wind power, biomass, and solar power. In terms of stated goals, all these projects failed, and financial support was terminated during the 1980s. However, interest in these measures has revived in the current episode of high prices of petroleum products.

A prime example of these technical misadventures was the Synthetic Fuels Corporation (SFC), established to find an economical means for converting North Dakota lignite, a very low-grade coal, into pipeline quality synthetic gas. The plant for this enterprise was conceived in the early 1970s, but no construction took place until 1980. The plant was built at a cost of $2 billion, of which 25 percent was financed by a consortium of five energy companies and 75 percent by Department of Energy (DOE) loan guarantees. The first synthetic gas sales commenced in 1984. By then, the political picture had changed radically, and the price of natural gas had fallen. DOE refused a request of the private consortium for additional funding, whereupon the consortium defaulted on the government-guaranteed loan and DOE repossessed the plant. In 1988, DOE sold the plant to a local power cooperative for $78 million, or pennies on the dollar from the standpoint of the taxpayer.[12] The buyer was Basin Electric Cooperative of Bismark, ND.

Ironically, the plant, known as the Dakota Gasification Co. (DGC) has produced some useful technology under its present ownership. It is apparently profitable on its present capital structure when natural gas prices are high.[13] By-products account for about half of the plant revenues. These include xenon and krypton, which go into the lighting market (3.6 million liters/yr), anhydrous ammonia fertilizer (1000 tons/day), phenol for the resins market (35 million lb/yr.), liquid nitrogen (varying amounts), cresylic acid for the chemical industry (33 million lb./yr), and carbon dioxide, which is going to the oil industry for enhanced oil recovery.[14] DGC has pioneered an ammonia-based flue gas desulfurization (FGD) system that has been studied with interest by a number of utility technical people from around the country and is believed to be superior to some of the existing FGD systems based on lime and limestone. Present management's strategy includes aggressive development of by-products and their markets to reduce dependence on natural gas prices for profitability.

National Institutes of Health (NIH)

The NIH is an agency of the Department of Health and Human Services under the Public Health Service. It is the largest single supporter of biomedical research in the country. NIH is composed of twenty-five specialized institutes corresponding to various fields of health and disease: diabetes and digestive and kidney diseases; allergy and infectious diseases; child health and human development, dental and craniofacial research; environmental health sciences, general medical sciences, neurological disorders and stroke, eye, aging, and arthritis and musculoskeletal and skin diseases; and others. Almost all of the research funded by the NIH is conducted in medical schools, universities, and other nonfederal institutions. The primary form of funding is the research grant.[15]

Industrial Policy and Technology

The term "industrial policy" has come to mean government regulatory intervention into the allocation of capital to different industries in order to further some economic growth objective. A nation's industrial policy can take any of a number of forms, ranging from a formal policy administered by a cabinet-level organization, such as Japan's Ministry of Industries and Trade (MITI), to a scatter of regulations whose effect is to channel capital into selected industries, but not under the guidance of any systematic body of coordinated thought. By this standard, there is variety of measures that could be interpreted as industrial policy, including various protectionist measures in international trade. The US has never resorted to the most formal form of industrial policy. Rather, it has relied on *ad hoc* legislation and regulation, some of whose effects resemble those of industrial policy.

The historical argument for industrial policies has revolved around the protection of infant industries in developing countries. Whether or not highly protectionist policies have actually benefited nascent industries in different countries is open to debate. It is widely believed that US growth to industrial power in the nineteenth century benefited from high tariff protection, but the twentieth century affords numerous examples of situations wherein high protection relieved so-called infant industries from any real pressure to operate efficiently. For India and Argentina, the result of protection was domestic industries that were incapable of surviving under competitive pressures when they were eventually exposed to foreign competition. In the worst example, many countries attempted to protect domestic employment by means of high tariff walls during the 1930s. Competitive protectionism is credited with the massive collapse of world trade during that decade that exacerbated depression conditions. It was in reaction to this collapse of trade that postwar arrangements for encouraging trade were formed, including the General Agreement on Tariffs and Trade (GATT). This has been the worldwide vehicle by which nations have undertaken in coordinated fashion to ease restrictions on trade in the postwar period.

Today industrial policy, including protectionism, is attracting growing interest, because diminishing barriers to trade in the last twenty years, while contributing positively to the general increase of wealth, especially in the developed world, have caused economic dislocations. Many people and groups of people that have been negatively affected use their political influence to arrest or retard the progress of free trade that the GATT agreement seeks. Moreover, even those countries that are overall most interested in fomenting free trade seem to be unwilling to expose various internal well-organized groups to the full force of international competition. This certainly includes the US, which, like the European Community and Japan, favors domestic agriculture with substantial income support measures that have implications for world trade.

The nexus between measures that can be construed as industrial policies and technology shows up in governmental interventions in energy markets. Recent years have seen an upsurge in the US and other governments' fostering of energy alternatives to the traditional (oil, gas, and coal) by means of subsidies. These measures have, to some extent, been political responses to the high gasoline prices starting in 2005, but the 1970s saw a number of legislative and regulatory efforts in response to the surge in energy prices, especially motor gasoline prices, which occurred in that decade. In the US, measures have been hastily enacted and implemented, and for that reason are of suspect wisdom. In many cases, the technology involved would *increase* the costs of motor fuels to the consumer relative even to the high levels they attained in 2005. There is no better example than US encouragement of the use of corn-based fuel ethanol to extend gasoline. Using corn ethanol this way has been compared with using sirloin to extend hamburger.

One proposal being pushed officially is to make fuel ethanol from various organic wastes instead of corn. The appeals of this are that the feedstock for the process would be far cheaper than corn, and production of large volumes of fuel ethanol would not have as much impact on food prices as with corn ethanol. However, the process to alcohol using these feeds is at present sufficiently more costly than that from corn as to overcome any advantage from using cheaper feeds. Other specific encouragements to domestic energy production have included a tax credit for dollars expended on tertiary oil recovery, and laws forcing utilities to buy electric power from cogeneration projects. There are other examples. These are measures that might be expected to be included in a formal industrial policy, but to label them as collectively such a policy would be to infer more organized purpose than was probably behind their implementation. Rather, the entire effort appears to be a politically motivated scattergun approach intended to give the appearance of responding to high energy prices that have upset many voters.

The US has sought to encourage the use of corn-based ethanol and soybean-based diesel fuel with tax subsidies. Subsidies combined with high gasoline prices have made corn ethanol almost competitive with gasoline. However, applying subsidies to these two alternative transportation fuels has had several deleterious effects. First, the subsidies favor a present generation of biofuels over possible alternatives that would use less costly feedstocks and be less costly to distribute, for the subsidy support does not support possible next-generation renewable fuels, such as cellulosics-based ethanol or butanol. While this discrimination in favor or corn ethanol and biodiesel from soybeans has not fully discouraged research into potentially superior renewables, the subsidy policy seems shortsighted. Corn ethanol is not an efficient way to energy independence (from oil exporters) because the premium it delivers over the energy cost of its production is small. It is not an efficient way to cleaner air because it cannot be used in most automotive motors other than in the form of admixtures in gasoline of 15 percent or less. At best, it can be defended as an interim stage between present technology and future development of renewable fuels with higher energy premia, higher energy density, less costly distribution, and less impact on food prices. As of 2006, the corn ethanol industry was experiencing a boom that was supported on thin advantages of energy independence and reduced pollution.

A formal organization of industrial policy in the form of a cabinet department (or ministry) empowered to intervene in the private sector's allocation of capital has the appeal that this seems to be a way of systemizing what might otherwise be a random scattering of legislated and regulatory measures that collectively appear to lack a central purpose. This is the model that the Japanese have taken. The approach has at least two weaknesses. First, the people who make the decisions within such an organization are human and are subject to the same limitations as are all other people: their thinking is heavily reflective of their experience, and this gives the decision process a strong backward-looking bias. This is not desirable in a body charged with seeing *future* opportunities. Second, such a government department could hardly be expected to be free of political influence stemming from a host of motivations, most of which would have only the thinnest possible relation to the goals of the industrial policy organization. Much of these sorts of problems can be found in the Japanese experience. For example, at the time when several Japanese industrial organizations were beginning to explore the possibilities of consumer electronics, MITI was pushing steelmaking, shipbuilding, and automobiles! One might deplore the US style of *ad hoc* ventures into industrial policy, but the US had demonstrated some capacity for abandoning projects (eventually) that prove hopelessly uneconomic in spite of the research resources they absorb.

Government Technology Encouragement: Critique

Successes and Failures

Generally, the government has done very well with encouraging technology when it has attempted only narrow objectives, such as possessing itself of weapons systems. This has been a successful pattern ever since the founding of the country. From the early efforts of Whitney and others to develop interchangeable parts through the Manhattan Project and the postwar support of risky research efforts, there have been commercial after-effects that have played crucial roles in underpinning the quality of life enjoyed by Americans. The absence of military-inspired research leading to prototype inventions during the interwar period of the twentieth century was at least an exacerbating factor in the Great Depression of the 1930s.

One of the less desirable results of the Manhattan Project was that it left a lasting impression among Americans and their elected representatives in Congress that anything can be achieved technologically, provided that sufficient talent and money supported the effort. One reason for that project's success was that war conditions created a political atmosphere favorable to huge public funding for a project that the public generally would not have understood even had it known the details. This war condition was probably what President Carter had in mind with his "moral equivalent of war" statement in 1979. This was the best he could do in the context of the aftermaths of the Iranian Revolution, but moral equivalence simply did not attain the level of real war in the minds of the electorate and its elected representatives when it came to tolerating expensive projects with no real end in sight. Many of the energy independence projects were either abandoned or severely modified with the change in political atmosphere that occurred in the 1980s.

The Manhattan Project had at least two other advantages over the energy independence attempts of the 1970s. First, its science was sound, for by the time the project was started in 1941, the theory of the atom was well advanced. From there to the spectacular culmination of the project, the challenge was difficult and hugely expensive, but the effort consisted essentially of engineering from understood scientific principle. There never was any dependence on new scientific breakthroughs. The same cannot be said for a project like the Synthetic Fuels effort, for its basic components, such as coal gasification, had been around for years but had never been able to compete effectively with traditional oil and gas on an economic basis. Success was dependent on scientific breakthroughs that would dramatically reduce costs, but that has not happened even to this day.

The second advantage was that the Manhattan Project had no implications for any commercial market that then existed. Its progress could not be rendered irrelevant by a fall in the price of some commodity, such as natural gas. The Synthetic fuels corporation's efforts were highly subject to commodity price risk. Whereas nuclear power eventually entered the commercial power market as an established technology in competition with other established technologies, synthetic gas never had that advantage. A third advantage for the wartime nuclear project is that the War Production Board had effectively suppressed other efforts based on novel designs, so that there was reduced competition for resources of scientific and engineering talent.

Resource Diversion

One criticism that is sometimes leveled at government encouragement of inventive activities is that it draws talent away from purely private sector activities, potentially crippling them. This criticism has at least a superficial appeal in light of the ubiquity of government support for invention, especially since 1940. However, like many such criticisms, this one misses the complexity of the inventive process and does not stand up to close examination. Again, the example frequently offered is the Manhattan Project, but that effort took place under circumstances especially favorable to it that are not likely to be replicated, especially in peacetime.

In the first place, the supply of engineering and scientific talent is elastic—at least over the time period needed to train people. If there is competition for talent between government-sponsored activities and strictly private inventive endeavors, this will show up in the form of increasing starting salaries. This condition will quickly become known among university faculties and students, and will attract increasing numbers of students into engineering and related disciplines. Moreover, technically trained people can be imported according to experience of recent times. Thus, an increasing demand for talent will elicit increasing supply after a delay of the time necessary to train talent. Of course, not every newly minted engineering or science graduate will be of the intellectual disposition to do creative inventive work, but increasing the size of the general pool of technically trained people increases the probability for finding real inventive talent.

In the second place, what motivates an inventor? It is far from clear that money plays a decisive role beyond satisfying the inventor's immediate need for income. What the inventive person does require is a work atmosphere that allows free rein to creativity, including a high tolerance for risk taking, and the would-be inventor will tend to seek out employers that offer this. If research shops deriving part or all of their support from government contracts are omnipresent in the field of job opportunities for this kind of applicant, then these shops will get much of the available talent. This is not the same as drawing talent away from the private sector, which is perfectly able to bid on the talent it needs. RAND was highly successful at creating the kind of working atmosphere that encouraged creative thinking. Xerox's PARC (Palo Alto Research Center), founded in 1970, was spectacularly successful in assembling an intellectually stellar group of minds, who proceeded to invent the personal computer, the laser printer, and a number of other devices that enabled the growth of personal computing in the 1980s and 1990s. PARC benefited from leadership with enough character to resist pressure from corporate headquarters that would slant research toward development of existing products. PARC is remembered for its owner's failure to capitalize on most of its inventions, but this failure was over what to do with the results of a highly successful research effort. PARC deserves to be a model of how to set up and run a research facility.

Assumption of Risk

Within the stream of innovation following a prototype invention, the first stage is by far the most risky as it requires the expenditure of money that may or may not result in a return. Among the risks is that some other organization is also working on the same problem and will get to market first. In some instances, an alternative technology in the hands of a competitive organization could prove to be superior in the market. As a general rule, professional managements are paid to minimize risk, and are loath to involve themselves in enterprises that are regarded as highly speculative, either as to technical success or to size of market. If the government shows a willingness to assume the risk involved in first-stage innovation, much of the private sector has an incentive to let the government do it.

Part of the ubiquity of government in the generation of prototype inventions can be attributed to the willingness of the Department of Defense and its predecessors to undertake and support research with little obvious relationship to commerce. In some ways, the government has an advantage in risky technical pursuits, for it does not face periodic performance judgment by financial markets. In some sense, government sponsored research may be vulnerable to impatience in the electorate as in a Manhattan-sized project, and this was at least part of the reason that some of the alternative energy projects of the 1970s were terminated; but in the cases of projects funded by agencies such as NIH, NSF, and DARPA, where there is often no detailed media attention, electorate impatience is less of a problem. Despite certain examples seemingly to the contrary, such as PARC, corporate research appears to respond to a need to stay close to the market. It has been estimated that corporate research and development spends 80 percent of its time and talent on "product improvements" and 20 percent on novel research.[16]

Patent

The US patent system is probably the earliest example of public policy encouragement of invention, the US Patent Office having been set up in 1797. The traditional role of the patent system has been to reduce the risk inherent in stage-one innovation by guaranteeing an inventor a term monopoly on commercial products resulting from his or her invention. The recent difficulties of the patent system have received much publicity. At least some of these problems arise from severe overloading of the Patent Office and have been aggravated by Congress' failure to define precisely what is and is not patentable in the world of information technology. The result has been intense litigation in the form of infringement suits.

Naturally we tend to be preoccupied with problems of our own time, but the present-day problems with the patent system are not new. Something resembling today's patent patterns seems to recur during periods of high technological flux. George Selden, an American, did not invent the automobile, but he was granted a US patent on it in 1895. He and associates used this patent to get themselves a handsome royalty income from the nascent automobile industry. On the face of it, it seems like the Patent Office ignored any requirement that the invention be truly novel. Henry Ford refused to pay tribute to Selden, took him to court, and at the appeals level succeeded in having the Selden patent set aside in 1911.

Notes

1. The Mexican national oil company, Petroleos Mexicana (PEMEX) has always had a reputation for high inefficiency when evaluated as a business enterprise. PEMEX, for most of its history, makes considerably more sense if viewed as a jobs-providing social welfare organization.
2. Blurring of the guns-butter dichotomy is not the same as eliminating it. Expenditures for military consumables is still purely "guns," and these expand dramatically in times of actual conflict.
3. Defense-related expenditures averaged 57.3 percent of total federal R&D expenditures between 1970 and 2004. The ratio stood at 57.5 percent in 2004, the latest year available. These figures do not include space expenditures. The defense-to-total ratio was somewhat higher from 1983 to1992, having ranged from 59 percent to 69 percent. These figures were computed from data on federal and nonfederal research and development spending compiled by National Science Foundation, Office of Science Resources Statistics.
4. In one example, the Electro-Motive Division of the General Motors Corp. had introduced a successful line of freight diesel-electric locomotives several years before the war, and these proved popular with railroad companies. Under the WPB, no expansion in productive capacity for these locomotives was permitted, and production was carefully rationed among customers. One batch was awarded to the Santa Fe system for use west of Winslow, Arizona, a region in which boiler water supply was a major problem. Other allocations depended on the customer's demonstration of some serious difficulty with the operation of steam power.
5. As a commercial airliner, the Boeing 707 was not the first to the world market but ended up dominating it. The "first" was the de Haviland Comet, which lost the advantage of its lead because of a tendency to explode in mid-air, a problem eventually traced to structural design defects. The Caravelle had a long and successful commercial life, but was a comparatively small aircraft adapted to relatively short hauls. The 707 was the first truly successful general-purpose airliner.
6. By 1975, nuclear power plants provided just over 20 percent of US electric power.
7. Recent designs are said to have the advantage of being much more tolerant of human errors than the light-water design. For a discussion of the circumstances of the selection of the light-water reactor over alternative reactor technologies, see Chapter 14 below.
8. Initially, the fuel tended to coal. The need for coaling stations for the world's navies was often a justification for the late-nineteenth-century acquisition of colonies by the US, UK, and other naval powers.
9. "Think Tank" was a new term at the time, having been imported from the British.
10. The term "packet switching" was coined by British physicist Donald Davies, who had independently developed the concept in the mid-1960s. Paul Baran had run into

trouble with his version by being several years ahead of its time: in 1960, decision makers both inside government and in the private sector had not yet learned to think in the requisite digital terms.

11. "Nodes" are the intersections within the redundant network and are computers.
12. Kent E. Janssen, "Lignite-to-Gas Plant Reveals Numerous Innovations," *Power,* (May-June 1997), pp. 79-82.
13. Profitability is helped by a low capital cost and presumably low debt, the DOE and the taxpayer having eaten the initial debt. This makes the plant irreproducible in a financial sense.
14. Kent E. Janssen, p.79.
15. The prohibition of federal funding of stem-cell research imposed by President George W. Bush has fairly transparent political origins. It does not prohibit the general area of research funded from other than federal sources, however. The bitterness of the reaction of the scientific community to the prohibition hints at the high level of dependence on federal funding by scientific research and development activities.
16. Anderson, Howard, "Why Big Business Can't Invent," *Technology Review,* Vol. 107, No. 4, (May 2004), 56-59.

7

Government Regulation and Technology

Normally one does not closely associate government regulation and the progress of technology or the lack of it. However, innovation does not take place in a vacuum, and, especially in its second stage, it will be cognizant of the *regulatory background.* Regulation of economic activity in the economy can be thought of as a constraint placed on production functions in the affected industry for the purpose of achieving a publically desirable result. In a steady state of regulation, the business activities affected will adapt to the regulation. Whether or not the regulation affects the level of innovative activity in the affected industry depends on the nature and quality of the regulation. It happens sometimes that the adaptation to the regulatory background diminishes managerial incentives to innovate.

The situation is much different when there is a major change in the regulatory background. Such a change acts like the addition or removal of constraints from the production function, and provokes a reaction that can strikingly resemble the impact of a disruptive prototype invention. Relating the impact of a change in the regulatory background to that of a prototype invention is not purely metaphoric. There is a historical tendency for a regulatory regime, once established, to persist even though the circumstances that may have justified it in the first place change radically. In cases wherein an established regulatory pattern that has been in place for many decades, such as with surface freight transportation and natural gas pricing, the market structure conditions under which the regulation was initially established had changed radically well before the time of deregulation. In both the transportation and natural gas cases, new competition was enabled by technological change. One can think of the removal of a regulatory regime that has been in place for a very long period as confronting the affected industry with the accumulated effects many years of technological change *all at once.* In effect, the regulation had prevented managements from adapting to the competitive changes affecting their markets as they occurred.

In some instances, adaptation to the changed regulatory situation can be achieved by modifications of currently used technology and practice, but in others, compliance by modification is sufficiently costly that the technology in use has to be replaced in its entirety. This chapter does not undertake to explore all possible examples to be found in the late twentieth and early twenty-first centuries, but it does offer outstanding examples by way of illustrating the above-stated principles: the impact of the deregulation of railway freight transportation, of motor carrier deregulation, of natural gas wellhead prices, the impacts of the clean air and clean water regulations that developed after 1970, and the unfolding regulation of important emerging parts of the telecommunications and electric utilities industries.

Freight Transportation Deregulation

Railroad Deregulation

The railroads were the first major industry that became subject to federal and state regulation. Federal regulation came by authority of federal legislation passed between 1880 and 1920. To illustrate the effects of regulation in the years prior to 1980, consider what a company management is basically called upon to do. Decisions have to cover what to produce, how to price the product, what factors of production to employ, and at what cost, and how to budget scarce capital to the furtherance of the business. Under regulation, authority over how to price the product was heavily compromised by rules whose effect was to make price changes so difficult in many instances as not to be worth pursuing. A proposed change in rail rates was subject to a waiting period of thirty days, during which time any interested party, including competitors, had a right to protest the change. Because a change in rates of any magnitude was certain to threaten some interest, opponents had an effective veto over the change, and as a result, even justifiable changes either never took place or were severely modified by the Interstate Commerce Commission (ICC), the regulatory body. Ratemaking ability later turned out to be crucial to the railroad companies in holding existing business and in gaining new business. No better example of this can be found than the case of fresh fruits and vegetables. Rail rates for these commodities were effectively deregulated about a year before general rate deregulation came with the passage of the Staggers Act of 1980. Prior to rail rate deregulation for fruits and vegetables, trucking companies would price to take the business for most of the year except in certain rush seasons, in which truck rates would be raised sharply. When this happened, the shippers would turn to the rails, which were unable to match the truckers' rate flexibility and consequently were in the position of maintaining an expensive capital structure solely as an overflow facility that was used only in the peak shipping season. After deregulation, railroads were able to compete for the business based on price, with the result that significant business that had been lost to highway carriage was attracted to rail haulage.

Decisions over what factors to employ were subject to veto by labor unions, which blocked a number of potentially productive work rules changes. Although not directly affected by regulation, the unions' position had been weakening in the 1970s, and this became sharply apparent with the liquidation of the Rock Island system. Under the assumption that the government would not let a major railroad be liquidated, the unions demanded the terms of a national pattern wage settlement, which the Rock Island demonstrated that it could not provide. The unions struck, shutting down the Rock Island, which was already in its tenth year in bankruptcy, and the result was the liquidation of the company and the loss of all its jobs. Pieces of the system survived as short-line companies, but these organizations generally were non-union.

The only things left for rail management under regulation was the capital budgeting and merger and acquisition functions.[1] Managements achieved significant cost reductions by such measures as replacing steam locomotives with diesel-electrics and industry consolidation through mergers, but the potential for fending off bankruptcy and possible nationalization by these measures proved limited. The cost-saving potential of these measures could not be realized without major changes in work rules, and these were successfully opposed by rail unions for many years. For example, many management decisions to replace steam with diesel power date to the late 1930s. Diesel locomotives do not require the services of firemen, but union power was sufficient to preserve firemen's jobs until the 1980s. Under regulation, the ability of rail managements to withdraw from loss activities, such as carrying passengers and branch lines that had long ceased to be profitable, was severely restricted. Deregulation changed this by making abandonments of routes and unprofitable services much easier to accomplish.

It would be difficult to exaggerate the profundity of the change that ratemaking deregulation brought to the railroads. In any principles of economics text, there is listed a number of conditions defining a perfectly competitive market, including a large number of choices available to the buying side of the market, ease of entering the business, and others. Under ICC regulation, all of these competitive criteria were forgotten except the one about large number of choices. To achieve large numbers, the ICC defined routing options between pairs of terminals and maintained rates to cover the costs of the least efficient carriers between each pair. Under this system, one major rail system enjoyed a profitable business hauling lumber from Oregon to St. Louis by way of Los Angeles and San Antonio. In another example, forty-five routing choices between Dallas and Kansas City were available to the shipper! Naturally, the carriers with the geographically most advantageous route structures had little incentive to provide capacity over their short routes that efficiency might otherwise justify. The obvious effect of this system was to eliminate cost efficiency as a determinant of routing. To deregulate ratemaking was to create the possibility for dramatic increases in the efficiency by which goods could move. Eventually, the effect of rate deregulation was to create actual price competition among carriers, and this eventually caused traffic to gravitate to the most cost-efficient routes. The implications of this for productivity in the business of moving goods should be obvious. It was an important result in a large country characterized by great distances.

How long did the adjustment to the competitive marketplace take? Does a management thoroughly accustomed to operating under one set of rules have the ability to take immediate advantage of opportunities created by a liberalization of the rules? Based on the actual reaction, the answer would seem to be "no," and this is not surprising. Under an established set of operating rules, members of management would have been recruited for skills in doing things that managements could do, and these were the skills that would have been honed through working experience. Where were the skills necessary to make profitable use of general ratemaking freedom[2] to come from? In many instances, the initial reaction to deregulation was denial and attempt at proceeding in the accustomed manner, followed by cautious experimenting with newly permitted options such as pricing strategies. It took time to determine which of these would result in new and profitable business.[3] All in all, the adaptation process took years, and there is at least some evidence that it was still being worked out two decades after deregulation. For example, some carriers experimented with contract rates—allowed under the new rules—in which the actual rate was concealed in the structure of the contract and could be known only to the parties to the contract between carrier and shipper. At least one major system appears to have abandoned this approach on the grounds that the rates embedded in a contract with one shipper would not be visible to competitive shippers, who would resent the possibility that their competitors were getting a better rate.

The effect of this deregulation was strikingly like the impact of a prototype invention, including the aspect of threat to established interests. The lag between deregulation and the successful taking advantage of it resembles a first-stage innovation period; only the reason is not a wait for enabling hardware technology, but a wait for human managers to retool themselves to understand their new situation.

Motor Freight Carriers

In the same year as the Staggers Act (1980), which deregulated railroads, Congress passed the Motor Carrier Act, which substantially deregulated motor carriers. Under ICC regulation, a motor carrier operated under authority granted in the form of a Certificate of Public Convenience[4] that spelled out precisely what the carrier could do; and limited the carrier's activity to the activities spelled out in the certificate. For example, the authority might cover haulage from Atlanta to St. Louis, but unless explicitly granted, this did not imply authority to haul from St. Louis to Atlanta. Needless to say, these rules produced many empty truck miles. Rates were set to cover the costs of empty back hauls. Under deregulation, rates were no longer imposed by the ICC but became competitively determined, and backhauls were permitted. It does not take much imagination to see that this latter change would have a profound effect. A backhaul in the hands of a competitor for your front haul could cover his out-of-pocket costs for the return trip, and enable him to undercut your front haul rate. Thus, in a competitive environment, the *right* to acquire a back haul becomes the *competitive necessity* for doing so. As with the railroads, trucking company managements did not react immediately to this situation. In 1985, truckers were still coping with the idea of securing back hauls.[5] As in the railroad case, managements had to confront the problems of operating in a competitive environment and adapt accordingly. Thus, the motor carrier case fits the pattern in which the abrupt change in the rules acts in the same way as a prototype invention, and the period of industry adaptation resembles a stage-one innovation period. The elimination of many miles of empty truck movements due to seeking backhauls was an obvious contributor to increased productivity in overland transportation and distribution of goods.

Natural Gas Price Deregulation

Viewing the termination of a long-standing regime of government regulation as the equivalent of a disruptive prototype invention focuses attention on the affected industry's reaction to the deregulation, and this is the primary interest in this study. However, it should be recognized that a deregulation that returns the affected industries to rule by the marketplace, is often a way of unwinding mischief that resulted from the imposition of regulation in the first place. In the case of the railroads, the imposition of regulation was a lengthy stepwise process that is well documented and will not be reviewed here. Regulation was imposed under very different competitive conditions than those existing in 1980, and even though it may have made economic sense when imposed before 1920, it was left in place far too long in the face of changing overall economic conditions. The result was the near financial failure of the industry, which was the crisis that induced Congress to end the regulation. The reason why deregulation had to await crisis conditions is that whenever a regulatory system is imposed on an industry and left in place for a substantial period of time, the adaptation to the regulated conditions creates powerful constituencies in favor of keeping things as they are both inside and outside of the industry. These unite and attempt to use the political process not only to oppose removal of the regulatory system, but also to oppose even incremental changes in the pattern that would make the regulation less inimical to the industry's commercial health. Change in the regulation comes only when some kind of crisis places extreme pressure on the regulated structure.

Another deregulation that is very much of interest at the time of the present writing was that of natural gas pricing, which was completed with Congress' passage of the Wellhead Decontrol Act of 1989 that took full effect on January 1, 1993.[6] This deregulation produced effects that were contrary to popular wisdom during the 1970s and 1980s, when it was anticipated that deregulation would lead to substantial increases in gas price and that there would be no positive supply response. This latter belief, widely held, rested on the notion that natural gas reserves were in a state of terminal depletion. The result that actually transpired was declining prices and increasing supplies to the interstate pipeline system. Subsequently, natural gas usage has expanded. Among all the large-volume hydrocarbon fuels, it offers the minimum production of carbon dioxide for the quantity of energy its combustion yields, and this makes it a preferred fuel in an era wherein carbon dioxide is widely thought to aggravate global warming and its attendant problems. In light of these positive effects, it is of interest to review briefly why the price of gas was regulated in the first place (the beginning of deregulation can be dated to 1978, and the process was complete only after 1992) and why the deregulation took so long in spite of the problems that regulation caused.[7]

The oldest part of today's natural gas industry was the local distribution companies (LDCs). These appeared in the US in the nineteenth century and had antecedents before that. Their business was to manufacture and deliver gas to customers for illumination and space heating, and their growth mirrored the decline of the wax candle as indoor light source. The gas they sold was a low-Btu product variously known as town gas or coal gas, and it was made by gasifying coal or oil by burning part of the fuel in an air-deficient environment. Its principal component was carbon monoxide, and it was low-Btu because of a heavy dilution with atmospheric nitrogen. The LDCs became subject to rate regulation at the state level, and under regulation, they enjoyed local monopolies. Replacement of town gas with natural gas (principally methane) started with the early development of the oil industry. Crude oil and gas are always produced jointly, but the proportions are extremely variable. In the early years of the petroleum industry, oil producers sold gas into local markets, but gas in surplus over this demand tended to be flared either at the well site or at a nearby refinery. One can speculate over how much useful energy was wasted by flaring. This began to change in the 1920s with the development of pipelines to deliver gas to larger markets, and with this development came the *interstate* pipeline. At first, the interstate gas pipelines were not regulated as to rates, having not been included in the regulation imposed by the Hepburn Act of 1906. This changed with the Natural Gas Act (NGA) of 1938, which placed the interstate gas lines under regulation by the Federal Power Commission.

The interstate gas transmission lines did not function as common carriers; rather, they operated as *merchant* carriers. By this, it is meant that the pipeline companies purchased the gas from producers and sold it to LDCs at the "city gates." Consequently, even though their transport charges became subject to regulation, the price of gas delivered to the LDCs continued to reflect the unregulated price of gas at the wellhead. This made regulation of the price of gas to the final consumer impossible from the states' standpoint. The language of the NGA did not clearly empower the FPC to regulate wellhead prices,[8] but the price of gas at the wellhead came under FPC regulation as a result of the Supreme Court's decision in *Phillips Petroleum Company vs. State of Wisconsin* (1954), in which the Court ruled that Congress had intended that price regulation be extended to the gas production industry.

At first, there was no obvious adverse impact on gas supply from the imposition of regulation. However, by the 1960s, there developed shortages of gas supplied to the interstate pipeline system, and shortages became increasingly acute in the 1970s. Price caps at the wellhead had initially been above perceived market value, but this condition reversed in the 1970s. As is common in regulatory regimes, there was a failure to recognize this change of conditions with increased price caps. Interestingly, the idea that the shortages were a logical result of placing gas under a price cap was resisted by many groups. One source of this resistance was the congressional delegations from states that imported gas, and that feared the expected political impact of an abrupt increase in gas prices. Such a fear presumed that an increase in gas prices would elicit no increase in supply. The 1970s saw the "energy crisis" in which the price of oil rose dramatically in response to a complex of domestic and international events. One idea that gained a popular following in the 1970s was that the world's oil reserves were in a terminal state of final depletion, and this idea was extended to gas reserves. Hence, what gas was being supplied was deemed in official circles to be what could be supplied and there was none beyond that. Therefore, how could there be a normal supply response to price increase when there was no more to be had?

The federal response to evident shortages at first grew out of the "terminal depletion" hypothesis and involved a number of attempts to deal with the problem through unconventional means, of which the principal three were encouraging the importation of liquid natural gas (LNG), unconventional gas from coal gasification (see discussion of the Great Plains Project in Chapter 6 above), and Alaska natural gas to be brought to markets via a costly new pipeline. The sum total of these efforts amounted to very little gas supply. The other major response to the shortages was the Natural Gas Policy Act of 1978 (NGPA), which began the process of wellhead price deregulation by exempting a number of categories of "new" gas supplies from regulation. Examples of new supplies included "deep gas (new wells to a depth of 10,000 feet or more)," tight gas (from geologic formations insufficiently porous to let gas to flow to production holes), and LNG imports. It was supposed that "old gas," or gas that was already being supplied to the interstate system, would not be affected by deregulation, so it continued to be regulated. Attempts to segment the gas by source induced some perverse results. The attraction of unregulated prices induced producers to drill for costly deep gas and ignore the possibilities for less costly tertiary recovery methods that could have increased the production of old gas. At one time, there were thirty-eight different categories of natural gas, and it goes without saying that administering this system was a massive headache.

With the passage of the NGPA, the Federal Energy Regulatory Commission (FERC, successor to the FPC) began a series of baby steps toward deregulation of wellhead gas prices, which culminated in total deregulation after 1992. During this span of years, there were increasing hints that deregulation would be a forceful cure to gas supply problems. First, shortages that affected interstate gas distribution never became problems in gas-producing states, Texas, Louisiana, and Oklahoma, in whose intrastate markets wellhead gas price was not under federal regulation. Second, during the early1980s, there was an unexpected increase in supply to the interstate pipeline system that was referred to as the "gas bubble," implying lack of official belief in the permanence of the emerging supply surplus. Belief in the terminal depletion hypothesis was still strong among various public officials, and died only slowly. Moreover, it was widely trumpeted by the popular media. For the most part, this evidence of experience accumulated into a strong argument in favor of deregulation and was eventually sufficient to overcome resistance of gas-importing regions with the result cited above.

As already noted, gas use has expanded under deregulation. At the time of this writing, a surge in gas prices has occurred, and it is fortunate that so far there has been no attempt at re-regulation and bureaucratically managed allocation of (questionably) diminishing supplies. It has become popular among energy analysts to predict that today's relatively elevated gas prices will persist for a considerable period of time into the future. Such forecasts typically are justified for a host of reasons, some seemingly long term and some not. Among the latter is the damage to gas production due to Hurricane Katrina. This kind of forecast implies a less-than-confident belief that the high level of prices will elicit a supply response that will result in a decline in gas price. However, there is ample evidence in the field that such a supply response is not only likely, but is in progress. There is intense exploration and development of what have been termed unconventional natural gas sources; such as gas shales, tar sands gas, and the Gulf of Mexico; that include new sources as well as restoration of production interrupted by Katrina damage. It remains to be seen whether this really represents the lessons of the 1970s having been absorbed.

Environmental Regulation

One of the standard teachings of modern economics is that a competitive environment that satisfies certain conditions will result in a superior social result. One of the assumptions underlying this conclusion is that prices for all goods and services fully cover the costs of their production. The fact that the actual world falls visibly short of meeting this condition was the rational basis for the

Clean Air Act (1970) and the Clean Water Act (1972).[9] The classic example of what economists call negative externalities is the factory whose exhaust products go up its chimney into the atmosphere and soil the neighbors' washing as it hangs on the clothesline. This example is not a bad parallel to coal-burning utilities exhausting tons of sulfur dioxide into the atmosphere, which falls out in the form of acid rain, and the operations of private automobiles whose collective exhaust streams are composed of carbon monoxide and nitrogen's oxides, the former toxic to humans and the latter a precursor of smog. In both cases, producers (of cars and power) are shifting the costs they forego in the form of avoided cleanup costs onto the general public, which pays in the form of reduced quality of health and life.

The modern environmental movement is generally recognized as having begun with the publication of Rachel Carson's *Silent Spring* in 1962. Since then, the public mandate for dealing with some very large environmental problems has grown to be irresistible politically, and this reality has driven the development of governmental regulatory intervention worldwide, as it has developed in the last three decades of the twentieth century. The Clean Air Act, which empowers the Environmental Protection Agency (EPA) to set and enforce national air quality standards, has been amended periodically, and the effect of the amendments has been to tighten regulation. This tightening regulation has had effects similar to that of freight transportation deregulation in acting like a series of prototype inventions. Adaptation to tightened air quality regulation has produced some innovation, including inventions.

Automotive Exhaust under the Clean Air Act

The EPA's first object of attack was automotive emissions. It called for a 90 percent reduction in auto emissions from 1968 levels, as well as elimination of most lead from all automotive gasoline by the 1975 model year. Underlying this directive was the assumption that the technology for achieving this goal existed and that all that was needed was a high standard backed by vigorous enforcement. William Ruckleshaus, the first administrator of the EPA, later admitted that the entire project was "presumptuous from the start . . ."[10] Achievement of these goals was left to private industry. In reality, the assumptions underlying the goals proved wildly optimistic, and the fact was that no one in 1970 could see how the goal was to be achieved. The fact that it was achieved on schedule was a true technological miracle, involving the automobile, glass, catalyst, and petroleum refining industries. In this case, it can be said that the government got away with one. This is not always the case, as evidenced in the alternative energy projects of the 1970s.

Immediate compliance focused on the automobile industry, which attempted to deal with the problem by redesigning engines to run on a much leaner air-fuel mixture, as this would encourage complete combustion of the fuel. This was a self-defeating approach in that adapting to a leaner air-fuel mixture required reducing the compression ratio of engines, thus reducing their thermal efficiency and reducing gasoline mileage.[11] Owners of cars manufactured in the 1971-1974 period will recognize other symptoms of the detuning: poor overall performance, especially from a cold start. The auto industry was resistant to the idea of a chemical solution, partly on the basis of cost, and continued to tinker with engines with the general goal of achieving more complete combustion.

The idea of oxidizing carbon monoxide to carbon dioxide—then regarded as a harmless gas—and unburned hydrocarbons to CO_2 and water in the presence of a catalyst had already undergone some development prior to 1970, most notably under the direction of Eugene Houdry, known for his catalytic cracking process of the late 1930s. Part of the problem was to develop a physical support system for the catalyst that would function in the severe temperature conditions in the automotive exhaust stream. For this, Corning Glass Works overcame huge problems in materials research, and backed that achievement with a new plant capable of producing the substrate in the quantities needed for automobiles. Englehard developed the catalysts after much experimenting with materials, finally settling on a platinum-based catalyst, and worked out a form of the catalyst that minimized the consumption of this precious metal.

In addition to oxidizing CO and unburned hydrocarbons to a more environmentally benign state, the additional problem of converting nitrogen oxides (NO_x) to nitrogen molecules also faced the designers of what became the catalytic converter. NO_x is a recognized precursor of smog. This treatment of NO_x called for a *reduction* reaction, as compared with an *oxidation* reaction called for by the treatment of CO. At first, researchers considered the possibility of a separate converter to deal with the NO_x, but this was rejected on the ground of additional cost. The solution involved the amazing feat of getting oxidation and reduction to take place simultaneously in the same converter housing. This accomplishment was enabled by computing and sensing technology that was just emerging at the time. It is also creditable to Englehard Corporation's development of its amazing three-way Catalyst, which ultimately resulted in the elimination of 97 percent of hydrocarbons, 96 percent of carbon monoxide, and 90 percent of nitrogen oxides.

As will be developed later, large corporations have demonstrated a certain reluctance to fund the kind of research leading to risky first-stage innovation. Against this background, the decision by Corning Glass to invest in a plant for mass producing its substrate for the catalytic converter, even before the technical details of the substrate had been worked out, stands out as an exception. The company faced the risk that some other solution to the overall problem of automotive exhaust cleanup would emerge, rendering its investment worthless. The key element in this was that Corning's senior management fully recognized the profit potential of success. All too often, corporate failure to exploit its own staff's invention can be laid to lack of senior management appreciation of the profit potential.

Part of the EPA's directive on automotive emissions required a supply of unleaded motor gasoline by the 1975 model year. Lead had been demonstrated to poison catalysts used in catalytic converters such that the catalysts would lose effectiveness after 10,000 miles at best, and the EPA standard called for full performance at least to 50,000 miles. A small admixture of tetraethyl lead had been found during the 1920s to be a cheap way to enhance octane rating. Therefore, the refining industry's challenge was to take the lead out of gasoline while maintaining or enhancing octane standards so that the catalytic converter that emerged in 1975 imposed no constraint on compression ratio.

Gasoline is a blend of a number of substances with the common characteristic of having a boiling point in a defined range.[12] Because gasoline boiling temperatures are generally higher than ambient temperatures, historically there was a problem of getting the gasoline into vapor form for mixture with air when the engine is started cold. This was solved for years by including in the blend something with a high octane rating and a low boiling temperature that would become a vapor at temperatures lower than engine operating temperature, and this role generally fell to butane.[13] The refiners' problem was made more difficult when the EPA eventually banned butane from gasoline because it would boil off at ambient temperatures on warm days. Anyone who has filled a tank with gasoline containing butane will probably recall having seen a plume of light distortion coming from the filler pipe on hot days. This was butane boiling off. Not only did the refiners have to replace the octane values lost by taking out the TEL, but also that lost from taking out the butane. The problem of getting the gasoline into vapor form was eventually solved by the fuel injection systems that became standard during the 1980s. These achieved vaporization with mechanical force. With carburetion, the maximum force that could be applied to vaporization was limited to that available from manifold and carburetor vacuum.

There were several tools with which the industry could address its problem. The first was catalytic cracking, which had been improved incrementally since the days of Houdry's process. Second, there was alkylation, a process that can be thought of as the opposite of cracking, as it involves forcing smaller molecules to combine into larger ones. Typically, butanes or pentanes and butylenes (a by-product of cracking) are forced together to form alkylate, a mixture containing substantial proportions of substances, such as isooctane, that have high octane ratings and low vapor pressures. Butanes, having been banned as direct components of the gasoline blend, became available for inputs to alkylation units.[14] A third tool was catalytic reforming, a process in which naphthas, having low octane ratings, are converted to isoparafins and aromatics that have considerably higher octane ratings. Not only did gasolines have to satisfy the octane requirements of modern engines, but they also had to satisfy vapor pressure requirements. With all the constraints on the blending problem, satisfactory solution depended on large computers able to handle some fairly large linear programming problems

Electric Power Generation and Environmental Regulation

In 1970, the United States drew its electric power needs from three main sources: thermal-electric plants, nuclear power plants, and hydroelectric plants. The first of these, the thermal plants, was a focus of interest under the clean air law and regulation. The majority of thermal-electric plants used coal as fuel.[15] Coal-fired plants posed several problems. First, most coals contain some sulfur, and coal from some regions, such as northern West Virginia, southeastern Ohio, and the Illinois Basin, have especially heavy sulfur loadings.[16] Sulfur was combining with oxygen under the boilers to form sulfur dioxide, which was being exhausted into the atmosphere. Sulfur dioxide (SO_2) is further oxidized in the atmosphere to form sulfuric acid and falls out as acid rain, causing much mischief environmentally. A second problem is that at high temperatures of combustion, NO_x forms and also is exhausted into the atmosphere. These problems have been attacked with a number of different technologies, and as of 2005, no complete solution yet exists. A third problem was particulate emissions, which were alleged to cause respiratory disorders in people.

In the original Clean Air Act (CAA), the principal emphasis was on automobile exhaust, and to some extent, the power industry drew a pass. Language in the act called for emissions control in new power plants, but older plants were effectively grandfathered. The assumption was that these old plants would gradually be retired, and the capacity that replaced it would be equipped with emissions control equipment. As with other assumptions of the 1970 CAA, this one proved naïve, for in practice it opened a large loophole in which utility companies could meet needs for increased output by *expanding* their old plants. Another option that utilities had under the CAA was that plants that the law required to reduce SO_2 emissions could start burning "compliance coal," or coal with a naturally lower sulfur loading than had previously been used. This led to the start of a boom for the parts of the coal industry with reserves of compliance coal: increasing demand in the east, coals from south of the Kanawha River and western coals from such areas as Wyoming's Powder River Basin,[17] and a corresponding diminution in demand for high-sulfur coals in Northern West Virginia, Illinois, and Ohio.

In 1977, the Congress addressed what it saw as the problems of the original CAA. These amendments introduced the concept of New Source Review (NSR), which requires older industrial facilities that owners want to expand to undergo an EPA assessment and install pollution control technologies if the planned expansion will produce significantly more emissions. This has led to considerable litigation, with utilities taking the position that certain plant activities are routine maintenance and the government taking the position that these activities are really modernizations that will raise capacity. Another aspect of the 1977 CAA amendments was the emphasis on "Best Available Control Technology" (BACT), and courts tended to interpret BACT as meaning lime or limestone scrubbers[18] to deal with SO_2 and electrostatic precipitators to deal with particulate emissions.

The CAA was again amended in 1990, and the new amendments targeted acid rain and the remaining presence of lead in gasoline. The 1990 amendments corrected the problem of the 1977 amendments in which the government effectively spelled out the technology for compliance and replaced this policy by allowing utilities a wide choice of means to achieve compliance. Flue gas desulfurization (scrubbers) became but one of allowable strategies, which included fuel switching. Additional flexibility was expected to decrease the cost of compliance. Between 1970 and 2000, there were sharp decreases in the emission of pollutants from stationary sources. For example, sulfur dioxide has been reduced 27 percent. Clearly, this leaves much SO_2 in the atmosphere, but as the elimination percentage increases, so does the cost of elimination.

A number of electric utilities took advantage of increased latitude of decision under the 1990 CAA amendments by building generating capacity based on natural gas. Natural gas has the advantage of being the cleanest burning of the fossil fuels available, and it contains no sulfur. Utilities used it in two different ways: as fuel under boilers (simple substitution for some other fuel, mostly coal) and as gas turbine fuel. Gas turbine plants took two forms: peaking plants and base-load plants. Gas turbines are well suited to the role of peaking power, for they can be brought from cold start to full power output within thirty minutes.[19] As base-load facilities, gas turbines can be used in combined cycle (CC) arrangements in which the waste heat from the gas turbine is used to produce steam, which is used in a steam turbine to produce more power. A modern-design CC plant can achieve up to 60 percent thermal efficiency, compared with a conventional thermal-electric plant that achieves a maximum of 35 percent efficiency. Growth in natural gas-based power capacity came to a halt after 2002 when the price of natural gas rose and many analysts became convinced that it would stay high.

Nonferrous Metals Smelting

Another field in which the Clean Air Act forced substantial change in the technology is nonferrous metal smelting. To illustrate the problem, consider a large-size copper smelter. Prior to the CAA, smelting took place in two steps: smelting and converting. Both of these were open in the sense that off gasses were vented to the atmosphere. A typical commercial copper ore contains, in approximately equal proportions, copper, sulfur, and iron. All the sulfur was vented in the form of sulfur dioxide, and the amount of sulfur vented in this form could be well over fifty times that vented by a large coal-burning power plant.[20] A copper smelter could, thus, be thought of as a very concentrated source of sulfur dioxide.

After several early efforts to mitigate the problem, such as tall smokestacks for wider dispersal of smelter exhaust and conversion of sulfur dioxide to sulfuric acid, there remained the problem that substantial quantities of sulfur dioxide were still escaping from smelters. The first major change in technology was the introduction of flash smelting to replace the traditional reverbratory furnace. In flash smelting, the concentrate feed is ground to a flour-like fineness and fed into a reactor in which it "flashes" in a manner similar to what aerosol flour does in the presence of a spark. Flash smelting is enclosed, and much of the SO_2 is captured and sent to an adjacent sulfuric acid plant and converted to acid, which can be marketed. Flash smelting was only a partial solution, for in moving the molten product of the flash smelter to the converter, significant SO_2 escaped into the atmosphere. Flash smelting satisfied the State Implementation Plan only into the mid-1980s.[21] The solution was to adapt the flash principle to the converter step. This allowed the entire process, from cold concentrate feed to finished anode copper to be enclosed, thus enabling the capture of virtually all the off-gasses, and that is the state of the art today.[22]

Regulation and Alternative Technologies

Regulation affects the application of technology to electric power production and distribution in two ways. First, the rekindled interest in some older technologies such as wind and solar power is directly attributable to present and anticipated future regulation under the Clean Air Act. Second, it often turns out that the application of these technologies can be affected, sometimes negatively, by various regulations at the state and local level. To complicate matters, the US is moving toward deregulation of the electric power industry. As traditionally applied, regulation of the electric utility industry worked under the assumption that the industry was a "natural monopoly." The term "natural monopoly" refers to the physical characteristic of an industry that if it were to achieve its lowest-cost configuration, its scale would be larger than any conceivable market demand. When the utility industry first emerged in the early twentieth century, it was realized that competing sellers would have to construct distribution lines that would be inherently underutilized and therefore high cost. It was widely felt that, left to market forces, these conditions would induce industry consolidation into monopoly sellers in order to get distribution costs down to its potential low point. Regulation was imposed in recognition of this supposed monopoly tendency for the purpose of preventing utilities from taking advantage of their monopoly by gouging consumers.

While it is not the purpose of this writing to expound on the progress of deregulation of the electric utility industry, certain aspects of it have induced revised thinking on the part of utility managements that are comparable with the kind of revisions induced by a disruptive prototype invention. It gradually became recognized that while the *distribution* activity of the utilities may have some natural monopoly elements, the *power production* element does not. Thus, the tendency of state deregulation measures has been to deregulate power production while keeping distribution under regulation.

Clean Coal Technologies

One technical strategy for eliminating undesirable emissions that appears to be emerging is gasification of coal in an integrated gasifier combined cycle plant (IGCC).[23] A more complete description of the technical and historical characteristics of IGCC is deferred to Chapter 12. The essential feature of gasification in the regulatory sense, apart from the fact that it utilizes coal (an abundant resource in the US) is that it converts virtually all the components of coal, including those that tend to form pollutants in conventional thermal-electric plants, into commercially marketable products. There are only minor waste streams that have to be disposed of. Even though an IGCC plant is still not fully cost competitive with a conventional thermal electric plant, the comparative disadvantage is shrinking, for the thrust of clean air regulation has been to increase the cost of dealing with the combustion products of the conventional plant, which would be avoided with an IGCC plant.[24]

IGCC applications have gained a market in petroleum refineries where they are able to process petroleum coke and other refinery wastes into synthesis gas that is then used to generate electric power for use in the refinery and elsewhere. The waste products used as feed to the gasifier had previously been considered hazardous wastes that, under established environmental regulations, would have had to be disposed of in an incineration process for which permitting was very costly in both time and money. Rather than treat refinery gasifiers as hazardous waste disposal facilities, however, state environmental regulators have generally opted to regard them as process units for converting the wastes to nonhazardous and useful products. This is a case regulators have recognized the environmental usefulness of gasifier technology and have removed a regulatory barrier to its use. The Federal Environmental Protection Agency has since implemented rules that treat refinery gasifiers and gasifiers in other types of plant as converters of the wastes to useful products. At least two major IGCC projects have been announced by coal-burning utilities and are in the planning stages now.[25]

Hydrogen

Another fuel technology that is presently receiving attention is hydrogen. Hydrogen's appeal is that combustion of it yields only pure water as a waste stream. While it is within the capability of today's engineering to build internal combustion engines and/or fuel cells that burn hydrogen, there is almost no infrastructure to support any level of reliance on this fuel. Providing such infrastructure would be hugely expensive, and any group of entities that commits major resources to this purpose takes on the very considerable risk that the technology future will take a different direction. Several companies formerly engaged in research and development of hydrogen-burning fuel cells have found the conventional internal combustion engine so well established that they have abandoned automotive research in favor of stationary applications for which there is a market.

There is, however, a line of development that could someday provide at least a start toward a wider hydrogen infrastructure. This is in the form of hydrogen plants built to provide hydrogen to petroleum refiners. The driving force behind this development is the regulated tightening of standards for sulfur in motor fuels. The traditional sources of hydrogen in the modern refinery, catalytic reforming units, have long since been oversubscribed, forcing the refiners to purchase hydrogen from the outside. The industrial gas companies have responded to this need and are building hydrogen plants and pipelines to distribute the hydrogen to consuming refineries. These facilities constitute the bases for any future hydrogen infrastructure for a consumer hydrogen market.

Nuclear Power

In the context of today's concern regarding emissions of carbon dioxide and other combustion products, there has been renewed interest nuclear power in energy circles in Washington. Nuclear power produces no noxious emissions, although it does present an ongoing problem of safe disposal of waste products of nuclear fission. Whether nuclear is a realistic option for future power supply is questionable, however, in light of its history. Following the success of the Manhattan Project, the US established the Atomic Energy Commission (AEC) with the mission of promoting nuclear energy as a source of electric power and of regulating the nuclear part of the utility industry. These two missions struck many as being in conflict, and in 1975 Congress responded to this concern by splitting the AEC into two agencies: the Nuclear Regulatory Commission (NRC) for safety regulation and the Energy Research and Development Administration (ERDA). In the meantime, the nuclear power industry had grown rapidly, and in terms of orders for new plants, had its heyday in the early 1970s. Not only were the utilities ordering numerous plants in this period, but the scale of the plants ordered had grown very large—large enough to present safety issues that had not really been perceived as problems with smaller plants.

As these safety issues were recognized with increased operating experience, there was a veritable avalanche of NRC regulations, one of whose effects was to increase the time needed to build a new nuclear power plant from 5.5 years in 1967 to 8.5 years in 1974. For a large conventional coal plant, the construction time held steady at five years in this same period. In addition, a high rate of inflation in the second half of the 1970s dramatically increased the costs of labor and materials in construction. Capital costs for nuclear energy grew high enough to offset the advantages of low-cost fuel as compared with coal. The result of all this was that utility executives completely lost enthusiasm for nuclear power, as it seemed economically not worth the trouble. By 1970, project cancellations had mostly killed the market for new nuclear plants.[26]

Then, in 1979, came the Three Mile Island (TMI) incident, followed in 1986 by the Cherbonyl plant's meltdown in the Soviet Union. The effect on public opinion was completely negative such that any new plant of the future will have immense difficulty with its siting and, absent drastic federal overrides of local control over plant siting, it will be a long time before any new nuclear plant will be started. Apart from the emotional reaction, what Three Mile Island exposed were serious flaws in the control and monitoring design that impaired operators' ability to deal with the situation as it developed. As for the Cherbonyl incident, the safety problems of design there were such as to make plants like Three Mile Island seem foolproof by comparison. However, TMI did not kill the nuclear power industry, for it previously had failed to live up to its early economic promise partly due to intensive regulation.[27]

On the positive side, control design and procedures have been immensely improved as a result of TMI. Also, the utility industry has undergone much consolidation, resulting in much larger and better-financed companies, offering at least the possibility for design standardization. Moreover, today's basic reactor designs are inherently safer than those of the 1970s. It remains to be seen whether emotional opposition to nuclear power can be overcome in the future. Perceived and anticipated energy supply difficulties and the absence of polluting products from operation of nuclear power plants have caused a revival of interest in this technology on the part of the US Department of Energy in recent years. The biggest unresolved problem is what to do with the spent nuclear fuel. The answer may lie in new reactor technology that would dramatically reduce the time period in which the waste products are highly dangerous.

Deregulation and Storage Technology

From its outset, the electric utility industry has faced a time profile of demand that is highly volatile on a daily, as well as seasonal basis. For example, in most of the US, demand for power in the summer greatly exceeds that of the winter because of, among other things, air conditioning demand. This same demand is highly volatile on a *daily* basis, for daytime demand for air conditioning greatly exceeds nighttime demand. Inasmuch as there never has been an economic way of storing electric power that can be applied in all parts of the country, the industry has always faced the problem of matching instantaneous demand and supply. The historical approach to this problem has been to build and maintain enough capacity, for both producing and distributing power, to cover the highest peak demand. Even effective economical day-to-day storage technology would have the benefits of relieving distribution lines of their peak loads, thus reducing the problems of brownouts due to their overloading, and eliminating the need for power production capacity previously needed to cover peak demand.

In very recent times, battery technology that could, with further development, provide at least day-to-day smoothing of the power production and distribution profile has begun to emerge in the form of very large batteries known as flow batteries. While the discussion of how these work is to be found in Chapter 12, an introduction is warranted here because of the regulatory problem that presently affects this technology. Flow batteries absorb power from the area grid during low-demand times and distribute it during high-demand times. They also serve as emergency stand-by power. A flow battery differs in principle from an ordinary battery in that the energy is stored in charged electrolytes *outside* the cell, and energy is released as charged electrolytes flow through the cell. There are two problems with flow batteries. The first of these is cost, and commentary is deferred to Chapter 12.

The second is regulation. Electric power deregulation in the US is in a state of flux. So far, the trend has been towards deregulating power production but keeping distribution under regulation. Where storage facilities fit in this picture is still controversial, as their historical nonexistence has kept them from being recognized as part of the regulatory problem until very recently. The precedent from natural gas deregulation says that storage is a part of distribution, but there is a school of thought that argues that batteries are effectively power producers during the discharge part of their cycle. If the decision is that flow batteries are part of distribution, then the power producers cannot very well be expected to invest in them. Such investment might come from the entities that end up owning and operating the power distribution systems, but until the regulatory question is resolved, neither kind of organization will be interested. The one significant investor in flow battery technology in the US now is the US Air Force, in a geographic area notorious for brownouts and blackouts. The Air Force's interest in flow battery technology comes from a desire for power supply reliability, not cost. The regional power supplier is TVA,[28] also a government entity. The regulatory limbo of flow batteries may persist until their cost characteristics make them attractive to private power, for when that happens, there will be significant political pressure on legislatures to decide.

Sludge Disposal

Coal burning power plants (the majority of US capacity) produce two kinds of pollutants: sulfur dioxide and a gritty soot known as fly ash. Prior to the Federal Clean Air Act (CAA), both of these were typically exhausted into the atmosphere. Under the CAA, the SO_2 was captured by flue gas scrubbers and landfilled, and the fly ash was captured by electrostatic precipitators and also landfilled. State public utility commissions approved inclusion of the costs of landfilling in the utilities' rate base. This was in spite of the fact that the landfills were recognized as having possible unpleasant long-term environmental consequences. The landfilled sludge from the scrubbers was chemically calcium sulfate, the same material used to make the most common of construction materials, gypsum wallboard. Wallboard companies have historically used mined gypsum, some from as far afield as British Columbia. However, when wallboard producers approached utility managements as proposed buyers of gypsum sludge, the utilities were not interested, for under regulation, any revenue they could realize from the sale of sludge would no more than offset the disadvantage of having the landfilling costs removed from their rate bases.

The various state utility deregulations have had the effect of moving power production out from under the regulatory umbrella thereby quickening power company managements' interest in cost control, and the idea of selling the sludge and avoiding the landfilling costs became attractive for the first time. As a result, a number of deals between power producers and wallboard manufacturers have been concluded. The other market for scrubber sludge is as a soil conditioner, and the power producers under deregulation have actively promoted its use among farmers. The fly ash has proved to be of interest to construction materials companies, which use it in lightweight cement. It is interesting to speculate on the benefits that the general economy has derived from some of these deregulations. Prior to deregulation of power production, wallboard was typically manufactured in the Pacific Northwest from raw materials imported from Canada, and distributed to the rest of the country, mostly by rail. In effect, deregulation has meant many fewer miles of transportation are presently needed to provide wallboard to the housing construction industry.

Regulation and Telecommunications

In the late twentieth century, there developed two basic approaches to supplying households with broadband communications capability: cable (provided by cable companies) and digital subscriber lines (DSL) from telephone companies.[29] The telephone companies have been regulated by various levels of government for many decades, and this pattern has been extended to DSL services: the companies are required by regulation to make their wirelines available to rival providers of broadband services at regulated rates. The idea behind this regulation is that a multiplicity of broadband suppliers at the retail level will foster competition while use of the basic wirelines by their owners plus tenant users will foster competition and low prices.

The cable companies have so far not been subject to this kind of regulation, and a recent Supreme Court decision appears to have affirmed this lack of regulation.[30] The market reaction to this regulatory dichotomy has been that consumers have awarded the closed access cable operators twice the market share of the open access DSL services of the telephone companies. Moreover, the cable companies have aggressively invested in their systems, preparing them for high-speed data transmission services. This appears to fly in the face of the theory that sees open access is pro-competitive.

To further affirm this conclusion, in February of 2003, the Federal Communications Commission relieved the telephone companies of some of their regulatory burden by making more difficult the terms under which Internet access providers could gain the use of telephone lines. The result was that the companies invested capital into their facilities in competition with the cable companies to a degree that they had not done before. One result was falling price of DSL service.

Taxes and Technology

The emphasis of this chapter has so far been on a major change in the regulatory climate, comparing the impact of such a change with the impact of a disruptive prototype invention. A change or difference in the structure of taxation can also have a profound impact of how technology develops. It is not the purpose here to develop either the regulation or the taxation impacts in all their complexities; an example suffices to demonstrate the taxation effect. It concerns the way in which the automotive engine developed.

Ever since the earliest days of automobiles, there has been a strong popular interest in their speed and power. Auto racing is almost as old as the auto itself, and given the evident interest in Indianapolis and NASCAR racing, this interest is undiminished today. Another thread of interest concerns fuel economy. The US has seen several surges in fuel economy interest since 1970, but it has been of interest in Europe for much longer.

Broadly speaking, there have been two engineering approaches to the speed and power issue. The first, which shall be referred to as the North American (NA) pattern, is based on comparatively large engine displacement (e.g., five liters and up), relatively low engine speeds, and high torque. By the 1950s, the typical NA engine was a V-8, with bore only slightly smaller than the stroke (sometimes larger, in what was known as an "over square" engine) and with a camshaft at the level of the crankcase driving valves via pushrods and rockers. This style of engine offered the advantage of a sense of effortless acceleration. Its disadvantage was that there were few driving situations outside of formal racing that utilized its full power potential. For most street and highway driving, the NA engine was understressed and, given reasonable care, was capable of several hundreds of thousands of miles of service or well in excess of the typical life of the vehicle of which it was a part. Another advantage of the NA engine is that it is relatively cheap to produce *en masse*.

The second pattern of engine development, to be referred to as the European (EU) pattern, as compared with the NA engine, emphasized smaller displacement and higher engine speeds and achieved speed and performance via gearing, turbocharging,[31] or a combination of the two. Camshafts were located in the cylinder head(s) and drove more-or-less directly on the valve stems. This arrangement eliminated the pushrods and rockers, major balance problems at higher engine speeds. The EU engine offered a better tailoring of power to power needs than the NA engine, but it is more complex and more costly to mass produce. The EU engine had the advantage that its smaller displacement offered superior opportunities for fuel-economic design.

Can one of these two approaches be justified as superior on purely engineering grounds? The answer is "probably not." The EU design had its genesis in the pattern of taxation in Europe, which included not only a stiff excise on gasoline, but also an excise on the car itself at the time of purchase, based on engine displacement. European tax policy also deliberately encouraged carbuyers to use diesel power, as a result of which the automotive diesel has been better developed there than in North America. These measures grew out of fears that oil supplies were not secure. Inasmuch as European customers wanted performance as did their North American counterparts, the EU design was aimed at achieving performance from a small displacement.[32]

By contrast, there has never been a displacement tax in North America, and gasoline has never been burdened with more than relatively modest taxation, even during episodes of high gasoline prices. Several of these occurred in the 1970s, but the effects of these experiences on consumer car-buying habits soon disappeared in subsequent periods of falling retail gasoline price. In the long history of North American motoring, gasoline has been relatively cheap. While recent high gasoline prices have definitely affected recent consumer buying decisions, it is not clear what will happen should gasoline prices recede.

Notes

1. The Staggers Act made railroad mergers easier, and a period of industry consolidation followed its enactment.
2. Under regulation, railroads created agreements covering the allocation of the revenues from interline shipments in organizations called rate bureaus. With deregulation, the rate bureaus came under the sway of the antitrust laws, and efforts to carry on "business as usual" brought rail managements face-to-face with their legal vulnerability, for the rate bureaus' activities had serious aspects of price fixing. Not surprisingly, these bureaus have mostly disappeared.
3. In the regulated environment, there was little incentive for railroads operating low-cost routes in heavily trafficked corridors to expand capacity. When traffic gravitated to the low-cost corridors under deregulation, bottlenecks tended to develop. For a while, certain circuitous routes in various markets survived as overflow facilities for the more direct routes. A large part of the pricing-strategy problem revolved around discovering how much shippers were able to pay for the carriers to eliminate the bottlenecks so that they could expedite shipments in ever-larger volume. This problem approached solution only in the late 1990s and early twenty-first century.
4. Regulated motor carriers carried their Certificates of Public Convenience as assets on their balance sheets. As far as the author knows, this was the only industry to follow this practice with its formal operating authority. One of the impacts of

deregulation was to render the certificates worthless; hence, deregulation had an immediate negative impact on these companies' profit and loss statements.

5.   In 1985, the author executed a consulting contract for a client in the business of cleaning chemical tank trucks after a haul. At this time, the chemical tanker business was beginning to cope with the implications of deregulation for back hauls, and it was anticipated that the tank cleaning business would expand rapidly in response to this change in chemical haulage circumstances. At the time, tank-cleaning technology was sufficiently developed that backhauls would not be contaminated by the previous load.

6.   Tussing, Arlon, and Tippee, Bob, *The Natural Gas Industry: Evolution, Structure, and Economics,* (Tulsa, OK: Pennwell Books, 1995), pp. 219-220.

7.   An excellent history of the origins of gas price regulation and the process of deregulation can be found in Tussing, and Tippee 125ff.

8.   Leaving an obviously controversial issue vague in the draft of a law is often a deliberate political tactic aimed at gathering support for the measure that would not be forthcoming if the language were clear and unequivocal.

9.   While noting the rational elements justifying these acts of Congress, it is useful also to recognize that the political impetus to these measures included a large amount of pure romanticism.

10.   Tim Palucka, "Doing the Impossible," *Invention and Technology,* vol. 19, no. 3, (Winter 2004), p. 24.

11.   As already noted, the efficiency of an internal combustion engine is proportional to its compression ratio.

12.   The gasoline boiling point range is approximately 90 degrees Fahrenheit to 220 degrees Fahrenheit.

13.   Normal butane boils at less than 80 degrees Fahrenheit.

14.   Alkylation was developed early in World War II and, along with catalytic cracking, produced high-octane aviation fuels that proved so valuable in aerial combat. In the postwar period, alkylates became less crucial as jet engines became the dominant power for both military and commercial aircraft. The refiners kept up their alkylation plants to serve the remaining market of general aviation, and these facilities attracted renewed interest when there arose a need to make up for octane values lost with the phasing out of TEL.

15.   Approximately 55 percent of thermal-electric capacity was coal burning in the early 1970s. The alternative fuels were residual furl oil and natural gas.

16.   Up to 5 percent by weight.

17.   The Powder River Basin coals are very low sulfur (close to 0 percent in some instances). While their heat value is at the low end of the scale for bituminous coals (approx. 8,000 Btu/ton), these coals exist in large deposits near the surface, so that they are relatively cheap to mine. This lets these coals bear the costs of a long rail haul to eastern utilities and still be available at the point of use for reasonable cost.

18.   This was a departure from the approach followed by the CAA concerning automotive emissions, in which EPA imposed the standard to meet it and left the means to the private sector. This aspect of the 1977 amendments has been referred to cynically as a "full-employment act for coal miners in high-sulfur coal districts."

19.   This compares with a number of hours with a conventional thermal-electric plant with coal fuel.

20.   A power plant consuming 10,000 tons of coal per day would emit roughly 100 tons of sulfur daily, assuming that the coal contained 1 percent sulfur by weight. A large smelter processing 25,000 tons of concentrate per day at a 25 percent sulfur loading

would emit roughly 6250 tons of sulfur daily, well over the emission of 50 of the example power plants.

21. Under the CAA, the EPA sets standards, but state governments have the duty to approve or disapprove of corrective strategies. This arrangement presupposes that state governments have greater familiarity with local conditions than the EPA staff.

22. The Japanese smelting industry provides a contrast to the US smelters' approach to the sulfur dioxide problem. The Japanese drastically modified the traditional technology whereas the Americans scrapped and replaced it.

23. There are several gasification processes today. Typically, coal or some other hydrocarbon feed is fed into the reactor in a water slurry along with a controlled quantity of oxygen. The principal product is synthesis gas, a mixture of H_2 and CO, which is then used as fuel by a gas turbine to generate power. Waste heat from the gas turbine is used to generate steam that is used in a steam turbine to produce more power (this is the combine cycle (CC) part of IGCC). Other products of the gasifier include steam that goes to the steam turbine, hydrogen sulfide, from which elemental sulfur is easily obtained, and nitrogen, used as a diluent in various processes.

24. As of 2005, it has been estimated that a conventional thermal-electric power plant of 1000-megawatt capacity can be built for $1.5 billion; an equivalent IGCC plant would cost closer to $2 billion.

25. The two are American Electric Power and Cinergy.

26. Heppenheimer, T.A., "Nuclear Power: What Went Wrong," *Invention and Technology,* Vol. 18, No. 2, (Fall 2002), p. 53.

27. Following the world oil crisis of 1974, France made a serious commitment to nuclear power and appears to have been successful for several reasons. First, it benefited from other countries' experience and settled on several standard plant designs, a measure which aided in holding construction costs close to planned levels. Second, it achieved a stable regulatory climate, something the US never has achieved. Nuclear provides around 80 percent of France's power demand as compared with 20 percent in the US See Heppenheimer, p. 52.

28. This facility, rated at 12Mw/120 Mwh, is being built at the Columbus Air Force Base in Mississippi.

29. Both of these can be viewed as lower-cost substitutes for extending fiber optic cables to the home. See the discussion of fiber optics in Chapter 8.

30. *National Cable and Telecommunications Ass'n v. Brand X.*

31. The turbocharger made use of the heat of the engine's exhaust for additional power in hard accelerations. It permitted the use of a smaller engine as the basic prime mover for the car than would have been necessary otherwise.

32. Interestingly, the only well-known European carmaker that historically relied on a NA-type design engine was Rolls-Royce. Considering the kind of market to which Rolls-Royce was selling, it is not surprising that the fuel economy possibilities of small displacement were neglected. Effects of high gasoline prices on carmakers in North America have been mixed. General Motors still relies on the NA pattern, while Ford has committed heavily to the EU pattern.

8

Computers and Other Electronics

While the antecedents of the modern digital computer go back much earlier, the ENIAC of the mid-1940s is commonly regarded as the progenitor of the modern electronic computer. ENIAC resulted from a wartime effort to mechanize the complex calculations underlying range tables for tube artillery and aerial bombardment. Today's computer resulted from several prototype inventions of the postwar period: the transistor and the integrated circuit. From a large machine operating in a centralized facility by a specialized staff, the computer has evolved into an extremely powerful analytical, process control, and communications tool in the hands of millions of individual, business, and government users. In centralized computing, the value of computer time was held to be greater than the value of any human user; but the massive decline in computing costs since 1970 has completely reversed that relationship. Computing costs have become sufficiently low as to have enabled extended applicability to industrial and business controls and to have underpinned an extended period of accelerated productivity growth in the US economy as a whole. It has made large-volume production less dependent on mass markets by introducing product flexibility into large-volume production.

Antecedents

Concept

The pioneer electronic computers of the 1940s benefited from several enabling streams. Arguably, the most important of these was the concept, attributed to Charles Babbage, an Englishman, in the 1820s and early 1830s. Babbage was the first to state the four main features that characterize the computer to this day: the mill (central processor), the store (memory and retrieval), the reader, and the printer. These four were the essential elements of Babbage's *Analytical Engine,* conceived as a general-purpose computing machine. Babbage's initial interest in this direction grew out of his *Difference Engine* project. The Difference

Engine was intended to create accurate entries in mathematical tables, such as Napierian logarithms, which were used in navigation and which Babbage had demonstrated to be full of errors, some not trivial. The general interest in the Difference Engine project was sufficient that Babbage was able to attract funding from the British Government to construct such an engine, and substantial parts of the difference engine design were actually built. The difference engine was entirely mechanical and called for large numbers of precisely machined gears and other parts. The machining technology of the time could provide quantity or precision but not both. The project was abandoned about 1833 when public funding ran out, for the Difference Engine was, in twentieth-century terms, well over budget in terms of both time and cost. The Analytical Engine was conceived as an improvement over the Difference Engine, which was not a general-purpose computer, but Babbage never was able to secure the kind of public funding that had supported the Difference Engine. Consequently, the Analytical Engine never had an existence other than as a detailed concept set out in Babbage's notes and drawings.

Punched Card Input

The data input technology that Babbage envisioned for the Analytical Engine already existed, as it had already been developed in the weaving industry in the form of the Jacquard Loom. Jacquard's system dates from 1804-1805, and used punched cards to regulate a loom to enable mistake-free production of highly complex patterns in fabrics, such as brocades. Punched card technology was improved in the 1890s by Hermann Hollerith, who adapted the approach to the tabulation of the US Census.

Vacuum Tubes

The triode, noted in Chapter 3, in addition to becoming the essential part of amplification circuits, was also an on/off switch in which a small current could control the state of the device. It was, thus, a natural element in the early digital machines. Babbage envisioned that the Analytical Engine would be entirely mechanical, driven by a steam engine. The triode enabled a fundamental break in how computers could be executed. In a wider context, the vacuum tube, of which the triode was the fundamental example, enabled the electronics industry. Prior to World War II, the electronics industry was almost entirely concerned with communications and the broadcast industry. As already mentioned in Chapter 3, the vacuum tube was the heart of the research that led to radar in the 1930s. Radar became a very important tool in the war, and its use in the Battles of Britain and of Midway are famous. The fragility of the vacuum tube was well known, and this weakness was especially threatening in wartime tactical situations.

Pioneer Electronic Computers

ENIAC

ENIAC (Electronic Numeric Integrator and Computer) was by far the most famous of the early electronic computing machines. It had several predecessors, however, of which the most important was the series of digital calculators developed at Harvard University under the direction of Howard Aiken. Aiken conceived of and oversaw the construction of a series of digital machines starting in 1937 and known as the Harvard Mark I through Mark IV. These machines made considerable use of vacuum tubes as switches with no moving parts, which, despite the fragility of the tubes, were a reliability improvement over the electromechanical switches that were the alternative. The Harvard Mark series machines were not the first programmable calculators, but they did make one important improvement over earlier machines: they utilized punched card readers to input data. The existing practice had been to use wire boards to program tasks, and the result of the punched card approach was that the Harvard machines were easier to program, not only in relation to prior practice but in relation to the later ENIAC. The Harvard machines were not small: the Mark I was fifty feet long and weighed five tons. In spite of its use of electronic circuits, these machines were largely mechanical and took between three and six seconds to add two numbers.

ENIAC itself was primarily a special purpose machine, but it embodied more general-purpose capability than any predecessor. Its genesis was in the early stages of the war, and it was financed by an army contract and executed under the direction of J. Presper Eckert and John W. Mauchly of the Moore School of Electrical Engineering at the University of Pennsylvania. It utilized about 18,000 vacuum tubes, which made reliability an issue, but it also made ENIAC approximately 1000 times faster than Harvard Mark I. Its primary purpose was the computation of range tables for artillery and for the atomic bomb.[1] ENIAC's weakness was in the size of its memory, for it used one triode per bit stored. Not only did this make for extreme bulkiness, but it also exacerbated the reliability problem, for to increase the number of tubes was to approach the point at which the machine would become useless as the probability that a tube would fail in any given minute moved toward unity.[2] The solution came in the form of the transistor and core storage, but these technologies did not become available until later. For the interim, there was at least one helpful invention, the delay line. This consisted of a mercury-filled tube with a quartz crystal at each end. Attributed to William Shockley of Bell Labs, the delay line could, in combination with ten vacuum tubes, store 1000 bits. The delay line was used in EDVAC (Electronic Discrete Variable Automatic Calculator), ENIAC's successor, and enabled a material reduction in the number of vacuum tubes. Also conceived at the Moore School, EDVAC resulted from a deliberate effort to create a more general-purpose machine than ENIAC.

Von Neumann's Concept

John von Neumann was a brilliant Hungarian-American mathematician who was the principal author of a 1946 paper entitled "Preliminary Discussion of the Logical Design of an Electronic Computing Instrument."[3] The most important concept advanced in this paper was that data and instruction should both be stored in a common memory unit. This meant that one program could be treated as data by another program. This idea paved the way for high-level programming languages and made possible much of the software advances of the following fifty years. Stored-program computers became known as von Neumann Machines, or IAS Machines, because of von Neuman's affiliation with the Institute for Advanced Studies at Princeton University.

Early Stored-Program Computers

The British were ahead of the American engineers in building the first Stored-program computer: the Manchester Mark I of 1949. This machine was attributed to Maxwell H.A. Newman, Thomas Kilburn, and Frederick C. Williams at the University of Manchester. Progress was advanced in the US with the invention of core storage by Jay Forrester of MIT and Jan Aleksander Rajchman of the Radio Corporation of America. Core storage enabled a material increase in processing speed and enabled MIT to build a computer with seemingly real-time potential, called the Whirlwind.

UNIVAC (Universal Automatic Computer) was designed as a stored-program computer by Eckert and Mauchly, already credited with ENIAC. The partners delivered the first UNIVAC to the Census Bureau in 1951. Univac used mercury delay tubes and thereby reduced the number of vacuum tubes to about 5,000. This improvement resulted in a main processor that *only* occupied 14.5' x 7.5' x 9'. UNIVAC was intended to replace punched-card accounting machines of the day. Having been designed as a business machine, UNIVAC represented the convergence of capabilities of the scientific computer with the needs of office automation, a development already well under way. The basic rights to the UNIVAC family of computers were acquired by the Remington-Rand Corporation.

The Transistor

It would be difficult to cite a more important prototype invention than the transistor. Its innovation streams enabled the enormous world of information technology that emerged in the closing decades of the twentieth century. Its invention was attributed to John Bardeen, Walter Brattain, and William Shockley of Bell Labs.[4] The first examples were made from the element germanium, which in its pure crystalline form is a near-perfect insulator. When "doped" with an impurity, however, germanium becomes a weak conductor, or semiconductor.

The element with which germanium is doped determines whether it has a surplus of electrons (n-type) or protons (p-type). It was found that layered doped germanium in either p-n-p or n-p-n pattern could serve as an off-on switch—for when a small current is applied to the middle layer, a larger current is permitted through the overall device.

First-Stage Innovation

In classic fashion, the very first transistors did little other than making their technical point. The management of Bell Labs recognized this, and, on the assumption that the concept would be more rapidly developed to a state of commercial usefulness if more people and organizations were involved in innovation, offered to license the transistor to anyone interested for $25,000. One of the licensees under this arrangement, Texas Instruments Corp. (TI), with the collaboration IDEA (an obscure maker of antennas), set out to develop a compact transistor radio. The result was the Regency (IDEA's brand) TR1 of 1954. As stage-one innovations went, this was not long by historical standards but much was accomplished in six years.

The first problem to be overcome was how to manufacture the transistor itself. Early techniques tended to have a high rejection rate: one process tried at Bell Labs produced one working transistor out of twenty.[5] TI's engineers were able to overcome this obstacle. A second challenge was in making all components smaller. One of the most attractive promises of the transistor was that an amplifier circuit could be physically much smaller than anything possible with vacuum tubes, but realizing this promise required attaining significant size reductions in the other necessary components, such as capacitors and transformers. This also was accomplished. The first transistors were based on germanium, but this posed a problem in that germanium tended to fail at elevated temperatures. This severely limited the usefulness of the germanium transistor in many military and industrial applications. Silicon was known to be a much more heat-tolerant material than germanium. TI took the approach that the consumer radio was an application where germanium's lack of heat tolerance would not be a problem, and that experience gained in developing the germanium transistor for the radio would apply to silicon transistor production.

It would be difficult to exaggerate the importance of the decision to pursue a large consumer market for the subsequent development of the computer and information technology industries, for that decision committed the company to learn how to produce transistors *en masse* and at a reasonable cost. TI priced its transistors to IDEA not at the cost of their production, but as the estimated cost assuming mass production. The TR1 radio used four transistors, which TI priced at $10.00 per set when the cost of production was actually in excess of this amount. IDEA priced the TR1 at $50.00, more than what AM radios with tubes cost, on the supposition that the compactness and technical sophistication of the TR1 would appeal widely. It did. However, the TR1 was never profitable for either TI or IDEA, with the result that both companies eventually withdrew from the consumer market.[6] TI did break even on its bold pricing policy for transistors, and the experience positioned it to achieve prosperity in the manufacture of transistors and, later, integrated circuits. As a benefit from its radio venture, TI found itself with a significant advantage over its competition in these fields. Moreover, the pattern of creating a mass market for information technology devices was to be repeated often in subsequent years, by TI and by other companies. Credit for TI's bold policy goes to Pat Haggerty, a TI vice-president. This kind of costly risk taking was to become characteristic of small, ambitious technology companies.

Early Computer Applications

The UNIVAC Model 80 of 1958 was the first completely transistorized computer. IBM was urging its engineers to design around the transistor in the mid- to late 1950s, and some of the designs of that period were modified around the transistor. The Model 1620 appeared around 1960 and was completely transistorized. In stark contrast with early computers, such as the UNIVAC, whose central processor occupied an entire room, this machine fit comfortably on a table about the size of a card table. For input and output, it used either a typewriter or punched cards, or a chain printer. Memory was in the form of core storage units. Consequently, the peripheral equipment completed the occupancy of a large room, but the size reduction from the earlier models was still dramatic.

The Integrated Circuit (IC)

An integrated circuit is an assembly of transistors, diodes, capacitors, and resisters placed on a base called a substrate and manufactured as a unit. The IC was independently invented in 1958 by TI's Jack Kilby and by Robert Noyce of Fairchild Semiconductor at about the same time. TI became highly successful at the design and sale of ICs. The IC became a crucial part of computers and any number of other consumer, military, and industrial electronic devices in subsequent years. It represented an ongoing reduction in the physical bulk of circuits and was the inspiration for what has become known as Moore's Law: that the number of devices that can be placed on a given area of substrate will approximately double each eighteen months.[7]

The IC became the crucial element in the development of increasingly powerful computers in increasingly small packages. Because manufacturing cost has proved roughly proportional to volume, the use of increasingly small features has meant a significant reduction in the cost of each individual device and an increase in circuit performance. The number of devices placed on a single small chip has increased from a handful to over 100,000,000. This increase represented an extreme increase in power in computers and other devices but was achieved only over a period of time.

The earliest integrated circuits contained transistors numbering in the tens and are now referred to as "small scale integration" (SSI) chips. The SSI chips figured importantly in early aerospace projects, which provided most of the market for them. Both the Minuteman Missile and Apollo programs required lightweight digital computers as components of their inertial guidance systems. The Apollo guidance system motivated the improvement of integrated circuit technology, while the Minuteman Program impelled the chips into mass production. One result was that the production costs of the chips fell from $1000 per circuit in 1960 to $25 per circuit in 1963. From SSI chips, the next phase was MMI (medium-scale integration) chips, which were attractive by costing little more than the SSI chips. The MMI chips enabled many devices using hundreds of transistors. The next step was the LSI (large-scale integration) chips of the mid-1970s, with tens of thousands of transistors. These were the bases of the pocket calculator and the pioneering personal computers.

The 1980s saw the introduction of VSI (very large-scale integration) chips with numbers of transistors up to and exceeding a million. These and even larger versions have been the enablers of the powerful central processing units that increasingly inhabited computers of the 1990s. To reflect increasing complexity, the term USI (ultra large-scale integration) has been proposed to refer to chips with more than a million transistors, but the use of this term seems to have been reserved for cases in which it is desired to emphasize the power and complexity of the chip. The VSI chips have found application, besides computers, in cellular telephones.

Second-Generation Computers

Big Iron

The mid-1950s saw the emergence of a business market for computers. The IBM Model 650 was a von Neuman Machine that was introduced in 1954 as a specifically business machine, and it was sufficiently popular that IBM sold about 1800 examples. The 650 cost approximately $200,000, as compared with the scientific model, the Model 701, whose sticker price was $1 million. In this era, IBM offered machines at sharp discounts to universities having computer science instruction programs. This practice had the effect of creating a generation of engineering graduates with familiarity and preference for IBM equipment. One of the most clearly successful of the IBM models of the late 1950s was the Model 1401, enabled by inventions by IBM: the high-speed chain printer and the magnetic disk storage. The 1401 sold over 12,000 copies. This computer and all its peripheral equipment were solid-state.

High-End Languages

The utility of the early computers was severely hampered by the necessity that each job be custom programmed in machine language. This led to considerable inefficiency because a computer required a corps of trained programmers between the actual users of the computers (the mathematicians, physicists, and others) and the machines themselves. One key to making computers more accessible and thereby to expand the market for them was the development of languages oriented to the users. This was recognized in the IBM organization in the early 1950s, and one result was the development of Fortran for the Model 704. Fortran and its compiler were developed by a team led by John Backus. Another language, COBOL, was developed for the more business-oriented models, such as the 650.

Operating Systems

The software programs that later became known as operating systems (OS) posed another serious problem for the first generation of computers. The need for some kind of supervisor program to run in parallel with the application program was recognized early. Such a program would have to monitor the actions of the application and take control when necessary. One of the more daunting challenges was to get the supervisor program to stop a running program while retaining the state of the application in all registers so that when the problem was solved, the work could pick up where it left off. There were some early successes. The UNIVAC 1103A had a single interrupt facility, and the Lincoln Labs TX2 generalized the interrupt facility so that various interrupt conditions could be set up under software control.

The IBM 360

The IBM Corporation in 1960 was fast emerging as the dominant player in the manufacture of computers. It solidified its dominance of the market with the 360 System, introduced in 1965. The 360 System was a family of computers having a common architecture and a common operating system. Not only were the 360s fully transistorized, but they employed a significant number of integrated circuits.

Time-Sharing

The best of the computers of the mid-1960s were very expensive and their time was deemed to be more valuable than that of the scientists and engineers who were the real users of the computer. In the batch-processing mode, the standard at the time, the user set up a problem and carried it to the computer center in the form of a deck of punched cards. The results would be available to the user with a delay of at least an hour and, in some situations, longer. The limitations of batch mode were causing considerable distress among users of the computers. One of the principal spokesmen for this distress was J.C.R. Licklider, who argued that the highest usefulness that computing could attain was for the computer to be a part of the user's thinking in the development of a problem.[8] In batch mode, the time from submission of the job to the computer center to availability of results was an *interruption* to the thinking on a problem—far from a part of it. Because of the economics of computing, however, the dedication of a machine to one user for any period of time was generally deemed to be out of the question. Time-sharing seemed to offer at least part of a solution to this problem. In it, a number of users could submit jobs to the computer from a remote terminal. The machine would then process the jobs with computer speed and, ideally, each user would see results within minutes of submission. While not perfectly real-time, it could seem that way under favorable conditions, and it came far closer to the ideal of having the computer as a tool of thinking.

Time-sharing required its own design of an operating system. For example, one of the requirements was that the system has the ability to queue jobs according to user priority and other considerations for situations in which the machine's immediate capacity becomes over-subscribed by users. The efforts to develop an operating system compatible with time-sharing ultimately resulted in the development of UNIX at Bell Labs. Bell Labs followed the policy of distributing the source codes for UNIX to universities and colleges but did not provide technical support for it. This resulted in a rapid but somewhat variegated development of UNIX. The strong multitasking and security features of this system have made it widely used by academic institutions and for World Wide Web servers. Time-sharing led to its own characteristic style of programming, for with UNIX, it was possible to pass results from one program to another as input. With this feature, small, single-purpose programs could be used together to achieve a more complex object than would be possible with any of the individual programs. Several new languages were developed to support this capability, most notably BASIC (Beginners' All-purpose Symbolic Instruction Code).

Minicomputers

Minicomputers came into use in about the same time period as time-sharing. These machines had all the generality of the mainframe machines of the time, but were stripped down, and cost approximately one-tenth the cost of the mainframes. Minicomputers, for example, did not support the higher languages that had been developed for mainframes, and they were programmed in assembly language. These machines were commonly used to control laboratory instruments and embodied simplified architecture and fast transistors. The first market for minicomputers was scientific laboratories, but the market became more general as time passed. The Digital Equipment Corp. (DEC), organized in 1957 by Kenneth Olsen and Harlan Anderson, became a large player in the minicomputer industry and produced a series of machines that expanded the usefulness of the mini. The DEC PDP-8 of 1965 was designed to be programmed in assembly language and was easy to attach to a variety of input/output devices and scientific instruments. It was popular among university scientists and engineers because its comparatively low cost put it within reach of departmental budgets and saved the bother of carrying jobs to the central university computing center and the wait for results. While it had limited memory, its architecture was simple enough that it could be reconfigured for a specific job by a clever graduate student.[9] A subsequent model, the PDP-11 somewhat generalized the minicomputer concept by providing much expanded memory and supporting higher languages.

Besides DEC, other players in the minicomputer industry included TI, Hewlett-Packard, Data General, and Honeywell. The last company was almost the only one of the established mainframe manufacturers to have made a significant penetration into the mini market. It is interesting that companies like IBM never made a meaningful penetration into the mini market. This possibly was because their managements never fully appreciated the appeal of the mini or the difference between the minicomputer and the small computers that they were producing.

Personal Computing

Xerox PARC

In the early 1970s, the Xerox Corporation had successfully brought the Chester Carlson invention to a highly effective office copying machine and was enjoying a period of high prosperity. Senior management, under the leadership of C. Peter McColough, was looking for a next big move, and, inasmuch as copier technology was analog, the idea of developing digital technology appeared to be the direction to take. To this end, Xerox founded the Palo Alto Research Center (PARC), and under the direction of George Pake, brought together a research staff of stellar talents. PARC was not intended to duplicate the efforts of Xerox' existing R & D laboratory; it was intended to work with basic physics to develop new inventions, of whose commercialization Xerox would be the first beneficiary. This was the plan, at least. PARC researchers then developed the Alto, the first personal computer, and a number of peripheral inventions that made the Alto a highly effective research tool.

The genesis of Alto came as a result of the failure of time-sharing to live up to its promise. During a normal working day, the number of time-sharing submissions could be sufficiently large to create a massive queuing problem, in which the wait for results could be no less than under batch computing. This problem varied according to time of day, for at night, traffic would lighten up to the point at which users could realize approximately real-time computing. This led PARC researchers to work all night, and this was inconvenient in terms of normal living. The idea which led to Alto was to build a computer that was powerful enough to handle the problems of the PARC researchers but cheap enough to manufacture that each researcher could have one. In other words, the Alto Project was revolutionary in that it proposed to reverse the established economics of computing that had led to time-sharing in the first place.

Alto's development was due to the efforts of two individuals principally: Butler Lampson and Chuck Thacker. Up until this time, computers had gained speed by providing the various functions; such as input/output devices, disk drives, the user's program, etc.; with their own memory capacity. This had the effect of relieving the central processing unit of operating these functions except for a general supervisory function. All these extra memory capacities ran up the hardware bill, and to reproduce these arrangements with Alto would defeat the purpose of building a low-cost machine that individual researchers could use one-on-one. The solution to this problem occurred to one of the principal developers of Alto, Chuck Thacker, and took the form of multitasking, in which the different functions would share the central processor (CPU) according to a preset rank order. If the CPU received a request, it would process it unless already under the control of a function with superior rank. This idea dramatically reduced the hardware requirement for Alto and, with other measures, brought its cost to about $10,000, or about 60 percent of the cost of a minicomputer of the period.[10] The Alto's first assignment was to serve the computing needs of the PARC research staff. By the summer of 1974, the lab had forty copies of Alto. These machines were tied together by the PARC-designed local area network known as Ethernet. Ethernet went on to become a widely used standard computer interconnection method. Other supporting (innovation) inventions included the laser printer and the graphic user interface.

Of all these inventions that have subsequently proved to be essential integral parts of personal computing, Xerox managed to wrest significant profit from only one: the laser printer. It neglected the others, including the Alto, and other organizations enjoyed their profit potential. Xerox management simply could not envision a large market for Altos. A discussion of the reasons for this performance shall remain for Chapter 13, which deals with the large issue of where might future prototype inventions be expected to be produced.

The Mass Market

It was left for other companies to find the way to profit from personal computing. Two of the entrants to the personal computing industry in the late 1970s were Tandy Corp. and Commodore Business Machines, Inc. The former had an established chain of retail outlets, the Radio Shack stores, and the latter was an established producer of electronic calculating machines that had working distribution channels in place. Persons who remember computer advertisements of the late 1970s will recall some of the come-ons used to appeal to the consumer audience: you can use these machines to store recipes. Also, you could play computer games with them. Prior to this, games had been playable on large dedicated machines to be found in hotel lobbies and in airport waiting areas. Also, there was the issue of price: one could have a copy of Tandy's first machine, the TRS-80, for $399. These appeals elicited a certain amount of scoffing from veteran academic and other professional computer users, but they had some success with the public. Tandy's TRS-80 Model, introduced in 1977, was sufficiently successful to justify a successor, the TRS-80 Model II, in 1979. While the original TRS-80 had been relatively stripped down, its Model II had more capabilities and was plausibly marketed as a small business computer.

The most significant of the early personal computer makers was Apple. Founded in 1976 by Steven G. Wozniak and Steven P. Jobs, Apple was instrumental in popularizing the personal computer through a combination of price, promotion, and design for ease of operation. Apple was a classic Silicon Valley startup. The first product was strictly for the hobby market and consisted of a circuit board but no case. The first order for it came from an electronics dealer, and the contract was executed in Jobs' parents' garage. After the success on this first contract, Apple organized itself with the aid of some venture capital funding and a professional formal business plan. The company succeeded with its Apple II Model.

Apple introduced the Lisa in 1982. Lisa was a relatively powerful machine, selling for approximately $10,000. It embodied two features that contributed materially to ease of use: a graphic user interface and a mouse.[11] Lisa was not successful in the market, but by the time it was withdrawn, Apple was ready with a somewhat stripped-down model, the McIntosh, which proved to be extremely successful. McIntosh was a thoughtful design, featuring an attractive case and an internal architecture that made it easy to attach peripherals, such as printers. It was well equipped with a user manual, and Apple published a professional manual that eased the job of those who would develop application software for the McIntosh.

Enter IBM

Prior to 1980, the major computer producers showed little interest in the emerging market for personal computers. The first sellers of these were startups, such as Apple, or companies established in distantly related or unrelated fields. This began to change around 1980, for that was the year that Hewlett-Packard (HP) brought out its HP-85, which was aimed at the small business market. It was more costly than the competing models of its time, with a price of $3250. It used a cassette for input at a time when its competitors had adopted disc drives and its proprietary architecture created problems for anyone who would develop application software for it.

IBM's entry into the personal computer market, as did Apple's, greatly encouraged expansion of the market. Once it had made its decision, IBM took several approaches that encouraged a rapid expansion of the market. It used open architecture, thereby encouraging the writing of software applications by third parties. This decision promoted speed of development. It contracted with third-party suppliers to provide hardware components. It contracted with software providers, and as a result was able to offer some standard applications, such as a word processor and a spreadsheet. The open architecture also allowed other computer makers to copy the PC. The heart of the PC was the Intel 8088 Processor.

For an operating system, it contracted with a small startup known as Microsoft, founded by Bill Gates and Paul Allen. This system, known as PC DOS, MS-DOS, or simply DOS, became the standard for the IBM PC and its clones. The PC was somewhat more powerful than its competitors at the time, having 16 kilobytes of memory, expandable to 256 K. This led to an interesting competition between IBM machines and the Apple McIntosh. The latter's operating system and graphic user interface used proportionally more memory than the IBM machines, which as a result had proportionately much more processing memory available than the McIntosh. The market overall responded to this tradeoff in favor of the IBM machines. These required at least some knowledge of the DOS language, but this did not prove to be a daunting problem. This comparison became resolved more and more in the IBM machines' favor with the development of the Microsoft Windows program. At first, Windows was little more than a shell working over DOS, but it gradually developed into a full operating system. It worked over DOS, but, to the user, it was the operating system. By the mid-1990s, Windows was effectively reproducing the ease of operation which had been Apple's big selling argument.

More Powerful Processors and Memories

Intel's 8088 Microprocessor, which IBM chose as the CPU of its PC, was the world's first general-purpose microprocessor. As such, the 8088 became incorporated into a variety of devices that became familiar in the 1990s, such as the Automatic Teller Machine, cash registers, and a variety of consumer devices.[12] With the success of the PC, the names of Intel and Microsoft became associated with personal computer development, with most PC-type machines with an Intel CPU and a Microsoft Windows operating system. Intel continued to develop its processors to ever-increasing performance capabilities. By the turn of the century, Intel's top-of-the line processor, the Pentium 4 had achieved a very large circuit density, having 42 million transistors on a single chip. This massive expansion of hardware capability enabled a number of developments in both the physical characteristics of computers and in software. Physically, the computer could be made small enough to be fitted, first into a line of highly portable computers, laptops, which have proved highly popular. Following the laptop's development came the palm-sized computer, which is a growing market as of this writing.

Microprocessors Everywhere

Different versions of the microprocessors, which are the analytical hearts of modern personal computers have become used in a variety of electronic devices found in the household. The typical American upper middle-class household has up to forty, regulating such things as toys, microwave ovens, other household appliances, and other devices.

Computer Data Storage

The power of the modern computer depends on its ability to store data and the programs for manipulating data. The modern computer has, broadly speaking, two kinds of memory: primary and secondary. Primary storage, commonly known as Random Access Memory (RAM) offers the advantage of very rapid retrieval of data, but it suffers the disadvantages of relatively high cost and inability to hold data unless the power is on (it is volatile).[13] In today's computers, RAM is based on large numbers of transistors integrated with retrieval circuitry. The transitory property of RAM storage is the reason why it is necessary for a computer to "boot up" in order to be useful. Some of the prospective computer technologies, such as quantum, nanotube, etc., promise the advantage of rapid retrieval and permanent storage. This would eliminate the necessity for booting up and might simplify overall computer architecture by eliminating the need for both volatile and nonvolatile storage.

Secondary storage, or mass storage, offers the advantages of non-volatility and comparatively low cost, but data retrieval from it is far slower than with RAM. The basic storage technology today is magnetic. In the earliest days of electronic computing, mass storage media took the form of punched cards and/or paper tape. In the days of "big iron," this took the form of magnetic tape and, later, tape cartridges. Today, the standard secondary storage device is the hard disk, which has been improved to the point of providing multiple gigabytes of capacity in a small volume. Alternative mass storage options include flash memories, external hard drives, zip drives, and other devices.

Software

Software and the Market for Personal Computers

The computer, at any stage of its development, is largely useless without software applications. In the early days of computing, the operators of computers found it necessary to employ programmers to support the problems presented by computer users. This custom programming quickly evolved into the availability of standardized packages that could be adapted to the requirements of a whole class of problems, such as a statistical time series analysis package or a discounted cash flow/rate of return analysis package. The lack of good software application packages hindered the early development of the market for the personal computer. Of course, these machines had their operating systems, game packages, and several other categories of applications, and the BASIC language was adapted to the PC for those customers who were able and willing to learn to program. None of these applications were sufficient to lift the personal computer out of the hobbyist market, however.

Two developments changed this and created a large and expanding business market for personal computers: the electronic spreadsheet and the word processor. To appreciate the importance of the electronic spreadsheet, it is necessary to remember how analyses were done in the days when a spreadsheet was literally a large piece of paper with ruled lines and columns. An example will illustrate. During the 1960s and 1970s, many companies were interested in growing their businesses by means of acquiring other companies. To make this strategy pay, it was necessary to find a company that could be acquired for less than the would-be acquirer estimated to be its true value, and this kind of evaluation of a proposed deal required much analysis. To accomplish this, a company with an acquiring strategy would employ a staff of analysts equipped with MBA degrees. Each analyst would laboriously create a spreadsheet analysis of an assigned problem with no more sophisticated tool than an electronic calculator and plenty of pencil lead. Upon completing the spreadsheet, the analyst would submit it to the senior management individual who requested it, the client. This person would generally react by wondering how the result would be affected if one or more of the basic assumptions of the problem were changed. This would send the analyst back to redo the whole spreadsheet by hand with the altered assumptions.[14]

In the days of mainframe computing, there was little point in computerizing this process. Under those conditions, the analyst would have quickly changed a few inputs to the analysis and submitted the job to the computer center, and then both analyst and client would have waited for the job to be handled in the computer center. Most of the time, it was faster for the analyst to do it by hand. However, once the analyst gained the use of a personal computer, the result of changes could be had very quickly once the problem had been set up initially on the electronic spreadsheet. Thus, the PC would appeal to a management, first, because of the possibility for more-or-less instant turnaround in analyses and, second, because of the possibility for getting a given amount of work done with fewer analysts.

The word processor on PCs came in two distinct stages, for between the IBM Selectric typewriter and the PC-based word processor, there was the dedicated word processing machine. This last was a large machine operated by secretaries, but it did save the time of retyping an edited letter or other document. The coming of the PC-based word processor allowed the composers of documents to make their own editing and final copy, and reduced the need for as many secretaries. The PC-based word processor became increasingly useful as computing power increased in the 1980s and 1990s. As staff persons, such as financial analysts, came to be provided with PCs equipped with powerful spreadsheet software, word processing software made it possible for these same people to produce finished internal and external correspondence without the need of secretaries. As a result of the personal computer, the ranks of secretaries primarily devoted to typing diminished, and only the most senior management persons enjoyed the services of the traditional secretary.

Pricing Issues with Software

Computer software represents an extreme case of which modern pharmaceuticals is an example: costly to conceive and develop and relatively cheap to reproduce. With pharmaceuticals, reproduction at least requires access to laboratory and manufacturing facilities. Reproduction of software can require as little as a personal computer that is within the financial reach of a typical, middle-class household. This condition is a continuing threat to property rights of the creators of software. The response to this threat has produced a variety of measures, including write-protection features embedded in the coding of the various software packages. The success of these measures has been partial, for the computer age has produced a large corps of technically sophisticated people. Their practice of "hacking" aims at the defeat of all manner of software protective measures for the purpose of gaining free use of the software involved or for other purposes of questionable legitimacy.

One idea that has intrigued producers of software is to recognize the ease with which software can be reproduced by distributing it free and to charge for it on a use basis. This concept is nothing more than the basis for charging for electric power, water, and fuel gas. With software, there is a large problem of metering use, the same problem that plagued the early days of electric power before effective metering devices were invented. One solution that was talked about considerably during the 1990s was to have software application available on the Internet that could be downloaded in time blocks as users needed the facility. This would make use metering entirely feasible. Under this idea, an Internet-based facility would duplicate the function of the hard drive on local computers and make such facilities unnecessary, at least in the conception of the advocates of this plan. The full-featured PC with hard drive would be replaced by a much more limited and cheaper device. At least one major software producer for a while envisioned this kind of arrangement as the pattern of the future. This future has not yet materialized except in a very limited fashion. Part of the reason is probably that many businesses use such software applications as spreadsheets and word processors so intensely that the flat rate pay system in which the software is simply bought and owned is seen as more economical than any use-based pay system.

Use-based pay systems have appeared in limited fashion where the matter downloaded is information rather than analytical tools. This area presents its own pricing issues. In these arrangements, the downloaded material draws from a database of information. A prospective consumer of a piece of this resource is naturally interested in whether the information the he or she can download will be of use.[15] This suggests that the fee per download ought not to be so high as to discourage a potential user with doubts. The service can be seen as like the purchase of a newspaper, which offers the chance at useful information at a nominal charge. Alternately, a user might take advantage of a free summary of the information download contemplated. This and other software pricing problems are still under experiment.

Supercomputers

Getting the Most from Hardware

The term *supercomputer* has come to be applied for the largest and fastest computing machines of any given period. These machines tend to be distinguished from conventional computers in several important respects. First, they typically have more than one microprocessor, as compared with one in the conventional computer. Second, their internal architecture tends to be cleverly arranged to minimize the distances that electrons must travel. This design feature implied very dense packing of internal components, with consequent problems of disposing of heat. To achieve this, processor and memory circuits are immersed in fluids cooled to cryogenically low levels, at which temperatures the components work the most rapidly. A third distinguishing feature of supercomputers is the capability to do vector arithmetic: in which, for example, a 1 x n matrix is multiplied with an n x 1 matrix in the same time that a conventional computer takes to multiply two numbers together.

Super computers are dedicated to the most difficult problems arising in science and engineering. Many such problems arise in national defense-related work, but not all do. For example, it was long not thought possible to calculate the optimum routing for a telephone call with conventional computers fast enough to be of any use to management. Conceptually, this problem appears to be a straightforward application of linear programming, but working it conventionally yields a result only after weeks of computer time. This classic problem was solved with the aid of supercomputing and some very ingenious mathematical algorithms. Among the better-publicized uses for supercomputers is chess playing, in which the fastest machines of comparatively recent design have defeated a human chess master.

Background

Development of the supercomputer to its present state is associated with a number of people, but two names that stand out in this regard are Seymour Cray and W. Daniel Hillis. After a period of employment with Remington Rand Corp., Cray and Norris founded the Control Data Corp. (CDC) in 1957. At the time, IBM and Remington-Rand almost completely dominated the business computer market, and CDC decided to specialize in the *niche* market for high-speed scientific computers. The CDC Model 1604k, designed by Cray, was one of the first computers to dispense with vacuum tubes entirely and rely on transistors. In 1964, CDC's 6600 was recognized as the fastest computer on Earth. It could execute three million floating point operations (FLOPS) per second, and it was the first machine to be widely referred to as a supercomputer.

Cray left CDC to found Cray Research in 1972. The Cray-1 of 1976 introduced vector arithmetic capability. A subsequent model, the Cray-2 of 1985, used four microprocessors and was the first computer to be capable of one billion FLOPS. Cray moved on again in 1989 to found Cray Computer Corp. All the companies that Cray left continued to produce machines that he had designed. The Cray machines had multiple processors, but these were coordinated by a central processing unit. At the speeds contemplated by computer designers, this CPU was a bottleneck. It occurred to W. Daniel Hillis, a graduate student at MIT, that it might be possible to eliminate this problem by decentralizing the control function and effectively eliminate the CPU. Hillis' ideas were incorporated in a multiprocessor machine built by the Thinking Machines Corp., which Hillis cofounded. This machine of 1985 used over 65,000 one-bit processors grouped 16 to each chip to achieve a capability of 3-4 billion TERAFLOPS, equal to the fastest of Cray's machines and did this with inexpensive components that operated satisfactorily at normal room temperatures. These machines became known as massively parallel processing machines.

Moore's Law and the Future: Some Possibilities

Gordon Moore was a cofounder of Intel, but he may be best known for his observation in 1965 that the number of transistors that could be crowded onto a single silicon chip would approximately double every year. He later modified that "law" to have the number double every eighteen months. This was not a law in the sense of a relationship of nature, but simply an observation of the actual progress of the technology of microprocessor design and manufacture. It has been thought for several years as of this writing that Moore's Law would break down in the foreseeable future as the conventional process of integrated circuit design encountered various physical limitations that would oppose the trend toward ever-increasing transistor density.

It is not within the purpose of this book to enter into the popular game of speculating on the number of future microprocessor generations that will preserve Moore's Law other than to note the observation, credited to futurist Ray Kurzweil, that the rate of data processing acceleration described by Moore's Law extends into history much longer ago than the invention of the integrated circuit. The increase in processing speed can be dated from the mechanical calculating devices used in processing the 1890 US Census, through the relay-based equipment used to crack the Nazi "enigma" code, through the tube-based computers of the 1950s, to the integrated circuit. Kurzweil's point was that there has always been a new technology to replace older technologies that have been developed to their potential. Several possible technologies that could extend the Moore effect indefinitely have appeared: quantum computing, optical computing, and computing based on nanotube technology.

Nanotubes

Phaedon Avouris of IBM's nanoscience group oversaw the development of a laboratory example of an atomic-scale transistor pair based on nanotubes that is a hundred times smaller than the transistors now being placed on computer memory chips and into microprocessors. While the IBM device is but a crude laboratory prototype, it promises to have a number of interesting properties for future computers. For example, it offers nonvolatile memory, meaning that the working memory of the computer will not lose its information when the computer is turned off. With such a machine, it will no longer be necessary to "boot up."

It is impossible to know exactly how long it will take for nanotube transistors to achieve meaningful displacement of silicon in the computer market. People closely involved with this development have a way of tossing off "ten years." Perhaps closer to commercial reality is nanotube-based television screens. In contrast to today's familiar cathode ray picture tube, an array of nanotubes would beam electrons at the viewing screen, with each pixel served by its own nanotube. This would permit a much thinner television receiver, and theoretically could be cost competitive with cathode ray tubes. If so, this technology would compete most effectively with today's liquid crystal or plasma technology. A number of established consumer electronics companies have undertaken to develop this technology, including Samsung.

Quantum Computing

In 1959, Richard Feynman observed that as electronic components reached microscopic scale, there would be effects predicted by quantum mechanics that might be turned to advantage in computing. In classical computing, speed can be increased through parallel processing, using more than one processor. To achieve an exponential decrease in computing time requires an exponential increase in the physical space needed by the equipment. With quantum computing, an exponential decrease in time requires only a linear increase in the space occupied by the computer. This is the promise of quantum computing. The difficulty with constructing a working quantum computer comes with the need to extract information, for in quantum mechanics, an observation, or measurement, alters the quantum state of the system.[16]

To date, no one has built a quantum computer except as an experiment to demonstrate the principle: the prototype. This was achieved in 1998 by Isaac Chuang of the Los Alamos National Laboratory, and Mark Kubinec of the University of California at Berkeley, who created a quantum computer that could be loaded with data and from which an answer could be extracted. This pioneer system was coherent for only a few nanoseconds and trivial from the standpoint of solving meaningful problems. From here, quantum computing enters its stage-one innovation period, which could prove to be of very long duration.

Optical Computing

There are several ways in which today's computers seem susceptible to improvement, powerful as they are. One of the more interesting technologies that may be coming together at present is the replacement of electronic circuits with optical circuits, both within the central microprocessor unit (CPU) itself and in the transmission of information among the different microchips within the computer case. The problem besetting present-day electronic circuits in the CPU is that the circuits have become so tiny and so crowded that cross-interference between and among different circuits is becoming increasingly difficult to prevent. As for the transmission of information between and among integrated circuit chips, the speed of electronic transmission along metallic conductors is below that of the processors, necessitating local storage of information which backs up while the circuits catch up with the microprocessors. Displacement of electrons by photons in these functions would go far toward not only making all parts of the computer compatible in the matter of speed, but it would also help immensely with the problem of undesirable cross-currents. The problem of implementing a photon-based computer was, up until recently, the lack of a means of converting electrons to photons that could be mass-produced sufficiently cheaply. This problem appears to have taken a large step toward solution with the 2005 invention of the silicon laser by Intel scientists, among others.

Mathematical Algorithms

In mathematics, an algorithm is a finite set of well-defined instructions for accomplishing a task for which, given an initial state, will terminate in a recognizable end-state. The name derives from the Persian mathematician Al-Khwarizimi who lived in the ninth century A.D. The algorithm has been used to simplify computation for centuries, but interest in this branch of mathematics quickened as the electronic computer developed. The essential difference between the computer and the human mind is that the former cannot think creatively (or at all), but is superior to the mind in its ability to do repetitive arithmetic. An algorithm can be thought of as a system for reformulating a computational problem in such a way as to take advantage of the computer's ability to shoulder large quantities of repetitive computations.

In general, an algorithm does not find an exact solution to a problem. Rather, it converges on the solution in a finite number of repeated steps. The mathematical requirement is that the solution has been proven to exist. In computers, all numbers have a finite number of digits, including numbers that in theory have an infinite decimal expansion. Therefore, repetitive calculations will build up chaotic errors from operations with numbers whose length exceeds the largest size number the computer is designed to handle. To be meaningful, the algorithm requires some means of assessing the size of the error between the estimated value and the actual value that allows for the chaotic remainder. When this is accomplished, it is possible to achieve the objective of the algorithm, which is to estimate a number within an error of preset smallness.

At any stage of the electronic computer's development, its power could be extended by developing a more efficient algorithm, meaning one that converged in fewer iterations, or one using less working memory capacity (RAM memory). An example illustrates how this works. In the 1950s, it was not possible to compute the most economical routing for a telephone call. Conceptually, the problem could be stated in terms of a linear programming program, in which an objective function (for example cost) is minimized subject to a number of constraints expressed as linear inequalities. The so-called simplex algorithm had been developed for linear programming solutions and called for the computer's strength in repetitive arithmetic operations. The problem was that there was an enormous number of routing options for a long-distance call, and the numbers were so great that a simplex solution took almost a month of computing by the most powerful computers of the time. A report of the optimum routing could have been useful to management on a daily basis, but the speed that was possible then made the exercise useless. Then Bell Laboratories mathematicians developed an alternative algorithm, based on the mathematical field of projective geometry, which improved on simplex's speed by orders of magnitude.

Notes

1. Computer development was moving forward also in Great Britain. In common with ENIAC, the wartime British machines were special-purpose, focusing on the problems of code breaking.
2. Reliability was a sufficiently large problem that first-generation computers were often programmed to "dump" their memories onto punched cards every fifteen minutes. This allowed a job in progress to be restarted with no more than fifteen minutes lost.
3. The other two authors were Arthur Burks and Herman Goldstein. The authors were affiliated with the Institute for Advanced Studies of Princeton University, and the computers following the standards set out in the paper came to be referred to as "IAS" Machines.

4. William Shockley had already contributed to early computer development with his delay line, mentioned above. Shockley and his co-inventors of the transistor shared a Nobel Prize for this work.
5. Robert J. Simcoe, "The Revolution in your Pocket," *Invention and Technology*, vol. 20, no. 2, (Fall 2004), pp. 12-17.
6. The manufacture of consumer electronic devices gradually moved to Asian producers. This included not only radios, but television sets and a host of other devices. Had the TR1 and its successor models been priced at $60.00, this story could have been quite different, or so it is speculated.
7. In 1968, Noyce and Gordon E. Moore left Fairchild Semiconductor and later, after being joined by Andrew Grove, founded Intel Corp., which became the leading producer of microprocessors.
8. J.C.L. Licklider was a professor of psychology at MIT—at the time, he expressed these ideas in an article entitled "Man-Computer Symbiosis," published in 1960. Licklider later became the head of the Information Processing Techniques Office of ARPA, from which position he became an effective promoter of time-sharing and oversaw the development of the tools to implement it.
9. At the time of its release, the PDP-8 sold for around $20,000. By the late 1970s, this was down to $3,000.
10. Over the two decades following the development of Alto, the cost of memory fell dramatically as consequence of both large-scale production and technical success at placing an ever-growing number of devices of a single chip. With prices of hardware components of the early 1970s, the accomplishment of Alto seems all the more remarkable.
11. Xerox had made a half-hearted attempt to market the Alto with its Star, which it brought to market in 1981. As a commercial venture, Star was a failure, and it was soon withdrawn. With that withdrawal, Xerox seemed to lose interest in personal computer manufacturing and sales, and became receptive to others, such as Apple, which were interested in using features such as graphic user interface and mouse.
12. The Intel Corporation was founded in 1968 by Gordon Moore and Robert Noyce, who was a co-inventor of the integrated circuit.
13. RAM storage is more costly than secondary storage by an order of magnitude.
14. During the 1970s, the author worked in the corporate planning department of a company with an active acquisition strategy. The company headquarters was located in a small and sleepy town, and the senior management would take advantage of their access to company aircraft by spending the workweek in more interesting places. They would come in on Friday afternoons with a load of problems for the analysts. The spreadsheets that would result had to be altered for numerous changes of assumptions, and the analysts involved were sometimes lucky to get home before 10:00 P.M. on Fridays. Moreover, the work would continue on Saturday morning. The analysts tended to welcome the basketball season, for many of the seniors were avid fans of the University of Kentucky and other schools and would take Saturday afternoon off.
15. The use-fee database tends to occur in technically specialized areas of information. At least one chemical industry source for a while offered a free trial period of a month, which it abandoned on the ground that people were "abusing" the trial period. One can see how this might happen if a user takes advantage of the trial period to satisfy an information requirement for a specific project and decides he of she has no interest in the service beyond that project.
16. Rieffel, Eleanor and Wolfgang Polak, "An Introduction to Quantum Computing for Non-Physicists," (Palo Alto, CA: Xerox Palo Alto Laboratory, 2000), p. 2.

9

Light

As rich and variegated as twentieth-century technology was, it can be said to have been dominated by two main themes: electrons and photons. While progress in the former has evolved for most of the century, the technology by which light is created and controlled may be the dominating theme of the twenty-first century. Electrons, generally speaking, move in conducting media, such as metals, against resistance. Photons move literally at the speed of light, and resistance is not a problem. This chapter deals with the tools of light.

The Laser

The Technology

Laser is an acronym for Light Amplification by Stimulated Emission of Radiation. It is a device that uses quantum mechanical principles and stimulated emission to produce a coherent[1] and highly collimated (small angle of dispersion) beam of light that is primarily characterized by one wavelength (color). The color of the light depends on the medium (lasing medium) used to produce the beam. This medium has to be of highly controlled purity, size, and shape. A laser beam can be continuous at constant amplitude (CW) or pulsed. With a pulsed beam, much higher peak energy can be attained than with a CW beam. A laser medium can be used as an amplifier of a light signal when it is "seeded" with light from another source. This property is highly important in telecommunications applications, as discussed below (See Figure 9.1)

The physics of lasers rests on the idea of producing a "population inversion" in a laser medium, by supplying energy in the form of light or electricity.[2] Referring to Figure 9.1, this energy is supplied in the form of light from a powerful strobe light, labeled "flash lamp" in the figure. The lasing medium is a cylinder of ruby with one end fully silvered to form a mirror and the other end partially silvered in order to reflect and pass light at the same time.[3] The object is for the light amplification (gain) to exceed the loss from the tube acting as a cavity

Figure 9.1
Ruby Laser

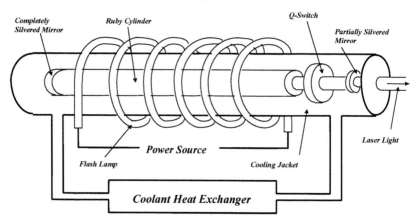

resonator. The equipment is so designed that the power of the circulating light can rise exponentially. If enough power is applied through the light source, the medium reaches an equilibrium capacity between stimulated amplification and loss from the resonator, and the light in excess of this capacity is emitted as a beam of laser light.

While the first working laser was built by Theodore H. Maiman of Hughes Laboratories of Malibu, CA in 1960, the genesis of the laser goes back well before that. The laser was preceded by the maser (similar to the laser, but with "microwave" substituted for "light"), which was invented by Charles Townes of Columbia University with graduate students J.P. Gordon and H.J. Zeiger in 1953. The research effort that resulted in the maser appears to have had its origin in the development of radar technology in World War II, for many of the scientists who had been involved in radar work became interested in the closely related microwave radiation field. Microwave research was aided by the availability of much laboratory equipment that came from the radar program.

Not long after the invention of the maser, scientists began to look into the possibility of simulated emission in other parts of the electromagnetic spectrum, with much interest in the optical and infrared parts of the spectrum. Townes worked with Arthur L. Schawlow on describing the physics of the optical maser (later renamed the laser) and published a paper in *Physical Review* in December of 1958, in which they proposed a means of actually constructing such a device. By this time, interest in an optical maser had become intense and had attracted considerable budget support from a number of large organizations, including Bell Laboratories, General Electric, Westinghouse, Siemens, and a variety of other players both large and small.

Maiman's prototype laser barely preceded a similar device by Schawlow and others at Bell Laboratories. Maiman used a synthetic ruby crystal to produce red laser light. Soon after his prototype, the Iranian physicist Ali Javan invented the gas laser. During the same general period, maser-laser research was proceeding in the Soviet Union, and, in 1964, Charles Townes shared a Nobel Prize in physics with two Soviet scientists, Nikolai Basov and Alexandr Prokhorov.

Maiman's achievement of winning what had become a race to the laser was amazing in several respects: first, because of the massive competition he faced and, second, his approach to the problem included several elements that had been rejected by other researchers, and this exposed him to ridicule from within his own profession. He was a junior employee of Hughes Aircraft at the time, and his employer questioned the utility of a success, should he attain it. Needless to say, his project was time-limited and not generously funded. Maiman's most controversial decision was to use ruby as a lasing medium, a material that had already been written off by the physics profession at large. His ability to succeed in the face of these difficulties can be attributed to his own confidence and a professional and personal background that was unusual for the times.[4]

At the time of its invention, the laser was so new and unprecedented that there was no large slate of existing applications which it could take over from earlier technologies, for there had been no prior strong source of coherent light. In view of this, it does not seem surprising that the research effort leading to the laser's invention should have received some criticism from the senior managements of the corporations that were funding the research. Against the background of doubting attitude on the part of senior management, the seemingly lavish funding of the research by a number of large corporate research organizations may seem strange. If there was no obvious large market for a laser, where was the research energy, as evidenced by the number of contenders and the size of their expenditures in the race to build the first working laser, coming from?

The answer may lie with the nature of corporate research shops in the postwar period. There is no question as to the level of enthusiasm for laser research in the physics community. The heads of corporate R & D divisions typically had risen through the research ranks or had been recruited from the outside based on educational and research experience qualifications, and were themselves fully susceptible to the enthusiasm of the research staff regarding the future of coherent light generators. The research heads were also very senior vice-presidents commanding substantial budgets and had much discretion over how these budgets were used. The successful R & D leaders also had the capacity for resisting senior corporate management's pressures, at least up to a point.[5] One is therefore tempted to the conclusion that the overall massive push to the laser by corporate research shops reflected the impetus of the scientific staffs aided by the power of their organizations within the corporate structure.

Perhaps the most important property of the laser is its ability to produce highly coherent and collimated light. Many of its subsequent applications depend on the former property especially, as it enables the design of controlled interference patterns. The following is a description of some of the major applications of laser light. Generally speaking, the important commercial and military applications began to appear after 1970, implying approximately ten years as the length of the stage-one innovation period for the laser. While the presence of direct research funding by government agencies in the research leading the invention of the laser is not evident, the government was a very active direct supporter of the first stage innovation process.

Telecommunications

Perhaps the best-known laser application is in long-distance telecommunications, enabled by the 1970s development of the optical fiber. In this application, a laser or LED (light-emitting diode, see below) is used to convert an electronic signal into a photonic signal using the optical fiber as a light-guide to transmit voice and data over long distances. In this application, the optical fiber directly replaces a traditional copper conductor and results in a number of significant advantages over copper wire transmission media, including a cleaner signal that has a low rate of attenuation. Most of the discussion of this technological combination is deferred to the discussion of optical fibers below, but one important contribution of the laser is its ability to emit a highly coherent beam of light into the end of the fiber. Also, the laser's capability for producing light of close to a single wavelength and amplitude has offered at least partial solution of problems related to the transmission of data over copper wires by increasing the rate of transmission to the speed of light. Not even a laser can produce a perfectly collimated beam of light due to the effects of diffraction;[6] but when laser light is put into a wave-guide, such as an optical fiber, the diffraction laws governing dispersion no longer apply and diffraction ceases to be a problem. As of the present writing, the laser/optical fiber combination has virtually taken over the long-distance voice and data market in the US and lacks only universal connection directly to houses to achieve literally complete takeover.

Navigation

Several devices for detecting changes in orientation and spin based on the laser that make use of interference of laser light were developed during the 1960s and 1970s. These have begun to replace inertial guidance systems and go under the name of *inertial reference systems*. Because they have no moving parts, they offer considerable improvement in navigational accuracy compared to mechanical inertial guidance systems, and are light and therefore sufficiently portable to be used in aircraft.

One example of such a device, the Ring Laser Gyroscope (RLG), was developed in the 1960s and 1970s for commercial, space, and military guidance. The RLG uses segments of transmission paths arranged in either a triangle or a square, with the segments connected by mirrors. One of the mirrors allows light to pass through as well as be reflected so that the light can reach the sensors. Laser light is introduced into this geometry in both directions, establishing a standing wave resonant with the length of the path. As the device rotates, light in the two directions travels different distances, such that light in one direction changes phase and resonant frequency relative to light in the other direction. The result is a changing interference pattern that the sensor reads. The angular position is read by counting the interference fringes. The RLG found its first major commercial application when the Boeing Co. selected Honeywell to supply inertial reference systems for its 757 aircraft of 1978. The Northrup-Grumman LTN-92 utilized three RLGs and three high-speed digital microprocessors to provide an all-attitude, worldwide navigation system offering up to five times the reliability of mechanical inertial navigation systems. The LTN-92 allows older aircraft designs, such as the L-1011, the DC-10 and Boeing 747 to conform with current regulatory requirements.[7]

Industry

When one considers that an ordinary 100-watt incandescent light bulb emits about 95 percent of its energy as heat in all directions, it is easy to appreciate the laser's accomplishment of putting all of its output energy into one highly collimated and coherent beam. Indeed, a high-intensity laser beam can be very hot and is capable of melting some metals with high melting temperatures. This is the basis for an important laser application in industry: cutting and welding. The laser is the basis for the welding robots that have come to be used extensively on automotive and other production lines. The resulting precise control of welding operation thereby obtained has been the basis for substantial improvement in quality uniformity over previous welding methods.

Another coming laser application in industry is three-dimensional measurement by means of coherent laser radar. This technology was developed with government funding, and enables non-contact measurement of objects that through sheer size or awkward positioning would be costly to measure by more traditional means. The device is portable and will perform under any light conditions and will work with any object having 1 percent reflectivity or greater. Uses include quality assurance applications, such as comparing dimensions of parts as programmed into computer-aided design (CAD) to as-built parts and monitoring wear on automated machine tools.

Medicine

One of the better-known consumer uses of lasers is in LASIK surgery. LASIK is an acronym for Laser-Assisted In-Situ Keratomileusis, a procedure used by ophthalmologists to reshape eyeballs in such a way as to correct vision, thereby reducing a patient's dependence on glasses or contact lenses. After a procedure to determine the exact shape of the eyeball in order to measure the precise dimension of the needed alteration, an excimer laser is used to remodel the corneal stroma. Excimer laser ablates tissue without damaging adjacent stroma.[8]

Other surgical applications of the laser include the surgical welding of a detached retina and the staunching of bleeding in the gastrointestinal tract caused by a peptic ulcer. Laser light properties of being absorbed by pigmented lesions enable its use to treat pigmented tumors and remove tattoos. Laser surgery has also been found useful in treating superficial bladder cancer and can be combined with ultra sonography for transurethral ultrasound-guided, laser-induced prostatectomy. Other uses include the treatment of glaucoma and lesions of the cervix and vulva, including carcinoma and genital warts.

Data Recording

Laser technology has become central to the recording of digital data in industry and in the consumer recording industry. In this technology, digital data are recorded by burning a series of microscopic holes by means of a laser, commonly knows as pits, in a thin metallic surface of a small disk. In the read mode, the data are exposed to low-intensity laser light played on the disk surface, and the reflected light is "read" by light-sensitive diodes. The intensity of light received by the diodes varies according to the presence or absence of pits, and the diode circuits convert the information for either screen readout or into sound readout. A laser can record gigabytes of information in the microscopic pits of a DVD.

The Silicon Laser

Technological development has produced increasing speeds in all the components of the personal computer but at different rates. The resulting speed differential has already become a problem. Central processing units (CPUs), for example, have become orders of magnitude faster than the capacity of their copper connections with memory and other units within the computer. As a result, the fastest CPUs can spend as much as 75 percent of their time idle, waiting for the electronic traffic jam in the copper connections to other circuits to clear.[9] There is widespread agreement that a technical solution to this problem consists of converting the internal information transmission within the computer from an electronic basis to a photonic one. Until recently, however, the devices needed to do this, a class of lasers known as III-V lasers (in which the lasing medium is some combination of an element from the third column of the periodic table with an element from the fifth column), which are difficult to manufacture and are therefore too expensive for personal computer application.

The silicon laser was invented in 2004-2005 by a research team led by Mario Paniccia at the Intel Corp. Its promise is the possibility of converting information transmission from electronic to photonic both within the silicon environment of the personal computer and in local area networks by cheap to manufacture. If this potential can be achieved, it will mean a massive increase in the speed and power of the personal computer. Moore's law will be extended for decades. Paniccia's achievement was a *tour de force* because silicon had not previously been considered as a good lasing medium.

As things stand now, further progress in this direction using copper conductor appears increasingly problematic. Even if engineers at Intel and other organizations succeed in increasing the speed and power of central processing units (CPU), these units will be able to produce electron flows that are too fast for the means of conducting these flows from chip to chip within the computer and within any wire-connected computer network. Moreover, speed increases within CPUs are expected to be hamstrung by such problems as heat buildup and cross talk between adjacent copper conductors. Converting from an electronic to a photonic basis promises to eliminate these and other problems.[10]

Light-Emitting Diodes

The Technology

A light-emitting diode (LED) is a solid-state electronic device that emits incoherent, narrow-spectrum light when electrically biased in the forward direction. The color of the light emitted depends on the chemical composition of the semiconducting materials of which the diode is made. Color possibilities range from infrared to visible spectrum to ultraviolet. Credit for invention of the visible-spectrum LED goes to Nick Holonyak, Jr., of the University of Illinois at Champaign in 1962. LEDs became generally available in the early 1970s.

While LEDs have not penetrated the general illumination market for reasons discussed below, they have a number of advantages over conventional lighting, both incandescent and fluorescent. They have an extremely long lifespan, over twice that of the best fluorescent tubes and over ten times that of the best incandescent bulbs. They are built in solid cases, which render them not easily subject to breakage. They are capable of yielding light of any of a range of colors without the aid of filters that are necessary with incandescent bulbs. Also, the shape of the LED package allows the light to focus, unlike incandescent or fluorescent lighting, which require external reflectors in order to direct their light usefully.

If the emissive layer of an LED is an organic compound, it known as an organic LED (OLED). Compared with regular LEDs, OLEDs are lighter and have the added feature of being flexible. Whereas an ordinary LED gives off a point of bright light, the OLED appears as a patch of glowing plastic. The potential of OLEDs comes from the possible ease and low-cost of their production, and because they require no costly chip-fabrication facilities, which make ordinary

LEDs expensive. The present impetus for the development of OLEDs comes from companies seeking backlighting for cheaper, brighter displays.

Niche markets

So far, cost and other considerations have kept LEDs out of the market for general illumination, but they have moved into a number of markets where their physical advantages justify their cost. One of the early uses was in calculator displays and readouts in digital watches and other timepieces. LEDs have largely taken over the market for automobile dashboards and other displays. Anyone who has undergone the frustration of replacing dashboard lighting, or the expense of having it done, will appreciate this use.

Local and state governments have taken advantage of LEDs' long service life by using them in traffic signal lights. In traditional traffic lights, a 150-watt incandescent bulb is mounted behind a colored lens, and has to be replaced after one year of service. By contrast, an array of LEDs behind a clear lens absorbs about 15 watts of power and lasts five years or longer. California's experience indicates the potential of this new form of lighting. As of 2002, the California Department of Transportation reported that replacement of traffic lights—red, yellow, and green—had shaved at least $10 million from the state's electric bill.[11] LED traffic lights, not surprisingly, are rapidly becoming commonplace. Another traffic-related application of LEDs is in traffic information signs.

The LED moves to roil the advertising world by means of the giant electronic billboard. These LED-display devices provide an animated picture and can change from one advertising message to another almost instantly. The advertisements are sold in blocks of time. Investors in these signs can promise captive audiences of motorists stuck in traffic jams, an audience that no other ad medium can easily match. This is causing concern in traditional advertising media such as newspapers.

Blue Light

The commercially viable blue-emitting LED was invented in 1993 by Shuji Nakamura, who at the time was working for the Nichia Corporation of Japan. Blue LEDs became widely available in the late 1990s. The significance of the blue LED was that, with LEDs of the other two primary colors, red and green, a white LED was possible. While some researchers have actually made white LEDs by combining the primary colors blue, red, and green, most white LEDs today use a blue gallium nitride (GaN) LED covered by a yellowish phosphor consisting of cerium-doped yttrium aluminum garnet (YAlG:Ce) crystals that have been powdered and bound in a viscous adhesive. Part of the blue light from the LED is converted to yellow by the phosphor, and the yellow light stimulates the red and green receptors of the eye. The mixture of blue and yellow appears white to the eye. White LEDs have found their way into several markets, such as flashlights, traffic signs (walk/don't walk), and head-mounted hiking lamps.

White Light and General Lighting

The real prize awaiting the developers of white LEDs is the general lighting market: offices, factories, and homes. This is why white LED development is being pursued by a variety of organizations worldwide, ranging from startups to established firms such as General Electric. Given the LEDs' theoretical ability to convert 100 percent of the input energy into light, replacement of existing incandescent and fluorescent lighting by white LEDs could save up to 10 percent of the world's power bill and obviate the need for $50 billion worth of new power plants by 2025.[12] The problem with the pioneer white LEDs for general lighting was threefold: lack of sufficient brightness, an incomplete light spectrum that yields light that is not pleasing to the eye, and cost. Progress on all three of these issues has been made, and the probability of ultimate success for the white LED in general lighting would seem enhanced by the sheer number of research organizations pursuing the goal.

Fiber Optics

Fiber Optics in Telecommunications

An *optical fiber* is a transparent, thin fiber, most often made of glass, which is a highly effective light guide for transmitting light over long distances. The combination of the optical fiber with the laser has enabled a virtual revolution in the technology of long-distance communication. Optical fibers have been available since the first useful fiber was invented in 1970 by researchers at Corning Glass Works, Zimar, Maurer, Keck, and Schultz. The physical fiber was preceded by a 1966 doctoral dissertation by Charles Kao, in which the author predicted that in order to be useful in telecommunications, a glass fiber had to have an optic signal attenuation of less than 20 decibels (dB) per kilometer. The first fibers from Corning had an attenuation rate of 17 dB per kilometer.[13]

In telecommunications, a laser-emitted infrared light beam is introduced into the end of an optical fiber. The glass fiber is coated ("doped") with some material which forms a cladding that has a slightly greater refractive index than the glass of the fiber itself. Light in the fiber that strikes the cladding is reflected back within the fiber. This property eliminates the problem of diffusion of laser light, and allows the light signal to travel around curves. It also eliminates "crosstalk" from one fiber to another, so that fibers can be arranged in bundles or cables. Fibers used in telecommunications typically have a diameter of 125 micrometers (μm). Fibers used in telecommunications are known as single-mode because they have only one strong propagation mode. These fibers are used in pairs, with one fiber of the pair carrying a signal in each direction.

The advantages of fiber optic telecommunication systems have permitted these to displace older systems based on copper conductors. Of these advantages, perhaps the most important, is the low signal attenuation rate: less than 0.3 dB per kilometer. This means that signals can be transmitted over several hundreds of miles without need for boosting in repeater stations, compared with twenty to twenty-five miles for copper. The complete absence of signal cross-talk between fibers, plus immunity from electromagnetic radiation, results in a much cleaner signal than was possible with copper. Also, as photons move at speeds approaching that of light in a vacuum and electrons encounter resistance in copper, fiber optic systems do a far better job of transmitting data over long distances. Another significant advantage gained in the conversion from copper to fiber is the much greater bandwidth potential of fiber.

With these advantages, it was not too surprising that the emerging fiber optics industry attracted huge amounts of capital in the 1980s and 1990s. Fiber cables were laid everywhere in the US and internationally under an assumption that the huge communications capacity would soon be filled with demanded communication services. This optimism proved largely unjustified, and the overbuilding that became apparent with the post-2000 downturn contributed to a very sharp recession that solidly impacted telecommunications investment. The magnitude of the miscalculation can be seen in the estimate, circa 2004, that over 98 percent of the fiber optic cable laid to that time remained "dark," i.e., no demand for it had ever emerged. As of the end of 2005, some of this "dark" fiber was coming into use, but the potential capacity of unused fiber remains large.

To date, fiber optic systems have become sufficiently widespread as to be within one mile of the individual home. During the later stages of what has been termed the telecommunications "bubble" of the late 1990s, telephone companies laid fiber optic cables into many residential neighborhoods, but conditions of the telecom recession that started in 2002 cooled these companies' ardor for making the final costly connection with individual homes. As of this writing, the delivery of broadband communication services over this distance is the object of contention between the cable operators and the local telephone companies with their digital subscriber line (DSL) services. It is not presently clear how this contest will work out, as the result may be heavily influenced by the form government regulation takes (see discussion in Chapter 6 above).

It is also not clear under what circumstances the last mile will be filled in with glass fiber. The cost of greenfield residential installation of fiber has fallen almost to the level of the cost of copper installation, and few communities have fiber-to-the-home technology, through which the residents receive a variety of ultra high-speed internet, television, and telephone services.[14] Cable and DSL can be viewed as lower-cost alternatives that may or may not turn out to have been stopgaps short of the general extension of fiber optics to individual residences. Yet another technology fits into this same category: packet switching. As will be remembered from Chapter 6, this is the heart of the Internet technology and consists of cutting a transmission into a number of packets, each of which has an address encoded and each can move to its ultimate destination by the most economical route, there it will be recombined into the entire message as sent. Until recently, packet switching has been plagued with quality problems in voice communications, but these have been overcome. The advantage from packet-switched networks is the significantly lower operation costs. The reason is that with present networks, a caller, once he has dialed a number, reserves a line that is exclusively devoted to that call for its duration. With packet switching, no line is tied up for the duration of the call. Lines are tied up only for the brief intervals in which packets are moving over them, and they are otherwise open for other traffic. As of this writing, Verizon is the telecommunications firm that has made the largest commitment to packet-switched networks.[15]

There is a powerful economic incentive to put off extending fiber to households: the unanswered question of just how much of the implied broadband capacity the consumer will pay for. The services that can be delivered with these technologies, including voice, data, high-definition video, and a large menu of other potential services, go well beyond what the typical customer is used to, and the demand for many of these is speculative.

Other Applications

While optical fibers used in telecommunications are exclusively of glass, several plastics have been used to make fibers useful for shorter distance applications. These include medical uses, in which light needs to be brought to bear on a target for which there is no line-of-sight path. Bundles of fibers are used in combination with lenses to view objects through a small hole, and these are called endoscopes. Medical endoscopes are used for exploratory or surgical procedures. Industrial endoscopes are used to inspect anything hard to reach, such as jet engine interiors. Optical fibers have entered the consumer market in such forms as Christmas tree lighting.

Holography

Holography is the science of the hologram. A hologram is created when two beams of coherent light, one reflected off the object to be depicted, and the other a reference beam, meet at a photosensitive plate. Because the two beams have to travel unequal distances, the two beams interfere with each other, and the plate records the interference patterns between them. When the recorded image is exposed to ordinary white light, the result is a three-dimensional image (See Figure 9.2).

In today's world, the obvious source of coherent light is the laser, but the invention of the hologram and the physical explanation of it predate the la-

Figure 9.2
Holograph

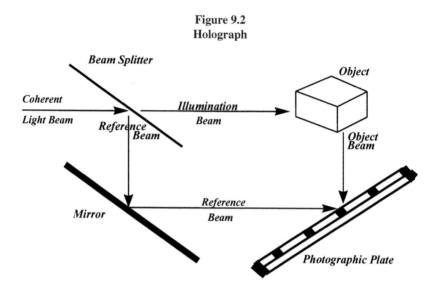

ser. In the first hologram experiments, coherent light was approximated by a combination of filter and slits. When laser technology appeared in the 1960s, it enabled the hologram to be applied in a number of important uses and form the holographic industry of today. In what can be considered the first stage of innovation, from its first discovery in 1947 to the introduction of laser light in the early 1960s, work on the hologram proceeded mainly from the impetus of scientific interest.

There were three different independent inventions of the hologram. The first, was by Dennis Gabor, a Hungarian physicist, to whom the idea occurred while waiting for a tennis court in Rugby, England, in 1947. The idea came in the course of Gabor's work on improving the resolving power of the electron microscope such that it could be used to observe an individual atom. Gabor was working for the British Thompson-Houston Company (BTH) and had the use of its laboratory facilities. He was able to cobble together apparatus to test his idea, producing coherent light in the manner mentioned above. His result was a two-dimensional hologram of a test transparency. His published paper of 1948 mentioned, almost in passing, that the technique could be used for three-dimensional imagery, but his attempts to experiment in this direction were frustrated by the weakness of the coherent light that he could produce with filters and slits.[16]

The second discovery of holography was by physicist Emmet Leith in 1954. Leith was working for the Willow Run Laboratory of the University of Michigan on an Air Force contract that sought a means of radar imagery as sharply resolved as conventional aerial photography. Because radar waves are 10,000 times as long as visible light waves, Leith and his colleagues were concerned with how to store the data that could be collected in bits and pieces by a surveillance aircraft as it flew over the ground to be surveyed; it was an enormous amount of data. In attempting to gather the data optically, Leith discovered that when light was shined on the optical data transparency, one could see a miniature image of the scanned object. Leith then worked out the physics of this phenomenon, but several months later he discovered Gabor's earlier paper. One can imagine his mixed feelings: disappointment at not being the first discoverer, and elation at having independently duplicated Gabor's discovery. Gabor was by then a luminary of the profession.[17]

Yuri Denisyuk, of the Soviet Union, became the third independent discoverer of holography in 1958. Denisyuk's variant of the hologram was the first that could be seen with ordinary white light, not the coherent light with which the exposure was made. Denisyuk discovered the work of Gabor and Leith only in the 1960s and seems to have run into some trouble getting his research published in the Soviet Union due possibly to professional jealousy or a political problem. It was only after these problems were sorted out that he received the Lenin Prize in 1970.

The possibility of three-dimensional imagery became real with the availability of the laser, and a change in technique, due to Leith. The laser meant a powerful source of coherent light, and became the basis for the commercialization of the hologram. Before this could happen, certain problems with laser light had to be worked out, for it was powerful enough to have exposed some sloppiness in experimental procedures that had not presented problems using the weak light sources available before the laser. The other improvement was moving the reference beam away from the line of the image beam. Before, when the two beams had been on almost the same axis, this arrangement produced a double image on the photosensitive plate.[18]

Arguably, the most important application of holography is in security: the holograms that are a feature of credit cards and some currency to prevent alteration or counterfeiting. The underlying technology is due to Stephen Benton, who, while working for the Polaroid Corporation was attempting to develop 3-D moving pictures that did not require special glasses to view. Industrially, the most important use is probably a technique for nondestructive detection of defects. Because of the extremely fine detail that a hologram can record, great care has to be taken to assure that the subject does not move. Failure to do this has produced some blurred holograms that once were simply thrown out. Karl Stetson and Robert Powell of the University of Michigan then discovered that the blurred holograms contained information about the nature of the movement that caused the blur. This discovery has led to a technique called holographic interferometry. As an example of the use of this technique, an automobile tire is first holographed at low pressure, and then holographed again at full pressure. The technique can distinguish between a sound tire, which has uniform expansion, and a defective tire, in which expansion is uneven.[19]

Liquid Crystals

The Technology

A *liquid crystal* is a substance that behaves as a liquid and a crystal at the same time. It acts as a crystalline solid in that the molecules have a consistent orientation, but are otherwise free to move around as in a liquid. The liquid crystal state is found in certain materials having the property of two different melting temperatures. At the lower of these, the solid state changes to a cloudy liquid, and at the higher, the cloudy liquid becomes a clear liquid. The liquid crystal state is found between these two melting temperatures. Mechanically, a substance in its liquid crystal state behaves like a liquid, but optically, because of the systematic orientation of the molecules, it acts like a crystalline solid. Liquid crystals (LCs) possess a number of properties that have led to their various uses. First, their color is highly sensitive to changes in temperature. Second, their color is sensitive to an applied electric current. Third, LCs are sensitive to light, and display different colors in light. This property has become the basis for what has become the principal application of the LC: displays. If the application works without an artificial light source, as in the display of a digital watch, the power consumption is extremely low, making battery power a feasible optin.

Three basic types of liquid crystals have been identified. The first is nematic, in which all of the molecules line up in the same direction but are otherwise free to move around. The second is cholesteric, in which the molecules arrange themselves in layers and the molecules themselves are oriented parallel to the planes of the layers. The molecules are free to move around within their layers. The third is smectic, in which the molecules arrange themselves in layers, and the molecules are oriented perpendicular to the planes of the layers.

Figure 9.3
Liquid Crystal Display Structure

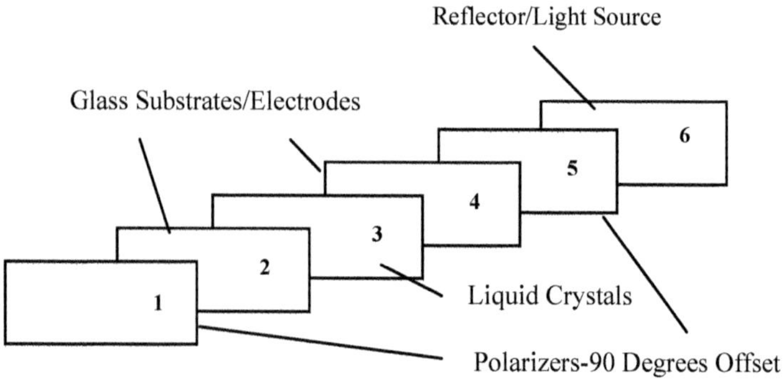

Liquid Crystal Displays

A liquid crystal display is a thin structure in which a column of a number of pixels is arranged in front of a reflector or a light source. In Figure 9.3, box 3 represents the layer of liquid crystals, and box 6 is either a reflector or light source. Layers 2 and 4 are glass substrates provided with transparent electrodes. Boxes 1 and 5 are polarizers, with polarizing orientations set at 90 degrees to one another. In the absence of anything in between them, light passed by one would be blocked by the other. This does not happen, however for when an electric current is applied to the electrodes in layers 2 and 5, the nematic liquid crystals in layer 3 are twisted such that light can pass through the entire system. The shapes of the electrodes in layer 2 determine where light can pass through the system. In applications where layer 6 can be a reflector, the display structure requires very low current to work. Where layer 6 is a light source, that source alone probably uses more current than the LCD itself.

Liquid crystal displays that rely on reflected light have become ubiquitous in the form of wristwatch, calculator, camera, and many other displays. The current consumption of these devices is so low that adequate power can sometimes be obtained from very small solar panels. The disadvantage of straight reflective LCDs is the difficulty of reading them in poor light. This is not a problem in cases such as modern gasoline pumps, which are operated in well-lit spaces. Also, in response to this problem, some LCDs work reflectively when light levels are high and transmissively in poor light with the aid of a low-level, supplemental light source. This type of arrangement is known as transflection. Where it is necessary to be able to read the display in low or no light, a light source is required. This type of arrangement is known as a transmissive LCD.

For high-resolution color display, such as for a computer monitor or television screen, the provision of contacts for each individual circuit, as in simple watch or calculator displays, becomes impossibly awkward and costly for use with large number of pixels. Instead, a matrix system is used, in which each individual pixel can be addressed independently. In personal organizer displays or older laptop monitors, each pixel has to maintain its state between refreshes in the absence of a steady current. This arrangement is known as passive matrix. In a modern color monitor, a matrix of thin-film transistors is added to the polarizing and color filters. Each pixel can be addressed independently. During a refresh, all pixels get addressed sequentially. The active matrix system offers considerably sharper images than the passive matrix and considerably faster response time.

LCD Background

Liquid crystal technology is one of several examples of technology that was discovered many decades before anyone found much use for it. It had its origin in the nineteenth century with the discovery of the phenomenon by Friedrich Reinitzer, an Austrian botanist in 1888, who found that the compound cholesteryl benzoate had two distinct melting points and was curious about the physical state of this material in between those two temperatures. Shortly after discovery, German physicist, Otto Lehmann became interested in LCs and managed to make a career of the topic.[20] The work of these pioneers established the characteristics of these materials, but their work was known only to a tiny subset of the scientific community and unknown industrially.

The first small break into the commercial product world came in the 1930s, when American physicist, John Dreyer discovered that liquid crystal molecules would align themselves along a surface that had been rubbed repeatedly in the same direction. This property could be used to make polarizers, and Dreyer used the technique to fabricate polarized lenses for sunglasses, and, until the 1950s, polarizers were the main use of LCs. In 1956, Westinghouse Electric assigned physicist James Ferguson to investigate the use of thermal imaging in the manufacture of electron tubes, and this work led Ferguson indirectly to observe the optical properties of liquid crystals in 1957. Ferguson went on to make a career of liquid crystal research, and it was his work and that of Glenn Brown that underpinned liquid crystals' success in the display field, which started out in the 1970s.

Ferguson did manage to obtain minimal support for his LC work at Westinghouse, but had serious difficulty in convincing his management of the products potential of the field; it was 1964 before he made significant progress on that front. As for product ideas, Ferguson was drawn to LCs' property of displaying a range of colors that were highly sensitive to temperature, and this suggested to him the field of medical imaging. The early imaging technique required that the LC material be painted on the patient's body where the colors that it assumed would constitute a "temperature map" of the patient. The procedure was a technical success; tumors would show in clear outline. Some doctors were interested as they believed LC imaging to be safer than X-rays. However, the technique seems not to have caught on because of its awkwardness, and because the same diagnostic information could be obtained with other imaging techniques, such as infrared thermal imaging.

Glenn Brown had become interested in LCs in the course of a search for research topics for his students while at the University of Cincinnati. He later moved to Kent State University, where he was able to generate enough interest in LC research to gain some support for founding the Liquid Crystal Institute, which is still the leading meeting forum for the liquid crystal research community. The Brown-Ferguson alliance at Kent State lasted until 1970, when Ferguson left to set up a company to sell LC display technology based on his patents, of which the most important was the twisted nematic technology that is at the heart of modern display devices. He soon faced the problem of massive competition, as his and Brown's success at interesting others in LC potential attracted a number of other firms into the display business, including RCA. Ferguson did not have the financial resources he needed to enforce his patents.

The first LC displays were in digital watches starting in the early 1970s. These were the culmination of much improvement of the devices, such as keeping moisture out. Products were expensive, and quality was highly uneven. Moreover, as typically happens with products based on new technology, prices fell, as did profits. A number of the companies that rushed into the business eventually withdrew almost as hastily. Ferguson was driven to sell his patents to Japanese interests, so that by the 1980s, the center of LC technology was Japan, and Japanese companies took most of the market for such products as computer monitors and LC television screens. This may be a monument to the shortsightedness of US corporate managements.

Photography

Conventional (Film) Photography

Since the early contributions of George Eastman and others there has been progress but it has been strictly incremental. Economically, the major change has been the evolution of photography from an art requiring high skill, and therefore largely limited to professionals, to an art highly accessible to consumers. In the US in 1890, the number of photographers was approximately equal to the number of accountants and dentists. Sales of photographic materials to amateurs were about 20 percent of the market, the balance being to professionals. As photography developed in the twentieth century, the amateur share rose by a factor of ten. One of the major improvements was the development of color photography for both the commercial market and for the consumer market. The first really successful color film, Kodachrome, was introduced in 1935.

Digital Photography

The technology that enabled digital photography at both the commercial and amateur level was the integrated circuit. The *charge-coupled device* (CCD), which is at the technical heart of digital cameras, consists of an integrated circuit and an array of linked capacitors and is used for converting light to an electronic image. When a photon strikes an atom, the collision can elevate the energy level of an electron sometimes releasing it. A released electron from a CCD gets placed in one of the capacitors. An image is projected on the capacitor array through a lens, and each capacitor accumulates a charge proportional to the amount of light that has fallen on it. This creates a grayscale image of how much light has fallen on each individual pixel. Both professional-level and consumer-level digital cameras that record color have become available.

Digital cameras, as of 2005, outsell removable film cameras.[21] Digital photography offers the advantage over film cameras of the capability for reviewing the results of an exposure immediately after taking and, if that is not satisfactory, of remaking the exposure to correct the problem. Digital images can be viewed, according to the system used, on a computer and printed with a low-cost inkjet printer or directly printed without the need for the computer. As of this writing, film retains the advantage of having considerably higher-resolution images, which makes it the preferred choice among those desiring large-format pictures, such as portraitists and artists. It suffers the disadvantage of the necessity for the significant processing necessary between exposure and finished image.

The commercially most meaningful competition between digital and film photography is in the huge consumer photography market, for which the convenience and accessibility of digital images have found favor. This can be viewed as the competition between a relatively mature technology and a very new one. Inasmuch as the technology of digital is probably still improving and the cost of this equipment is falling, it may be that film photography will be reduced to certain professional and art markets. The precedent for this can be found in developments that took place in the late nineteenth century. In the pioneering days of photography, the only way to obtain a large picture, say 8"x 10", was to use a camera with that size sensitive plate holder. This changed with the invention of the enlarger, in which a smaller negative image could be projected on easel in magnified form. This development enabled the portable camera, especially the 35mm format film. This did not produce the total demise of the large-format camera, however. Throughout the twentieth century, any need for large-format images of the highest resolution had to be met with large-format cameras. These became the domain of professionals and extremely advanced amateurs. The work of Ansel Adams was largely done with the 8x10-view camera. In the same way, it is unlikely that digital will replace film in all of its markets today.

Two areas in which digital imaging has demonstrated clear superiority are infrared photography and in night-vision devices. These applications rest on CCDs' sensitivity to infrared light. This contrasts with film photography, for to get infrared images, the film must be infrared sensitive, and the camera must be fitted with an infrared filter.

Xerography

Xerography is a plane paper process of dry photocopying in which the image of text or illustration is transferred by using the attractive forces of electric charges. It is alternatively known as electrophotography. In this process, a beam of light strikes the original material. The laser is the most common light source in modern copy machines. Light rays are reflected off the white areas onto a photosensitive plate over which electric charges have been deposited. Charges are neutralized in areas struck by the light. Because no light is reflected from black lettering, and only some is reflected from gray areas of illustrations, charges on the plate corresponding to the lettering and illustrations are retained on the plate. A sheet of paper is passed between the plate and another charged plate which draws the powder from the plate to the paper, which is then fixed to the paper with heat. Because the image on the charged plate (in the form of a cylindrical drum) is a projected one rather than a contact image, the xerography process is capable of yielding either enlarged or reduced images of the original material.

Chester Carlson, the inventor of Xerography, had been working for several years during the 1930s before he, with the aid of an assistant, devised the experiment that demonstrated that the principle was viable in 1938. Carlson received his first patent in 1942 and shopped his invention to a number of household-name corporations before finding some interest on the part of the Battelle Memorial Institute in 1944. Later, the Haloid Corporation of Rochester, NY became interested in developing xerography into a commercial copying machine. It carried out the development successfully and introduced its Model 912 in 1958. This adds up to a stage-one innovation of twenty years if measured from the 1938 experiment. During much of that time, Carlson enjoyed no financial support other than his own resources prior to 1944, and he stands out as a classic individual inventor during an era when the individual inventor appeared to be a disappearing breed. Haloid renamed itself the Xerox Corporation. Carlson himself ended up with a substantial block of Xerox stock during the years when that company's shares rose meteorically based on its success with its copy machines. One of the more interesting aspects of this entire stage-one experience is the lack of interest on the part of the numerous companies that Carlson approached in the 1940s. Battelle's interest arose from the anticipation that a successful dry copying process could replace technologies such as mimeograph, some offset printing, carbon paper, and photocopying. This insight was available to all, but only one saw it.

In addition to office copying, variations on the xerographic process have been applied in the form of the laser printer—a device that has found a huge market in offices and residences. It has also found application in the plain-paper facsimile machine. The laser printer story is another one of the ironic accounts of how the Xerox Corporation failed to exploit its in-house inventions. It was invented in Xerox's Palo Alto Research Center (PARC) and was the only one of PARC's inventions for which Xerox achieved significant profit.

Notes

1. Coherency is the property by which the photons emitted by the device have a fixed phase relationship to one another.
2. The term "population inversion" refers to the relationship this exists between N_1, atoms in the ground state with energy E_1, and N_2, atoms in the exited state with energy E_2. In a state of thermal equilibrium, N_1 will always be larger than N_2. The

effect of stimulating the lasing medium is to increase the ratio of excited-state atoms to ground-state atoms such that $N_2/N_1 > 1$, this condition is referred to as a population inversion. The difference between the energy levels, E, determines the rate of light amplification.

3. The ruby laser illustrated here roughly corresponds to the first laser as invented by Maiman. Since then a host of materials have been found useful as lasing media, both solid and gaseous.

4. Among other things, Maiman had a solid background in optics and electronics, an unusual combination. As a student, he had gained admittance to Stanford's physics program only after prior work in the Electronic Engineering Department. He was also a maverick, which helped him pursue his ideas in the face of professional criticism.

5. This had to be an important element in the puzzle, for the technical success of Xerox PARC clearly owed something to its leadership's success at resisting pressures from corporate headquarters to make something salable.

6. Even though perfect collimation is theoretically impossible, laser light comes close. A beam from a small laboratory laser such as a helium-neon laser has a diameter of approximately one mile on the surface of the moon.

7. http://en.wikipedia.org/wiki/Ring_laser_gyro

8. http://en.wikipedia.org/wiki/LASIK

9. Gibbs, W. Wayt, "Computing at the Speed of Light," *Scientific American,* Vol. 291, No. 5, (Feb. 2004), p. 81.

10. Service, Robert, "Intel's Breakthrough," *Technology Review,* Vol. 108, No. 7, (July 2005), pp. 54-55.

11. Talbot, David, "LEDs vs. the Lightbulb," *Technology Review,* Vol. 106, No. 4, (May 2003), p. 32.

12. *Ibid.* The incandescent bulb emits only 5 percent of its input energy as light, and the rest heat. The fluorescent tube does better, recovering up to 30 percent of its input as light.

13. See http://en.wikipedia.org/wiki/Fiber_optic.

14. Verizon has largely limited its placing of fiber to individual homes to new and upscale subdivisions. See Fitzgerald, Michael, "To Fight, Verizon Switches," *Technology Review,* Vol. 107, No. 10, (December 2004), pp. 46-52. SBC and Qwest are also reported to be considering greenfield fiber services to residences. See Wood, Lamont, "Copper for Fiber," *Scientific American,* Vol. 293, No. 1, (July 2005), p. 24.

15. Fitzgerald, Michael, pp. 46-52.

16. Palucka, Tim, "Holography: The Whole Picture," *Invention and Technology,* Vol. 18, No. 3, (Winter 2003), pp. 15-16.

17. Palucka, pp. 17-18.

18. Palucka, p. 19.

19. Palucka, p. 20.

20. Lehmann is credited with the name "liquid crystal."

21. Kodak ceased production of removable film cameras as of the end of 2004. This attracted some proclamations of the end of film photography even though Kodak for years has been but a minor player in the removable film camera market. Still, the move represented a not-surprising erosion of the removable-film camera market at its fringes.

10

Medical and Related Technology

Introduction

One of the more dramatic occurrences of the twentieth century has been the large increase in human life expectancy at birth. In the United States this number was forty-nine years at the turn of the nineteenth century; one hundred years later, it was seventy-seven years, an increase of 57 percent. This increase came about for several reasons, prominent among which are improved diet and improved health care, based in part on improved medical technology. Improved health care was underpinned by increasing knowledge of how the human body works, and of how to deal with its problems. Historically, the largest improvements in human longevity have resulted from diminishing the number of deaths among the very young. This is true generally in the more advanced economies of the world. Many researchers believe that the past emphasis in helping the young to mature may be changing to improving the health prospects of the oldest segment of the population. The hope is that extended lifespan will underpin more un-debilitated and productive years at the upper end of middle age. Thus, advancing medical technology influences the general economy not just by the investment opportunities that it creates. By improving health in middle age generally, it creates a more productive workforce. By extending life, it adds to the problem of how to support the part of the population no longer in the workforce. These are conditions applying not only to the US but also to Western Europe, Japan, and China.

Medical technology emerged from the nineteenth century with a solidly founded germ theory of disease and imaging tools based on the microscope and nascent X-ray technology (see discussion in Chapter 2 above). Of all the possible branches of technology, medical is arguably the one most admired by the general public. Interestingly, it is not easy to find examples of genuine prototype inventions in the medical area from the twentieth century. Much of modern medical technology is derived from adaptation of prototype inventions and discoveries that originally did not originate from work on medical problems.

For example, the sulfa drugs shared the same heritage as chemical dyes; both could trace their origins to the destructive distillation of coal tar. In the medical imaging field, ultrasound was adapted for medical diagnosis after having been successfully applied to detecting flaws in welds in steel structures such as ships, pipelines, and rails. Magnetic resonance imaging rests on principles of quantum mechanics, and has at its heart powerful computing, a descendant from the integrated circuit. Chemotherapy in the treatment of cancer was an adaptation of substances related to those first used as that most noxious of weapons, mustard gas. The outstanding example of a medical technology whose diagnosis possibilities were recognized from the time of discovery is the X-ray. The history of medical technology provides a lesson on the power of the prototype invention to affect fields of endeavor totally unforeseen by the early pioneers in the technology.

A fairly recent discovery in the field of biology that may bear importantly on the future of medicine is recombinant DNA, which is believed to promise the possibility for treatment of diseases heretofore not considered curable. Recombinant DNA may turn out to be a discovery of prototype dimension. In addition to its potential for the development of treatments for a number of genetic-related diseases, it has become the tool for creating seeds for plants with desirable characteristics. In this capacity, it may hold the key to the kind of food production increases needed to feed an expanding world population. Of late, it has also become the tool in the search for low-cost fermentation processes for producing alternative motor fuels.

Pharmaceuticals

The industry that grew out of knowledge of organic chemistry was chemicals with its subdivision, pharmaceuticals. Starting with the work of Paul Erlich prior to 1910, the search for medicines has been aided by an expanding understanding not only of the characteristics and structure of organic compounds, but of how these compounds react in the body in relation to specific diseases. A growing body of knowledge of how specific compounds act in the body gives researchers to focus their search for cures for specific ailments (see the discussion of the work of Pasteur, Lister, and others before 1900 in Chapter 2 above). It is needless to say that the search for new medicines is far from over, given the threat still posed by cancer and other ailments presently considered incurable. Nevertheless, the search has been productive to date, and it has produced an impressive list of Nobel Prizes in medicine. It would require far more than the space available in a reasonably sized book to detail all of this progress, but at least two achievements stand out as high points. They are illuminating the structure of DNA and the mapping of the human genome. Pharmaceutical progress certainly occurred prior to the gaining of full understanding of the structure of DNA and its genetic role. In many cases, understanding of DNA enabled understanding of previous results achieved experimentally. Moreover, it was the basis for the discovery of recombinant DNA, which may underpin the future of medical research.

Antibiotics

An antibiotic is a substance that kills or retards the growth of bacteria. Antibiotics are a subclass of the larger class of antimicrobials, which also includes antiviral and antiparasitic substances. Antibiotics have been highly successful in medical treatment because they are effective, have relatively few side effects, and are relatively harmless to the host organisms. The first antibiotic substances were the sulfonamides (sulfa drugs; see Chapter 3), but these are largely out of use because of the development of bacterial strains that are resistant to them. Today the term refers to substances, such as penicillin, a derivative of natural growth, and synthetically compounded substances (see discussion of discovery, Chapter 3 above). Modern antibiotics fall into two categories. The first is the bacteriocidal substance, which kills bacteria. The second is the bacteriostatic substance, which retards the growth of bacteria. The effectiveness of the latter rests on the intervention of the body's immune system to eliminate the reduced bacterial population left after the drug has worked.

It is fortunate that synthetic antibiotic substances have been relatively easy to compound and to manufacture, for new varieties appear from time to time. The impetus behind the development of new antibiotics is the appearance of antibiotics-resistant bacterial strains. When the object is to treat the ailment by killing the bacteria, the desired result is to kill as close to 100 percent of the bacteria as is possible. If some bacteria survive the treatment, then their progeny are likely to have a high level of resistance to the particular antibiotic substance that was used. From time to time, the medical profession has been accused of over prescribing antibiotics with the result of the spread of resistant strains of bacteria. To appreciate the magnitude, it was estimated that by 1984, half the people in the United States with active tuberculosis had a bacterial strain that was resistant to at least one antibiotic. Some hospitals and child-care facilities have become notorious as places where bacterial resistance is so high that none of the simpler, cheaper antibiotics have any effect at all. New antibiotics that are developed tend to be more costly than older drugs, and the race to develop drugs to stay ahead of bacterial evolution has contributed to the tendency for drugs to become increasingly expensive.

Vaccines

Vaccines were an important development of the pre-DNA pharmaceutical industry. A vaccine is an antigenic preparation designed to produce an active immunity to a disease. The process for administering vaccines is known as vaccination. In general, the immune system of a vaccinated person comes to recognize the vaccine agent as a foreign substance, destroys it, and remembers it. This experience means that the immune system will react to an attempted invasion of the body by a related virulent agent (disease), by (1) neutralizing the invading agent before it can enter into healthy cells of the body, and (2) by recognizing and destroying infected cells before their population can grow uncontrollably. To date, vaccines have succeeded in the virtual elimination of smallpox. Diseases such as rubella, polio, measles, mumps, chickenpox, and typhoid are far less common than they were even sixty years ago due to vaccines.

Vaccination in some form goes back thousands of years. The Chinese and Indians had vaccination procedures before 200 BC. The discoverer of the basic inoculation procedure for smallpox was Edward Jenner, an English country doctor. In 1796, Jenner was aware that milkmaids would occasionally contract cowpox, a disease related to smallpox, through their contact with cows' udders, but milkmaids who recovered from cowpox never seemed to come down with smallpox. In that year, he took infectious fluid from the hand of a milkmaid named Sarah Nelmes, and placed the fluid in a scratch in the arm of James Phipps, a healthy neighborhood boy of eight years. Phipps came down with cowpox but recovered with no serious permanent effects, and did not come down with smallpox when Jenner injected some smallpox-infected matter into Phipps' arm. Readers in the early twenty-first century who have any familiarity with the research that has to go into the development of a vaccine or any other drug before it can be tried on humans will appreciate the simplicity of procedures in the eighteenth century.

Modern vaccines have been divided into three categories. The first is "inactivated." These are previously virulent materials that have been killed with either chemicals or heat. This class includes vaccines against influenza, cholera, plague, and hepatitis A. There is a tendency among this group to have short-term immune responses, and booster shots are needed to maintain patient immunity. The second group is "live, attenuated." These agents are live microorganisms that have been cultivated under conditions designed to disable their virulent characteristics. Agents of this group typically provoke more durable immunities than those of the inactivated group, and include vaccines against yellow fever, measles, rubella, and mumps. The third group is "toxoids." These are inactivated toxic compounds from microorganisms in cases where these cause illness. Toxoid-based vaccines include tetanus and diphtheria.

Conjugate vaccines deal with certain bacteria that have polysaccharide outer coatings and are poorly immunogenic. By linking these outer coats to proteins (e.g., toxins), the immune system can be led to recognize the polysaccharide as though it were a protein antigen. Subunit vaccines create immune responses with a fragment of an inactivated or attenuated microorganism. Recombinant Vector vaccines combine the physiology of one microorganism and the DNA of another, and create immunity against diseases that have a complex infection process. DNA vaccination undertakes to create immunity by using an infectious agent's DNA rather than the agent itself. As of 2005, all these varieties of vaccines are in development and not in general use.

Modern vaccines are not without problems. Inasmuch as they work by giving the patient a mild or subclinical case of the disease against which they purport to protect, or a related milder disease, there will occasionally be instances in which symptoms that are supposed to be subclinical, or at worst mild, will be considerably more violent than that. This kind of event has produced lawsuits against the vaccine manufacturers to the extent that a number of drug companies have withdrawn from vaccine development and production. In the US, withdrawal has reduced the country's capacity for producing vaccines to the point where there is doubt as the adequacy of vaccine supply in the face of a really serious epidemic approaching the magnitude of the 1918 influenza epidemic that was spectacularly fatal worldwide. The concern now is over bird flu, an avian condition believed to be capable of affecting humans. This disease has appeared in rural China, where agriculture is in a semi-modern state at best, and people often live in very close proximity with their fowl.

Early Cancer Treatments

In the early 1940s, a diagnosis of cancer of any kind was tantamount to a death sentence. The known treatments consisted of surgery and radiation. The former involved the removal of significant healthy tissue along with the cancerous tissue, and the latter's effectiveness varied from patient to patient. The total effect of this battery of treatment options was the extension of life for approximately one patient in four. One of the few medical accomplishments of the World War II period was the addition of chemotherapy to the cancer treatment possibilities in the form of the drug mechlorethamine, which is of the family of nitrogen mustards. These chemicals first emerged in the public view as one of the most hideous of the poison gases to come out of World War I. The discoverers of mechlorethamine, Louis Goodman and Alfred Gilman, were engaged in wartime research on poison gases.[1] In their investigation of how these agents killed, they found that the nitrogen mustards inhibited cells' ability to reproduce, with their strongest effect against cells that reproduce rapidly. Inasmuch as cancer is characterized by very rapid cell reproduction, they got the idea that the nitrogen mustards might have some potential for arresting cancerous development without fatal damage to the patient. The Goodman-Gilman experimentation on animals and some human cancer patients led directly to modern cancer chemotherapy. Understanding of the mechanism of how these agents worked was one of the early results following the DNA revolution starting in the 1950s.

In order to appreciate the value of the discovery of the three-dimensional geometry of DNA, it helps to relate some research work that was in progress in 1944. Dr. George Hitchings, who was director of biochemistry research for the

Burroughs-Wellcome Company had formed the idea that a class of chemicals known as purines could be the basis for attacking a variety of diseases, especially some forms of cancer. At this point in history, the composition of nucleic acids was reasonably well understood, and it was known that these materials were involved in heredity. The purines were present in nucleic acids, ergo Hitchings' interest in them. Hitchings also knew that cancers involved exceptionally rapid rates of cell division and suspected that the purines were at the center of this phenomenon. His idea, and that of Gertrude Elion who became his research partner and later shared in the same Nobel Prize, was to attack the disease with a purine that was similar to one found naturally in the body, but which was just different enough that it would not function in the rapid cell division character-istic of cancers. Success came in 1951 when Elion and associates effected the substitution of a sulfur atom for an oxygen atom to form a substance known as 6-mercaptopurine (6-MP), which led to dramatically improved success in the treatment of leukemia. 6-MP was also successful as a suppressant of the immune system and was used to prevent rejection of organs in organ transplants.[2]

Elion and Hitchings are credited with an important change in the methodol-ogy of pharmaceutical research. Instead of relying completely on trial and error methods, they used differences in biochemistry between normal human cells and pathogens to design drugs that could kill or inhibit the growth of particular pathogens without harming the host cells. Elion is credited with (besides 6-mercaptopurine) azathioprine (the first immuno-suppressive agent),[3] allopurinol (for gout), pyrimethamine (for malaria), trimethoprin (for meningitis), acyclovir (for viral herpes), and, most recently, for the aids drug AZT, for the development of which she came out of retirement.

DNA

Since today's understanding of fundamental atomic structure emerged from the work of Bohr, deBroglie, and others, much of the basic work in chemistry has revolved around inferring the structure of organic and inorganic chemical compounds given knowledge of the substances under study and the constraints imposed by the known rules of basic atomic structure. Knowledge of how a substance is structured at the molecular level represents a substantial advance over prior knowledge about that substance, and serves as a way to improve understanding of how given substances relate to given diseases of the body.

DNA is the common acronym for deoxyribonucleic acid, the nucleic acid which contains the genetic instructions specifying the biological development of all cellular forms of life. DNA was first isolated in the nineteenth century in cell nuclei, mixed with RNA (ribonucleic acid). Both were recognized as polymers. Recognition that there were two different substances was the first real step in the discovery path, but at the outset, no connection between either DNA or RNA and genetics was suspected. As of 1940, it was generally accepted that chromosomes, very large molecules, were the overall agents bearing heritable traits, but it was not known which component parts of chromosomes were the heritable agents. In 1943, Oswald Theodore Avery of the Rockefeller Institute for Medical Research proposed DNA as the heritable agent in the chromosome based on his own experimental work, not protein as many had supposed before. In 1953, Alfred Hershey and Martha Chase performed experimental work that verified DNA as the heritable agent.

The three-dimensional structure of DNA was discovered in 1953 by James Watson and Francis Crick of Cambridge University. They determined that DNA is a double-helix polymer, a spiral consisting of two DNA strands wound around each other in the manner of a twisted ladder (see Figure 10.1). The rails of the ladder are composed of sugars and phosphates. The rungs are formed by bonded pairs of nitrogenous bases: adenine (A); guanine (G); cytosine (C); and thymine (T). An A on one chain bonds to a T on the other, forming an A—T ladder rung. A C on one chain bonds to a G on the other. If the bonds between the two chains are broken, the two chains unwind, and free nucleotides within the cell attach themselves to the exposed bases of the separated chains following the base-pairing rule: A—T or C—G. The result is two identical DNA molecules from one original and is the method by which hereditary information is passed from one generation of cells to another (see Figure 10.2).

The importance of the Watson and Crick model and their and others' subsequent research was the suggested relationship between DNA, RNA, proteins, and the genetic code, which was deciphered not long afterwards. The findings here described in summary collectively formed the basis for the science of molecular biology. [4]

The Human Genome

The genetic code is a set of rules that maps DNA sequences to proteins in the living cell and controls the process of protein synthesis. Genes are composed of DNA. The genome is the genetic information carried by an organism that is inscribed in one or more DNA molecules. The original human genome project undertook to map the entire human genome down to the nucleotide (or base pair) level and to identify all the genes present in it. The project originated in 1986 due to the efforts of Charles De Lisi, who was then director of the Health and Environmental Research Programs of the Department of Energy. The Human Genome Project (HGP) expected to cost $3 billion and to take fifteen years was launched officially in 1990. Progress was hastened by considerable foreign cooperation and was enabled by the massive expansion of computing power in the 1990s. A rough draft of the genome was complete by 2000, two years ahead of the original schedule. Eight years after the HGP was started, a private sector startup, Celera Genomics (founded by Craig Venter) also undertook to map the genome. Celera proceeded much more rapidly than the government effort through the use of a faster technique known as genome shotgun sequencing, with the result that Celera produced a genome map at about the time the HGP did and at a fraction of the cost—$300 million versus $3 billion. The private sector effort was animated by the prospect of patenting some of the genes, but the administration at the time took the position that genes could not be patented. In October of 2004, researchers in the HGP announced that the human genome contained an estimated 20,000 to 25,000 genes, well below the 100,000 that had been predicted at the outset of the project in 1990.

There are many benefits expected from the mapping of the human genome when the ongoing research is viewed as a logical continuation of the type of work started by Paul Erlich before 1910. For example, a researcher who has focused attention on one particular gene in the context of a cancer investigation can refer to the database that has resulted from the genome project and find all that has been previously written on that gene, including its three-dimensional structure. Much of the work in interpreting the genome data still lies ahead. The

Figure 10.1
DNA Structure

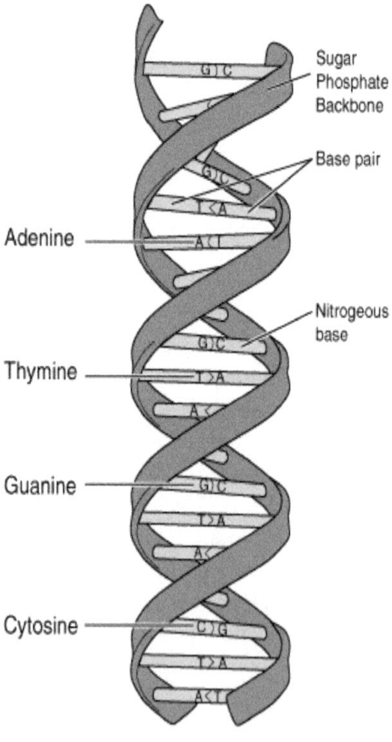

Source: Wikipedia, the free encyclopedia

Figure 10.2
Splitting DNA

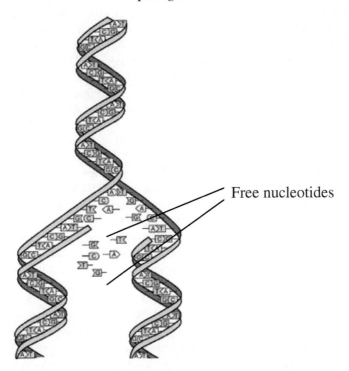

Free nucleotides

Source: Wikipedia, the free encyclopedia

ultimate hope is that, once the fund of knowledge of what function each gene performs in the body has grown sufficiently, gene therapy will prove to be a way to eradicate faulty genes. It might be possible to insert a normal copy of the gene and cure the disorder that was traced to the faulty gene. Present opinion differs on how successful genetics will be with cancer and other dread diseases.

Biotechnology

Biotechnology deals with understanding and using biological processes to develop or manufacture products for improving health, improving the reliability of the human food supply, and, of late, developing energy sources. People are used to thinking of biotechnology as a very recent development mainly because certain aspects of it have become controversial and have attracted the attention of popular news media. Actually, systematic activity that comes under the biotechnology heading goes back almost to the beginnings of human agriculture. The Romans, for example, practiced selective breeding for the purpose of improving farm animals by imparting superior traits such as better milk production, improved meat production, and others. Biotechnology in the animal husbandry field died out in medieval times when it was customary to graze animals in common fields, but in the seventeenth century, the practice revived, enabled by fencing of pastures so that animal breeding could be subject to control.[5] Selective breeding was applied to both plants and animals. During the eighteenth century, selective breeding gained a firmer scientific basis as a result of the work of Gregor Mendel, who theorized that an inherited characteristic is determined by the combination of a pair of hereditary units, later called genes.

Recombinant DNA

Part of the revolution triggered by the identification of the DNA and RNA molecules in the 1950s has consisted of the birth of the technology of recombinant DNA, which allows direct manipulation of genetic material. There are two kinds of alterations achievable by modern bioengineering: transgenic and intraspecies. Transgenic refers to transferring genes from one species to another, and intraspecies refers to gene transfers within the same species. Intraspecies does essentially the same thing as the ancient practice of selective breeding, but is far faster and more efficient, as researchers can tweak an organism by inserting genes to achieve desired traits. The origin of the recombinant DNA technology was largely in the work of two individuals, Stanley Cohen and Herbert Boyer. Boyer, leading a team at the University of California at San Francisco, isolated an enzyme that could be used to cut strings of DNA into segments that carry the code for a pre-determined protein that could be attached to other strands of DNA. This enabled the precise tailoring of DNA with traits from other strands of DNA. Cohen, an associate professor at Stanford University, had developed a method for introducing antibiotic-carrying plasmids[6] into certain bacteria as well as a method for isolating and cloning genes carried by the plasmids. The two met at a conference on bacterial plasmids in Hawaii in 1972, compared notes, and started a collaboration that eventually led to the founding of Genentech, the first of the biotechnology startup companies, incorporated in 1976. Genentech scientists achieved the synthesis of insulin in 1978 to the benefit of diabetes sufferers. The company has since become a major corporate player in the recombinant DNA field.

Recombinant DNA seems like a candidate for inclusion in the list of prototype discoveries by the standards set out in Chapter 1, and it may yet be so deemed. The broad reason for not labeling it as prototype here is that its applications to date fall far short of what was anticipated for it. Part of the problem with recombinant DNA progress so far has been that the tools for manipulating DNA fragments remain relatively crude. Recombinant DNA research has no counterpart to planar technology, which enabled design and volume production of low-cost integrated circuits starting in the 1960s. There is no public domain library of simple recombinant DNA results to which a design engineer working with DNA can refer, and, as a result, each recombinant DNA maneuver has the character of a laboratory experiment and is relatable only to the immediate research project. Moreover, there is no widely accepted body of manufacturing techniques and standards by which recombinant DNA products can be manufactured. It was design standards and efficient manufacturing techniques that supported the rapid growth of the integrated circuit after the 1960s. Lack of these has to contribute to the high prices commonly asked for such products as Procrit and Enbrel. At least part of this problem has been attributed to the practice of patenting each DNA result using what has been called extremely broad language.

Food and Agriculture

Bioengineering has altered crop seeds to produce desirable traits. The first genetically engineered food product came into the market in the early 1990s in the form of the Flavr Savr Tomato, which was engineered to taste fresh year around. In the ensuing ten years, the US approved more than forty bioengineered crops. These have included the well-publicized "Roundup Ready" seeds, which allow farmers to control weeds by using Monsanto's "Roundup," a wide spectrum herbicide, without damaging the crop. These seeds have enabled improved yields per acre with less overall reliance on chemical herbicides.

The seeds that have resulted from intraspecies genetic manipulation have become controversial over the issue of whether they are free of environmental threat. This has engaged the environmental movements in the industrial countries, and in Europe, the political force of these people has been sufficient to bring about legislation all but banning the sale of foods deriving from bioengineered crops in the European Union. The objection seems to be that there is no absolute proof that the bioengineered seeds will not cause unpredicted environmental harm.[7] One fear, that is often mentioned, is that bioengineered crop seeds will share their special genes with weeds with effects such as making the weeds resistant to herbicides; in other words, creating "superweeds." Unlike the bioengineered agricultural products, bioengineered medicines have not provoked controversy regarding environmental effects. Of the face of it, it seems strange that substances for direct human consumption that are medicinally useful have escaped the controversy that surrounds seeds that promise solution to problems of providing food to a growing world population. Resistance to the seed products of transgenic manipulation has so far not materialized in a serious form, for the relation to this form of seed development traditional selective breeding has come to be widely recognized. The speed with which seeds can be improved through selective breeding has dramatically increased as a result of genetic mapping and an expanding knowledge of how genes are related to plant traits. Using what is known as gene marker technology, the seed companies are able to determine whether a plant has a desired trait almost from the time it sprouts by determining if the appropriate gene is present. This represents a dramatic speedup, for previously it had been necessary to grow the plant, plant its seeds, and let the resulting plants grow to maturity in order to tell whether the desired trait was or was not present. With traditional selective breeding, it used to take ten years from the discovery of a plant with a desired trait to the commercial production of seeds with that trait. With gene marker technology, the duration has been reduced to less than five years.

Energy

Fermentation processes are ancient, but due to the recent increase in the price of energy products, fermentation ethyl alcohols (ethanol) have become marginally attractive as motor fuel supplements. At least one major country, Brazil, claims to have become energy self-sufficient on the basis of ethanol based on fermentation of sugar. Sugar's low price in the world market suggests a history of overproduction relative to demand. Inasmuch as Brazil is a major sugar producer, it makes sense for it to divert some of its sugar crop to fuel ethanol production. Conversion of sugar to ethanol is a simple process, as industrial processes go, and in Brazil, production of sugar is based on low-cost labor. US production of sugar is much smaller than that of Brazil, and most fuel ethanol in the US has so far been derived from corn.[8] Ethanol production is inherently more costly with corn as feedstock than with sugar, for it requires separation of the sugars in corn from the other components, an additional production step. Quickened interest in fuel ethanol has produced an almost "gold rush" atmosphere in which many companies are investing heavily in ethanol production facilities to use corn as feed. This development may prove a fad, for it is inherently limited by a finite supply of corn, and with present technology, does not offer more than a small energy premium over the oil and gas consumed in the production of the corn. There is also the matter of interference with the nation's food supply. Not surprisingly, the ethanol rush has resulted in a sharp increase in the price of corn. Recombinant DNA may offer at least a limited way to overcome these limitations in the form of bacteria that are more efficient in extracting sugars from corn than have been known in the past.

Recombinant DNA may enter the energy puzzle in another way. There is much interest in this country of using feeds other than corn to make fuel ethanol, such as corn wastes, switchgrass, and other cellulosic products that have essentially no present market value. An obvious advantage of this material as feed is that it would be costless except for the cost of harvesting and transportation to the ethanol facility. This is the origin of the interest of developing bacteria that are more efficient in extracting and converting the sugars in these alternative input materials. The object of some recombinant DNA research is to create bacteria that are efficient extractors of sugars from cellulosic materials than bacteria that have performed this role in the past. The economic logic of this research is that at some point the low costs of bio-wastes will offset any premium of production expense over that associated with corn, presumably a far more costly feed.

One variant of this pattern of research is an effort to develop a relatively cheap way to a butanol fermentation process.[9] If this can be done, it would offer some important advantages over fuel ethanol. Ethanol is not only costly to make (as compared with gasoline), but it is also more costly to transport to its point of use. The reason is that ethanol mixed with gasoline comes out of phase in the presence of water, and today's petroleum products pipelines are not waterproof. Therefore, ethanol has to be mixed with gasoline at or near the point of consumption, to which it has to be moved by means more costly than pipeline, such as truck. Butanol has no such problem with water. Also, butanol has greater energy density than ethanol, and a gasoline-butanol mixture does not require engine modification in order to be usable in car and truck engines. A fuel mix that has ethanol content much greater than 15 percent requires significant engine modification to be usable.

One might be tempted to ask: Why not use sugar to produce ethanol? The reason in the US is almost totally political. The reason that Brazil is able to use sugar as ethanol feed to such advantage is twofold: the sugar is valued at the low world price, and the conversion is considerably cheaper than with corn. In the US, sugar is price supported at 20¢ to 25¢ per pound as compared with 4¢ to 8¢ in the world market. Importation of sugar is tightly subject to quotas and importation of sugar-based ethanol is also tightly restricted. It would seem as though the nation's interest in low-cost fuel ethanol conflicts with the interests of sugar producers in the US. The sugar producers are solidly supported by corn producers, who see sugar at world prices as a threat to their corn sweetener market. It may be that the only way around this apparent impasse is a successful research effort to create bacteria to get sugar out of biomass efficiently.

Drug Pricing

Pharmaceutical Pricing Issues

In recent decades, the pharmaceutical industry has created a number of new drugs that have brought cure or at least control to a number of diseases previously thought to be beyond the ability of medical technology to influence. At least some of these new drugs have been very expensive and accessible to patients only with the help of various third party payers. The pharmaceutical industry is an example of a pattern that is becoming increasingly common in recent times: industries whose products are hugely expensive to bring through stage-one development but are relatively cheap and easy to reproduce. The high sticker prices on new medicines reflect a small amount to cover the cost of reproduction and a very large element to cover the costs of research, development, and testing that was necessary to bring the drug to market availability. Not everyone pays these high asking prices. Drugs are discounted to large-scale buyers, and payment of full asking price falls mainly to consumers who have inadequate health insurance.

The combination of high asking price and low reproduction cost of drugs deemed to be essential to life preservation has brought about a conflict between the pure commercial interests of the drug companies and governments that respond to political pressures for pricing that reflects marginal cost of producing the drugs. Marginal cost pricing of drugs would not cover the costs of development of new drugs, and failure of the pricing to cover these costs would remove the incentive to develop further new drugs. A possible extreme result of official forcing of marginal cost pricing on the drug industry would condemn patients to whatever drugs existed at the time of forcing whether these are adequate or not. By and large, the US government has recognized the difficulty that the companies face and has generally prohibited acquiring drugs at discount from retail for resale. Also, the patent system has generally been effective at preventing domestic low-cost generic reproduction for the duration of patent life in order to recover the pharmaceutical companies' costs of discovery and stage-one innovation. The companies themselves justify discounting to large buyers, such as Medicare, with the logic that what the volume buyer pays covers marginal cost of production plus a contribution to research and development, testing, and approval costs. The great difference between marginal cost and fully allocated costs leaves room for substantial contribution to development costs even after a large discount.

The weakness in this system arises from the export market for the new and costly drugs. Many countries have socialized health care systems, which in theory make a full range of health services available to their citizens free of direct cost to them. As with any valuable good or service that is popularly perceived as "free," the quantity of the good or service demanded will exceed the supply of it. The various national health care services are large enough that the cost of their operation has serious fiscal implications for the governments and, therefore, face budgetary constraints. The result is that health care has to be rationed in some way and in a typical pattern, many patients face very long queues in order to gain access to some services. At times, these long waits can apply to services that are not merely discretionary, but are essential to continued life. With budgetary restrictions, foreign national health care services are aggressive in using their buying power that comes from large-scale purchasing. Indeed, many of these services face caps on what they can pay that have been imposed by their governments. Countries that have a tradition of indifferent respect for intellectual property rights, such as India, have been tolerant of domestic companies that produce generic copies of drugs patented in other countries.[10]

One example of a socialized national health care system that comes to mind is that of Canada. Supposedly, sales of drugs to the Canadian Health Service are for Canadian consumption, but this only prohibits the legal import of discounted drugs into the US for resale. There are many cases in which private individuals journey to Canada and stock up on their prescriptions, and US customs officials have up until recently been tolerant to the point of not interfering with this contraband importation by individuals. Recent developments suggest that this leniency may be changing. US and European Union drug companies have faced pressures to make AIDs drugs available to third-world countries at low costs. This has produced some donations, but the value of these is compromised by some unintended side effects. Not surprisingly, the supply of such donations is limited, and the AIDs virus is infamous for its ability to mutate to resistant forms. Inconsistent supply and usage by African and other third-world sufferers creates favorable conditions in which the virus can mutate to resist drug effects. This opens the possibility that a limited distribution may create a worse virus than would have existed in the absence of the distribution.

Bioengineering and Rising Health Care Costs

Bioengineering has produced a number of medicines for the treatment of human diseases. Examples include Procrit (erythropoietin), found effective at stimulating red blood cell production in the body, and Enbrel, found effective in the treatment of rheumatoid arthritis. The year 1982 saw the approval by the Federal Drug Administration of synthetic insulin, which gave diabetics a more plentiful and affordable source of this vital substance. Other medicines resulting from recombinant DNA technology include factor VIII (for males suffering from hemophilia A), factor IX (for hemophilia B), human growth hormone, several types of interferons, lepton, and others. In the ensuing twenty years, more than seventy biotechnologically engineered medicines have been approved for such uses as treating cancer, heart disease, cystic fibrosis, and immune disorders.

One of the problems with bioengineered drugs has been their cost, which can range from several thousand dollars per year of treatment to several hundred thousand. This reflects the costly development and production process plus the lengthy approval process in the Federal Drug Administration (FDA). Inasmuch as much of the cost of these drugs is borne by private employers (through their employee health benefit programs) and government agencies such as Medicare and the Veterans Health Service, the paying entities have begun to struggle under the burden. The cost problem has become especially recognized in the Medicare and Medicaid programs. The latter has had heavy impact on state finances. At the time of this writing, some sort of political reaction is taking form, but it is unclear how this process will work out. It is not obvious that there is anything the government can do to reduce the costs of some of the bioengineered drugs.

When a new technology appears, the popular media and the stock market have a way of massively underestimating the time and expense of bringing laboratory results to a commercially viable stage. Bioengineering is a case in point. Several new companies were organized in the 1970s for the purpose of exploiting the new science of recombinant DNA to produce bioengineered drugs. After a comparatively brief investor rush into the shares of these new companies, their failure to produce early and spectacular results caused investors to lose interest. The reality behind these events was that bringing recombinant DNA laboratory results to viability as commercial products proved to be an expensive and lengthy process.[11] It is likely that the initial optimism regarding bioengineering possibilities rested on supposed simple one-to-one relations between diagnoses and abnormalities in one particular gene. This is reminiscent of Robert Koch's working hypothesis of a one-to-one relation between diagnosis and a particular microbe (Chapter 2). The problem arises from the often-recognized situation that a diagnosis often is associated with a genetic abnormality, which is only a necessary condition for the disease, but not a sufficient condition by itself. The disease arises when the abnormality exists in *addition* to other conditions, which can be other genetic conditions, lifestyle patterns, or other factors. When this happens, the complexity of the problem can be hugely increased as, correspondingly, is the cost of research.

One has to ask why managements of publically traded companies undertook this exercise in stage-one innovation, which was not only costly but, in the absence of managerial confidence of there being a market for some costly medications, highly risky. Put another way, who was to pay for this stage-one innovation exercise? One answer that was likely to have been in the minds of managements was that resulting drugs, such as Procrit and Enbrel that would undoubtedly have high list prices, could be sold to third-party payers of health care costs. These payers were the health insurance industry, of which the largest entity was the federal Medicare program.[12] The onset of Medicare program roughly coincided with the beginning of a major acceleration in the rate of expansion of overall expenditures for health care in the US[13] and a nearly simultaneous expansion of the health insurance industry in the US. This latter expansion was well under way in the decade of the 1970s when a number of biotechnology companies were formed, including Amgen and Genentech, both of which have become quite large. The presence of the insurance industry, especially Medicare, probably acted as a powerful mitigator of the risk in the stage-one innovation of bioengineered drugs. Given the political origin of Medicare, failure to approve payment for a drug to an elderly sick person on the grounds of expense could lead to problems, and the private component of the insurance industry naturally had to adopt an equally generous coverage.

The appearance of costly bioengineered drugs was certainly a proximate element in the rapid rise of health care costs in the US. It apparently fits with the widely held notion that a substantial part of this increase is attributable to new technology. An important basis for this belief was a study by the RAND Corporation, which concluded that only about 10 percent of the health care cost rise could be attributed to the expansion of insurance coverage led by Medicare, and that a much larger share was attributable to the cost of new technology.[14] While it is not the purpose here to dissect the RAND study critically, this result has to seem strange upon reflection. To blame a substantial share of rising health care costs on the rising costs of medical technology implies that this technology somehow has appeared full-blown from outside the economy. As pointed out in the preceding paragraph, however, the stage-one development of bioengineered medications was likely to have been financed by company managements that were confident that third-party payers would provide a market for resulting drugs that might be priced beyond the financial means of most patients. It has always been a salient trait of successful free market-based economies for entrepreneurs to see and take advantage of opportunities such as those presented by the appearance of a rapidly growing third-party payer industry for health care products. Stated another way, some of the increase in health care expenditures nurtured the birth and growth of a bioengineered drug industry.

The RAND study results suggest that only about 10 percent of the total increase in health care costs can be attributed to the expansion of the third-party payer system. This does not seem to leave much room for the hypothesis of the preceding paragraph. Fortunately for this hypothesis, however, some recent empirical results have pointed to the possibility that the share attributable to the expansion of the health insurance industry is far larger than that concluded in the RAND study. In particular, a recent study by Amy N. Finkelstein of the Massachusetts Institute of Technology (MIT) has concluded that at least 37 percent of the increase in total healthcare costs can be traced to the expansion of the insurance industry.[15] The Finkelstein study concentrated on the impact of Medicare, for that was largely the source of her data. If her results can logically be extended to the private component of the health insurance industry, the implication is that the increase of overall health care costs traceable to expansion of the insurance industry is even greater than her basic results suggest. The basic hypothesis of the Finkelstein study has been around for several decades, but has heretofore not benefited from serious empirical backing. The importance of the Finkelstein study lies in its provision of such backing. Not surprisingly, it has already attracted considerable attention from economists and others who study the rise in health care costs.

Medical Diagnosis Imaging

At the end of the nineteenth century, the principal established medical imaging device was the microscope, whose resolving power had been immensely improved since the days of Van Leeuwenhoek, especially in the 1870s. The discovery of X-rays was technically of the nineteenth century, but its innovation was part of the twentieth century story. The "big three" of twentieth century medical diagnosis imaging are X-ray based technology, ultrasonics, and magnetic resonance imaging. Interestingly, the first two of these had their roots as part of the nineteenth-century technology legacy and are applications of prototype inventions from before 1910. Magnetic resonance imaging dates from the 1930s, and its underlying basic science includes the development of quantum physics. Needless to say, all three technologies have undergone dramatic improvements and have benefited from a number of innovation inventions.

Limitations of X-rays

While many improvements in X-ray technology were made following the discovery of X-rays in 1895, the technology's potential in medical diagnosis was recognized from the time of discovery. The absorption of X-rays in tissue is roughly proportional to the density of the tissue, with bone being the most absorbent form of tissue. Consequently, bone appears light on a photographic plate, with softer appearing as progressively darker. Thus, while X-ray imaging was highly effective at diagnosing fractures of bones from an early date, it was severely limited in the ability to distinguish between soft organs having similar density that might be located adjacent to each other. This problem was approached by the development of various substances that are relatively opaque to X-rays and could be introduced into organs of diagnostic interest to provide a contrast in the photographic image. Different kinds of contrast media can be injected into body cavities, into the bloodstream, and otherwise introduced into areas of interest. Use of contrast media greatly enhanced X-ray technology's ability to explore different parts of the body.

Tomography

Tomography is an X-ray based technology by which images of structures deep within the body can be secured by focusing X-rays on a single plane. In its simplest form, linear tomography, the X-ray tube moves in one direction and the sensitive plate moves in the opposite direction in a parallel plane. This produces a focused image of the plane that is parallel to the planes of motion of the plate and tube that passes through the pivot point of a ray from tube to plate. All tissue not in this plane is out of focus. In a more sophisticated version, plate and tube move synchronously in a circular or elliptical path to produce an image of objects in the focal plane that is sharper than with linear tomography. The tomographic techniques were developed to overcome simple X-rays' inability to distinguish between tissue masses that are of approximately equal density that are adjacent to each other. For example, the technique has been used extensively to study the kidneys, which are found in an area of objects of similar density.

Computerized Axial Tomography (CAT)

This is the most sophisticated of all the forms of X-ray technology and is widely known as the CAT-scan or CT-scanning technology. Here, a large number of tomographic images are made and digitally processed into a sharply focused 3-dimensional image of structures within the body. CT technology was invented independently in the early 1970s by Godfrey Newbold Hounsfield of EMI Central Research Laboratories and Allan McLeod Cormack of Tufts University. The two shared the Nobel Prize in medicine in 1979. The first EMI scanner required use of a water-containing device that enclosed the patient's head, and its use was limited to brain scans. The first CT-scanning system that could create images of any part of the body was the ACTA scanner designed by Robert S. Ledley, DDS, at Georgetown University.

The improvements in CT-scanning equipment since the 1970s have been enabled largely by microprocessor speed and capacity improvements and detector improvements. The original EMI scanner required several hours to collect the data and several days to produce the images. By dramatic contrast, modern CT systems can achieve a scan of the chest in less time than one breath-hold, and the computed images are produced in something approaching real time. Three-dimensional images are based on several hundreds of cross-sectional images in contrast with a dozen or fewer in early versions.

Magnetic Resonance Imaging

Magnetic Resonance Imaging (MRI) is a medical diagnostic technique that produces high-contrast photographic images of soft tissue that are more detailed and revealing than X-ray images. MRI became available in hospitals and clinics starting in the second half of the 1980s. Its physical bases go back to the 1930s, to the work of American physicist Isidor Isaac Rabi, who discovered a means of measuring molecules. Rabi discovered that when molecules are placed in a magnetic field and are simultaneously subject to radiation at radio-frequencies (RF), their nuclei emit a radio signal at the same frequency as the signal they receive. Different molecules have this property at different radio frequencies; that is, molecules of different substances *resonate* at different radio frequencies. Consequently, the researcher can gain information about the molecular structure of a given substance by applying a fixed magnetic field and searching through a range of radio frequencies or, alternatively, holding the radio frequency constant and searching through a range of magnetic field strengths. His work earned Rabi the Nobel Prize in physics in 1944.

What was first known as nuclear magnetic resonance (NMR) was first described independently by Felix Bloch and Edward Mills Purcell in 1946. Purcell had worked in radar research during the war at MIT's radiation laboratory in the area of production and detection of radio waves, and it appears that this experience reinforced his interest and understanding that led to the development of NMR. The tie with radar research also was present in the postwar development of the maser and laser. Bloch and Purcell shared the Nobel Prize in physics in 1952. When the technology began to be used in medical diagnosis, the term NMR was dropped in favor of MRI in order to avoid patient concern that might be provoked by the use of the word "nuclear."

The searching of the range of radio frequencies has been speeded up by adding the mathematical technique of Fourier analysis. In this technique, pioneered by Richard R. Ernst, the sample is held in a static magnetic field and simultaneously subjected to a short square pulse of radio frequency energy that contains all the frequencies thought to be of interest. The resulting data are then analyzed by Fourier series analysis. This technique produces a huge amount of data, and the need for processing it as quickly as possible makes MRI a candidate for early editions of super fast all-optical computers. Today's MRI equipment has also been enabled by superconducting materials in their magnets.

NMR has found a number of uses in analysis outside of medicine. In a comparatively recent example, it was used to determine the configuration of the buckminsterfullerene (buckey ball) at Rice University in 1985. Inasmuch as the molecule of interest contained no hydrogen, it was necessary to use the resonance frequency of carbon 13, which is technically more difficult than using hydrogen, the standard in medical analysis. After some difficulty, the spectrum was obtained by Richard Smalley and Robert Curl, for which the two shared a Nobel Prize in chemistry.

Ultrasonics

The medical adaptation of the principles of ultrasonics (See discussion in Chapter 3) is commonly known as ultrasound analysis and is widely used in medical diagnoses. An ultrasound examination consists of moving a probe, containing one or more acoustic transducers, in contact with the skin over the organ to be examined, such as a kidney. Faithful contact between probe and skin is aided by a water-based gel. The probe emanates sound waves, generally in the range of 1-10 megahertz (MHz), which are reflected when the waves encounter a material of different impedance, and the reflected echoes are read by the probe. The sound data are converted to an electric current that, when processed by a computer, produces an image on a screen for the examiner to read. The images can be recorded as a series of photographic images to be reported to the prescribing physician. An ultrasound examination can produce a two-dimensional image by means of an operator's rotating the probe. A 3-D image can be obtained with special equipment. Ultrasound can be focused in a narrow beam from a fixed source of sound waves, and the readout will show a spike whenever the sound waves encounter a change in impedance. Ultrasonography can be enhanced by taking Doppler measurements, which can reveal whether blood is flowing towards or away from an examined area.

Ultrasonography generally does an excellent job of imaging soft tissue. The equipment is portable, and examinations can be conducted at bedside. There are no long-term threats to patient health from the use of the technique, as is the case with X-ray imaging. Moreover, the procedure is low-cost in comparison with alternatives such as MRI scanning. The disadvantages of Ultrasonography include not being able to penetrate bone and wave range is short, i.e., the sound waves do not penetrate deeply into the body. Also, the method is highly dependent on the skill of the operator.

The medical use of Ultrasonography in the west was originated in Sweden by cardiologists Inge Edler and Carl Helmuth Hertz at Lund University in 1953. The two conducted the first ultrasonic examination of heart activity using equipment borrowed from a local shipyard. They published their results in 1954. There was a parallel development in Scotland, by Professor Ian Donald and colleagues at the Glasgow Royal Maternity Hospital. By coincidence, Donald and colleagues resorted to using equipment borrowed from the local Babcock & Wilcox marine boiler factory—Glasgow is also a shipbuilding center. With the collaboration of Dr. James Wilcox, Donald successfully adapted their techniques to the point of being able to study foeti during the entire course of a pregnancy. Together with collaborators Tom Brown (a medical physicist) and Dr. John MacVicar, Donald published the results in the *Lancet* in 1958. The technique has been successful in aiding the detection of fetal abnormalities and has been adapted to the needs of many other branches of medicine.

Electrocardiography

One of the publically best known of the medical diagnostic techniques is the electrocardiogram (EKG). An EKG examination produces a recording of the of the electrical activity of the heart in the form of a strip graph, which is used to monitor heart function and to diagnose abnormal heart conditions, such as skipped heartbeats; heart damage from heart attacks; detection of potassium, calcium, magnesium; and other electrolytic disturbances. The EKG examination is carried out by measuring electrical potential differences between various parts of the body. Twelve sensors are placed on the body, arranged in pairs, and measure potential difference along six standardized directions. These measure heart electrical activity from six vantage points. A normal heartbeat produces a wave that is divided into a number of distinctive parts, and diagnosis of problems proceeds from recognition of variations from the standard wave form.

The history of the EKG goes back to the late nineteenth century when it became recognized that the heart produced electricity. The first systematic study of this electric activity was by Augustus Waller who worked at St. Mary's Hospital in Paddington, London in 1913. Waller, however, failed to see any diagnostic potential resulting from his investigations. Willem Einthoven, working in the Netherlands, helped his investigations of heart-generated electricity by inventing the string galvanometer, which produced a better-defined readout than the capillary galvanometer that Waller had used. It was Einthoven who first parsed the normal heart wave form into its parts that have been part of the analysis ever since. Einthoven received the 1924 Nobel Prize in medicine as a result of this effort.

Electroencephalography

The electroencephalogram (EEG) is a cousin of the EKG in the sense that it also is a procedure for measuring electrical activity, but it focuses on the brain. Apart from the similarity, the EEG also has a different historical heritage from the EKG. The EEG examination is carried out by placing electrodes on the scalp, and the readout is used to assess brain damage, epilepsy, and other conditions. Pioneers of EEG analysis include Dr. Richard Caton who, while practicing medicine in Liverpool, presented findings regarding electrical phenomena of the exposed cerebral hemispheres of rabbits and monkeys in 1875, and Russian physiologist Vladimir Vladimirovich Pravdich-Neminsky, who published the first EEG (of a dog). German physiologist Hans Berger began studies of human EEGs in 1920 and is sometimes called the inventor of the process, although others were working in the area at about the same time.

Other Medical Hardware and Treatments

Pacemaker

The artificial pacemaker is an electronic cardiac-support device that regulates the heartbeat in patients suffering from types of heart disease in which the natural pacemaker has faltered or failed. The artificial pacemaker dates from 1950 and was a large device that had to be used externally. It was designed by Canadian electrical engineer John Hopps. The first pacemaker implantation in a human took place in Sweden in 1958. The pioneers of the process were Rune Elmqvist and Åke Senning. The first pacemaker implant was short-lived, but the patient lived until 2000; during his lifetime he received twenty-two pacemakers. Early implanted pacemakers were highly dependent on batteries, and required an operation every twenty-four months for the replacement of batteries. In the mid-1960s, transvenous electrical leads permitted placement of pacemakers without opening the thoracic cavity and without a general anesthesia.

NASA and the Applied Physics Laboratory of John Hopkins University adapted technology initially designed for two-way communication with satellites for communicating with and reprogramming pacemakers without need for surgery. This technology became generally available in 1997. Pacemaker design and ease of use benefited from miniaturization coming from space research and integrated circuit development. Improvements included rechargeable, long-lived batteries and the range of control functions that pacemakers are able to deliver. The pioneering examples controlled only the heart's two largest chambers, the left and right ventricles. The latest pacemakers not only control these, but they also control the atria. They have the ability to sense the patient's level of physical activity and respond appropriately by varying the rate of pacing using rate response algorithms.

Dialysis

One of the functions of a healthy kidney is to filter substances out of the blood that adversely affect the general health of the body. Recognition of kidney failure goes back to ancient times and was known as "uremia," a Greek word whose literal translation is "urine in the blood." The present-day treatment of this condition is known as dialysis, or hemodialysis, which is an extracorporeal procedure for filtering uremic substances out of the blood, after which the blood is returned to the body. In dialysis, two liquids separated by a porous membrane, exchange components whose molecules are small enough to pass through the pores of the membrane. Blood is introduced to one side of the membrane and a sterile liquid is introduced on the other. Dissolved substances in the blood, such as urea and inorganic salts, diffuse through the membrane. Red and white blood cells and platelets are too large to fit through the pores. To minimize the loss of substances needed by the body, such as sugars and amino acids, these substances are added to the sterile liquid such that flow of these substances across the membrane in one direction is offset by the flow in the other direction.

The underlying physical processes for dialysis are osmosis and diffusion. These two processes were first described qualitatively by Scottish chemist Thomas Graham, who has been called the "father of dialysis." The first fully scientific description of the laws of diffusion is due to Albert Einstein, who based his work on his own theory of Brownian molecular motion in his 1905 paper.

The pioneers of dialysis faced two basic problems: materials for the porous membrane and a way to suppress the blood's tendency to clot, which would be encountered while handling the blood outside the body. In some of the earliest experiments, on dogs, the membrane material was collodion, a material derived from cellulose. The anticoagulant of the early experiments was Hirudin, a substance that was first identified in 1880 and is the active anticoagulant in the saliva of leeches.

Some of the earliest attempted dialysis treatments on humans took place in the 1920s under the direction of Dr. Georg Haas of Giessen in Germany. He used collodion as the membrane and hirudin and the anticoagulant. None of the patients/subjects survived the treatment for the possible reasons of the advanced state of their condition when they were treated or to imperfect purification of the Hurudin. In the last of these experiments, Hirudin was replaced by Heparin, which is the universal anticoagulant in mammals. It was found that Heparin caused far fewer unfortunate side effects than Hirudin, and Heparin remains the preferred anticoagulant to this day.

Willem Kolff of the Netherlands achieved the first successful dialysis on a human patient in 1945 using a rotating drum dialyzer. The patient, a woman aged sixty-seven at the time of treatment, died at age seventy-three for causes unrelated to her kidney failure. The Kolff dialyzer represented some significant improvements over earlier designs and used cellophane as its membrane material. Kolff's dialyzer design was imported into the US after the war and was distributed to a number of hospitals in this country.

The next improvement originated in Sweden under the direction of Nils Alwell. The Alwell design combined the functions of dialysis of the Kolff lung with the function of ultrafiltration for removing surplus water from the body. This is a function of a healthy kidney and has to be replaced during dialysis. A number of further improvements in technique and hardware of dialysis are attributable to Belding Scribner of Seattle. These improvements included better ways of accessing blood from the body for treatment and hollow-fiber dialyzers, which have greatly eased treatment of patients with chronic renal failure.

Stent

A stent is an expandable wire mesh tube that can be inserted into a hollow passage within the body for the purpose of keeping it open. The best-known use of stents is in the coronary arteries, but they are used in a number of other applications in the body, including other arteries, veins, trachea, bile ducts, esophagus, ureters, and urethra. The stent is applied in collapsed form and expanded to contact with the walls of the structure in which it is placed by means of a balloon, which is subsequently deflated and withdrawn.

Stents have offered a less risky alternative (for the patient) to surgery in many instances. One of the historical drawbacks to stents has been the formation of scar tissue in the area of stent application, which in some patients can be sufficiently severe as to reclose the passage that the stent was intended to maintain. This condition, known as restenosis, historically affected about 30 percent of patients and was perceived as a problem sufficient to have attracted considerable research effort to find a cure. The modern response to the restenosis problem has been in the form of the drug eluting stent (DES), which is a stent coated with a drug that prevents or slows the formation of scar tissue for up to six months, after which scar tissue formation becomes a much smaller problem. The development of the DES faced the serious challenge of engineering the rate at which the drug is paid out into the body, but this problem has been solved satisfactorily. The DES is credited with reducing the restenosis rate to less than 10 percent.[16] The stent takes its name from the English dentist, Charles Stent

(1845-1901), who is credited with inventing it. Stent is credited also with other innovative contributions to medical treatment.

Notes

1.   One of the military lessons from World War I was that gas warfare conferred no decisive tactical advantage when used against a technically sophisticated enemy. The US and other warring nations carried out precautionary research on war gases during World War II against the possibility that gases might be used by enemy belligerents. Poison gas never was used tactically in World War II. The military uses of these gases were strictly in situations in which the defense had no protective masks. For example, Italy used mustard gas against Ethiopians in the 1930s, and the Germans used poison gases in their death camps. More recent has been the use of war gasses by the Iraqi army against the Iranians in the 1980s, and against Kurdish rebels in Iraq.

2.   Shagam, Janet Yogoda, "Designing Drugs," *Invention and Technology,* Vol. 21, No. 2, pp. 21-22.

3.   Used for organ transplants.

4.   The discovery of DNA's structure was aided materially by the work of others working in the field at about the same time. The helical structure first came from work by Linus Pauling, whose team at Cal Tech found this structure in many proteins. The team led by Maurice Wilkins and Rosalind Franklin at Kings College, London, provided some X-ray diffraction images from which Watson and Crick were able to measure the diameter and pitch of the DNA helix. This was a controversial matter, for Watson and Crick saw parts of Franklin's work without her knowledge of permission. Edwin Chargoff was also a valuable contributor to Watson and Crick.

5.   This was the enclosing of the commons that landowners learned was necessary to carry out selective breeding, the profitability of which they understood. The enclosures deprived large numbers of people of at least part of their livelihoods and triggered a gradual mass migration to cities of peasants displaced from the land. Some historians have said that this population enabled the early industrial revolution by providing a labor force.

6.   Plasmids are molecules of DNA that are found in bacteria separate from the bacterial chromosome. They (1) are small (a few thousand base pairs), (2) usually carry only one or a few genes, (3) are circular, and (4) have a single origin of replication.

7.   This standard is almost impossible to meet. Nothing that is new can be guaranteed in advance to be absolutely free of potential for environmental harm, however, that may be defined.

8.   The federal income support program for sugar producers appears to be in sharp conflict with an expanding domestic demand for fuel ethanol. The program has long been based on severe restrictions on sugar imports. There is also a stiff tariff on imports of ethanol into the US. It would seem to make sense to use a much-expanded domestic demand for fuel ethanol as a basis for encouraging domestic sugar production and importation of ethanol based on low-cost foreign sugar production. This would enable increased production of sugar-based ethanol in competition with more costly corn-based ethanol. One result of the sugar support program historically has been to drive the largest US sugar consumers (e.g., Coca Cola and Pepsi) to the use of sugar substitute high fructose corn syrup. Both the sugar producers and the corn interest oppose any change in the present support programs.

9.   Butanol is produced industrially by the Oxo Process, which is based on petroleum-based feeds. Butanol from the Oxo Process is much too costly to be considered

for use as a large-volume motor fuel; hence, the interest is a low-cost fermentation process.

10. As it evolved to higher levels of industrial sophistication, India seems to be becoming more respectful of intellectual property rights. The perception in India that that country might have something to lose from past attitudes toward intellectual property rights appears to catching on slowly.

11. There is no better example of this than the record of gene therapy, whose promise appeared great in the early 1970s. In the ensuing years up to the present time, however, the record of gene therapy research has been one of consistent failure. In a study by the Recombinant DNA Advisory Committee (RAC) of the NIH in 1995, it was noted that large investments in gene therapy research by NIH and private firms had not resulted in a single case of success in a human in a refereed journal. This very difficult first-stage innovation continues to the present. See Judson, Horace Freeland, "The Glimmering Promise of Gene Therapy," *Technology Review*, vol. 109, no. 5, (Nov.-Dec. 2006), pp. 40-47.

12. Medicare was enacted by Congress on July 1, 1965 and implemented July 1, 1966.

13. Total health care costs in the US rose from about 5 percent of GDP in 1960 to approximately 16 percent in 2004.

14. Manning, Willard; Joseph Newhouse; Naihua Duan; Emmett Keeler; Arleen Leibowitz; and Susan Marquis, "Health Insurance and the Demand for Medical Care: Evidence from a Randomized Experiment," *American Economic Review*. Vol. 77, No. 3, (1987), pp. 283-295.

15. Finkelstein, Amy N., "The Aggregate Effects of Health Insurance: Evidence from the Introduction of Medicare," (April 2006), www.nber-afinkels/papers/Finkelstein_Medicare_April06.pdf

16. Kling, Jim, "The Lucrative Elution," *Technology Review*, Vol. 108, No. 10. (October, 2005), p. 32.

11

On Technology and the Postwar Economy

Prototype Inventions since 1945

A list of the inventions from this period could be very long; but there was definitely a "big four" prototype inventions from which most of the others were ultimately derived. They were the transistor (1946-1950), the integrated circuit (1960), the laser (1959-1960), and optical fiber (1970). These four became the bases for a large amount of capital investment. Figure 11.1 shows a partial postwar technology heritage. The principal impacts of these four did not begin to show in the general economic statistics until the 1980s for several reasons. First, the transistor took a number of years to work its way into the computer, and this did not happen until the 1960s. In the form of the integrated circuit, the transistor underpinned the personal computer industry, which did not really get under way until the decade of the 1980s. It was the industrial mass market for the personal computer that fomented an acceleration of productivity growth.

Second, fiber optics technology, the basis for revolutionary changes in the telecommunications industry, was based on a marriage of the laser and the computer in the form of the router. This suggests that the impact on investment occurred between the recession of 1981-1982 and the recession following 2000. Even though optical fiber was available in the 1970s, it did not begin to be applied widely until the 1980s. Prior to 1981, the period following the 1974-1975 recession is remembered as the "stagflation" period in which there was simultaneous high inflation and low growth. The recession of 1981-1982 is usually attributed to the Federal Reserve's determination to bring the inflation under control through tight money policies.

The 1990s saw relatively high rates of investment and productivity growth in the national economy. Many of these changes were based on computer technology, in particular, on the rapid improvement in computing speed and capacity and on the falling cost of computing capacity. Examples of increased efficiency due to computer applications were not confined to manufacturing, but came from distribution, transportation, and inventory control. The link between the

Figure 11.1
Partial Postwar Technology Heritage

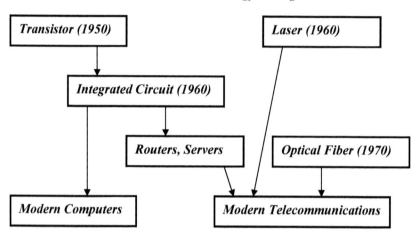

innovation streams in question and the investment flows of the 1990s is well supported by circumstantial and anecdotal evidence. The primary link between innovation streams and the condition of the general economy is the investment activity that an innovation stream generates. Indeed, the fundamental reason for identifying a subset of inventions as prototype is that these inventions are of the first magnitude among all inventions in their potential for creating investment opportunities. There is little question that the postwar's active technological innovation activity has been a support to the more-or-less general prosperity of the period, especially after the 1982 recession. However, the available data generally do not support a precise parsing of all the contributors to the postwar, economic, cyclical, and growth experience sufficient to measure the role of innovation in isolation. One way by which an idea of the general power of an innovation stream and its induced investment can be gained is indirect: an examination of the effects of failure of an investment stream. From this, one can indirectly infer the power of the innovation that generated the investment activity that supported the preceding cyclic upturn.

The Recessions of 1981 and 2000

The US economy underwent a number of recessions in the postwar era. Among these, the two that can be related, at least circumstantially, with technological innovation, or lack thereof, are that of 1981-1982 and that of 2000. The earlier of these occurred about the time of a low point in national productivity growth (see Figure 11.6 below, which appears in the discussion of productivity evidence). This recession is usually attributed to restrictive monetary policy which was designed to bring high inflation under control. While discussion of the full implications of the low productivity condition is deferred until the next section of this chapter, it should be remarked here that low productivity growth reflected maturity of the investment opportunities created by a number of prototype inventions that appeared in the 1930s and 1940s. This suggests that the severity of the early 1980s recession was aggravated by the weakness of technology-induced investment. Moreover, the second half of the 1970s is remembered as a stagflation episode in which relatively low growth and high inflation coincided. Low growth is consistent with the weak technology growth foundation, while high inflation is consistent with an attempt to stimulate growth by means of an easy money policy in the absence of a strong technology foundation.

The recession of 2000 grew out of an investment failure in the telecommunications industry that was of sufficient magnitude to have been at the heart of a downturn in investment in the overall nonresidential investment sector. Thus, the recession emanated from an investment failure in an industry wherein there had been heavy investment driven by stage-two technological innovation. Figure 11.2 shows the nonresidential concentration of the recession, for, by 2003, investment in nonresidential construction was the least recovered of the major sectors of private fixed investment. The pattern of investment failure in an industry whose recent growth had been generated by investment opportunities created in the course of a stage-two innovation process is reminiscent of the pattern of the late 1920s and early 1930s. The enormous difference was the far greater industrial diversity in the economy of 2000 as compared with that of 1930. In 2000, there was no approximately simultaneous failure of investment across a wide spectrum of industry; the failure did not spread to other stage-two innovation processes.

Figure 11.2
Private Fixed Investment, 2003, as Percent of 2000

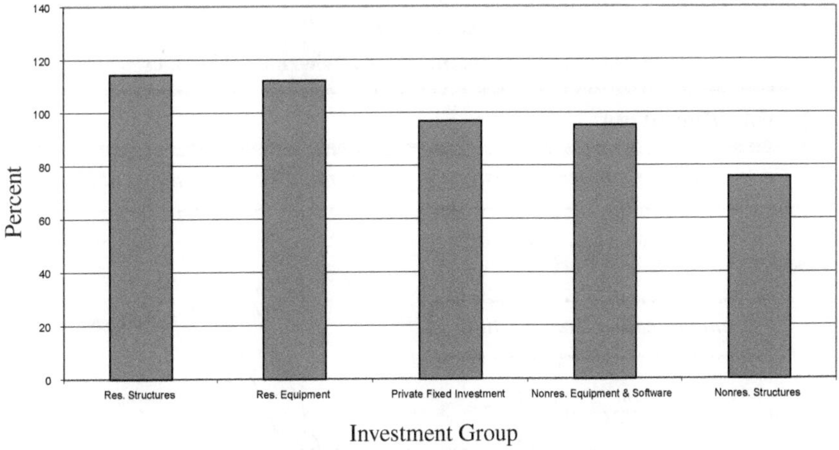

Source: US Department of Commerce, Bureau of Economic Analysis

The government's statistics identify certain subsets of the investment totals that are closely associated with innovation-induced growth. Two such data series that are broken out in the investment statistics are investment in communications structures and in communications equipment. These industries are suppliers of equipment and structures used by the telecommunications industry. Figure 11.3 maps out the crucial relationships. When the telecommunications industry engages in construction, say to lay optical fiber lines, it buys a flow of construction services from the communication structures industry. This flow of services is offset by a reverse flow of revenues to the communications construction industry. Similarly, the telecommunications industry's investment in equipment generates a flow of equipment into itself offset by an opposite flow of revenue to the communications equipment industry. Figure 11.3 illustrates how a failure of demand affecting the telecommunications industry can result in a failure of demand for both the construction machinery industry (indirectly via its effect on the communications construction industry) and the communication equipment industry, both of which are subsets of the total manufacturing industry.

Figure 11.4 compares the level of nonresidential construction activity for 2003 with the level in the peak year 2000 by way of indicating the severity of the recession for nonresidential construction overall and for the industry groups of which this total is composed. Overall nonresidential construction was down to 76 percent of its 2000 level, and the only one of its components to have been at a higher level of activity is mining and exploration, a group which includes oil and gas exploration, a very active industry during this period. At the low extreme were manufacturing and communications construction. While clearly overall manufacturing construction suffered a deeper decline than communications construction, both were down sharply. Prior to 2000, there had been heavy investment in communications construction in the form of laying fiber

Figure 11.3
Relations among Manufacturing and Telecommunications

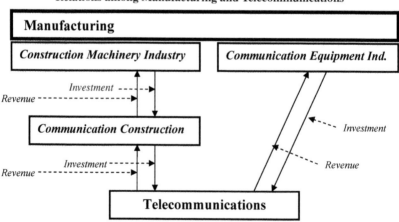

Figure 11.4
Nonresidential Construction Activity, 2003 as Percent of 2000

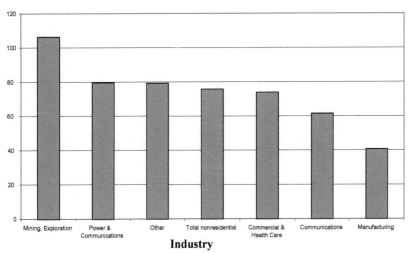

optic cable. Obviously, the sharp drop in manufacturing construction activity cannot be blamed fully on the setback in the telecommunications sector, for other capital-goods components of manufacturing were also set back sharply by recessionary conditions, and these also contributed to the setback in total manufacturing construction activity. These included the engines and turbines and the metalworking machinery groups. Still, the point is that the failure of one industry, telecommunications, appears to have been at the heart of the overall setback in manufacturing construction.

Figure 11.5 provides a comparison between growth in investment in communications construction and communications equipment. The equipment series underwent strong growth from 1993 to a peak in 2000. This investment stream included purchases of such items as routers and amplifying lasers used in a growing fiber optics network. The rate of increase in communications construction was less rapid up to 1998, but then it jumped abruptly to 1999. This reflected a boom in the laying of fiber optic cable. As it turned out, this boom was based on some incredibly optimistic forecasts of the future demand for the immense broadband capacity that would result from this investment. The unrealism of the underlying demand forecasts became obvious with time, with the result of a collapse of this line of investment after 2001. Much of the fiber that was laid has not been used to this day. This collapse affected the demand for communications equipment, for at least some of the late 1990s.

Both of these investment streams were driven by the stage-two innovations based on optical fibers, the laser, and computer-related facilities needed to operate the fiber optic systems. The post-2002 collapse represented the temporary

Figure 11.5
Investment in Communications Equipment and Construction

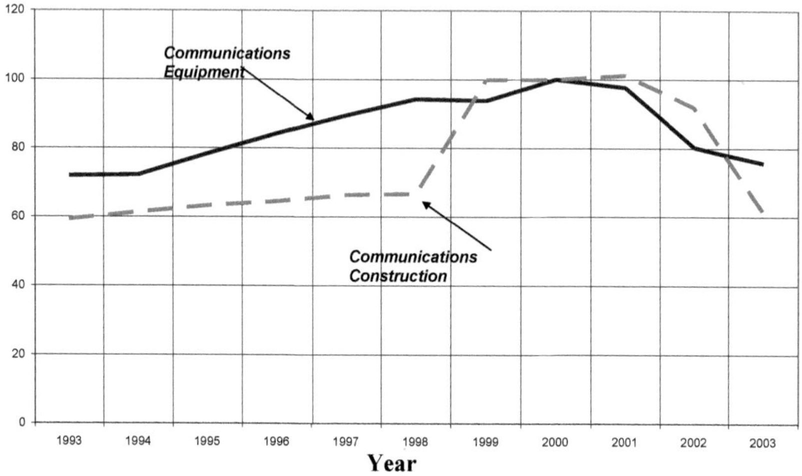

Source: US Department of Commerce, Bureau of Economic Analysis

cessation of the investment opportunities growing out of these two innovation streams. It should be clear from this example that there is no automatic market or other mechanism that signals precisely when innovation-driven investment has run beyond the market's ability to absorb the production level implied by the resulting capacity. What seems to have happened is that a number of firms convinced themselves that all fiber that could be laid would find a market. Firms were laying optic cable competitively, with the result that far more was laid than the market could absorb in the immediately foreseeable future. The investment spree ended only when reality finally intruded on irrational optimism. In a very broad sense, the psychology of investors in communications facilities in the 1990s fell into the same pattern as did the investors in automotive-related facilities in the mid- 1920s.

It is difficult to ignore the suspicion of bubble conditions in the latter stages of the pre-2000 telecommunications boom. The bubble earmarks include a very rapid growth in investment in fiber optic capacity and an emerging disconnect between the created capacity and existing and foreseeable demand for this capacity. There is an apparent similarity between this situation and what happened in the latter 1920s when automotive-related capacity was pushed beyond the needs of market demand. The telecommunications bubble of the late 1990s was far less harmful to the general economy because it involved a much smaller proportion of the economy than did the automotive sector of the 1920s, but the pattern seems to have repeated. When telecommunications investment collapsed, there was a follow-on collapse in share values of the companies involved, and this has been also recognized as a bubble and has attracted significant attention from academic scholars. The 2000 recession also involved a collapse in the valuations of the shares of dot-com startup companies, also with bubble earmarks, which may have been but was not necessarily a direct consequence of what happened in the telecommunications industry.

The Productivity Record

Indexes of productivity are ratios of measures of output to the quantity of productive factors consumed in producing the output. In a technologically static world, there would be little variation in the productivity ratio over time. The reason is that when the replacing capital is technically equivalent to that replaced, a constant rate of capital input to production will support no more than a constant rate of production expansion. In such a world, growth in production can be achieved only by expanding the stock of capital. Technological progress creates the possibility of replacing older capital plant with new capital that is more productive than that replaced. Even if the producer has decided to do no more than maintain production constant, this can be achieved with *less* capital than was necessary before the introduction of the new capital. The availability of capital plant of productivity superior to that of the existing plant shifts the decision emphasis of the capital-using firm from that of maintaining/increasing/decreasing the capital employed to that of maintaining/increasing/decreasing *productive capacity.*

In the long run, technological improvement is the source of growth for the productivity ratio. However, year-over-year change in productivity can be affected by any of a number of short-run cyclical events. For example, the onset of a deep recession or depression can cause a dip in a productivity index because various kinds of limits to producers' abilities to reduce factor inputs quickly means that inputs cannot be reduced as quickly as the business is falling off. In order to suppress short-term, cyclical influences in the productivity data and focus attention on movements that reflect technological improvement, the growth rates in Figure 11.6 are averaged compound rates of growth over nine-year intervals centered on the center year of the interval.

Figure 11.6 presents the productivity change record for most of the twentieth century. The underlying measure used is real GNP per worker hour. Ideally, a productivity ratio should reflect both labor and capital inputs to the production process. There are two reasons for the choice of the unit used here. First, available measures of capital have been compiled under the assumption that replacement capital is technologically equivalent to replaced capital. This is completely contrary to the idea that technological change enables replacement of worn-out capital with technologically superior capital. Second, productivity data extending as far back as 1911 are in terms of output per worker hour.[1]

A review of Figure 11.6 reveals two conspicuous dips in productivity growth, one centered in 1929 and another centered in 1976-1977. The earlier of these has been discussed in these pages (see Chapter 4 above). The relatively high values for 1921 and earlier reflect the growth of the mass-production industries in which capital replaced dramatically less productive capital (e.g., mass-production facilities replacing far less productive facilities). The general falloff of productivity growth into the 1920s does not reflect a lack of new capital investment, but only that the replacing capital did not result in as dramatically improved productivity than the replacing capital as in the 1910s. This, in all probability, reflected advancing exploitation of the investment opportunities created by the nineteenth-century technological legacy. Year-on-year productivity changes in 1920-1929, while positive, generally declined. Absolute levels of the productivity index declined from 1929 to 1934. The relatively high values of change that appear in and after the mid-1930s reflect at least two factors; one was reviving production with industrial recovery following the depression low year of 1933. There was a recessionary relapse in 1938. The other was reduction of costs relative to output. In the short-run, this result is generally not achievable because of the difficulty of shedding costs by means other than layoffs, but the longer depressed demand conditions persist, as in the Great Depression, the greater are opportunities for reducing all costs including capital.

The dip in productivity growth centered in 1976-1977 divides the postwar period into two broad phases from the standpoint of productivity growth. In the first, there was a postwar period of replacing worn-out capital left over from the depression and war years (1929-1945) accompanied by a productivity push resulting from a number of prototype inventions that had appeared in the 1930s and 1940s. These various forces behind productivity growth had run their course by the late 1970s. The subsequent productivity growth recovery, especially after 1990, reflects the massive investments that exploited the possibilities of post-1970 prototype inventions, especially integrated circuits, transistors, fiber optics, and the laser.

Postwar Technology and Productivity: Phase I (Pre-1978)

During roughly the first decade following World War II, there was considerable capital investment, which had a significant impact on output growth apart from any impact of new technology. The reason was that the 1930-1945 span of years was characterized by investment levels below what replacement of wear and tear on existing capital would have required under normal circumstances. In the depression, when profits were at historically low levels, there was a powerful incentive to keep old plant running as long as that was possible, even by solvent companies. With the onset of war-level production, there was adequate cash flow to support an adequate capital replacement program, but opportunities to

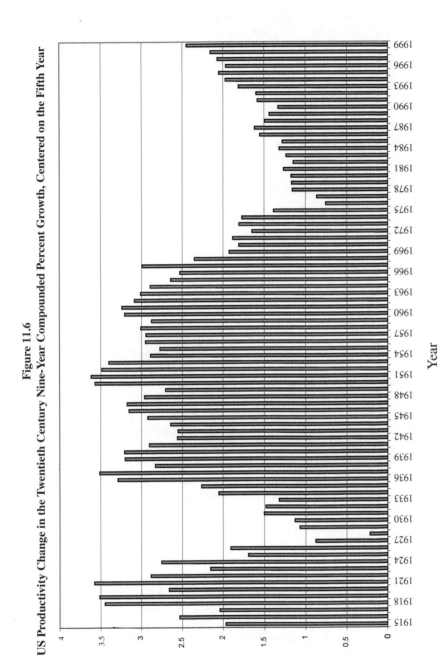

Figure 11.6

US Productivity Change in the Twentieth Century Nine-Year Compounded Percent Growth, Centered on the Fifth Year

Year

Source: Department of Commerce and OECD

do so were limited by the rulings of the War Production Board (see discussion in Chapter 6). The effect of the WPB was especially to enforce a static-like atmosphere, as it tended to discourage investment in civilian technology that it deemed remotely novel and not fully proven in past practice. The immediate postwar capital investment practices thus reflected adequate cash flows and relief from WPB restrictions.

Following the war's end, there was a productivity push from new technology with superior productivity characteristics. Railway transportation was an important example. Diesel locomotives replaced steam power. The result was very large manpower savings.[2] Most railroad companies had become persuaded to make this changeover just prior to the war but could not do so largely because the WPB would not approve any expansion in the capacity for producing diesel locomotives during the war. The actual changeover commenced in earnest in 1945 and was largely complete by 1955. In commercial aviation, the jet engine became available after 1955 and completely dominated the industry by 1965. The massive expansion of air transportation of passengers and freight was enabled by development of the radar-based air traffic control system—radar having emerged from wartime research as a highly usable tool. Highway freight transportation had its own diesel revolution before 1960 based on technical innovation during the 1930s.

In broadcasting, the new postwar development was television. This was the center of a growth industry until much of the manufacturing activity moved off-shore (see discussion in Chapter 3). Also, there was considerable improvement in household major appliances. These areas formed a substantial part of the very large postwar consumer goods expansion. Several 1930s products found greatly expanded markets in the postwar period, such as nylon and polyesters.

The contribution of technological innovation to the relatively high productivity growth in the first postwar decades weakened after 1965. It is not the purpose here to parse the causes of the problems of the 1970s—remembered as "stagflaton." However, productivity growth fell to levels well below those prevailing in the previous decades, and a likely major contributing reason was the full exploitation of the investment opportunities that grew out of stage-two innovations that began during and after the war.

Postwar Technology and Productivity: Phase 2 (Post 1978)

Productivity growth recovery through the 1980s and 1990s reflected the impact of the "big four" technologies of the postwar coming into stage-two innovation, especially that of the electronic computer. Why did the introduction of computers not affect overall productivity sooner than 1990? The answer lies in the history of the costs of computing. To affect overall productivity above a background rate of increase, it was necessary that computer applications be accepted widely into the management of and operations of industry. This had to wait on (1) the cost of computing power falling to the point wherein many applications became cost attractive that had not been so before and (2) for computing power to rise to the point of being able to engage the larger business management problems. These conditions really began to be met only during the late 1980s. Prior to then, computing power was costly and computers were accordingly confined to a limited number of applications that could economically justify the high cost, such as corporate and government accounting functions. Falling costs of computing power induced widening applications to design and production control, and ultimately, enterprise control.

The simultaneous decline in computing costs and growth in computing power resulted in a reversal of a basic condition of computing. If one looks backward before 1985, one finds that the value of a computer's time was held to be greater than that of any human being using the computer. In order to have a computer job executed, it was necessary to submit the job in the form of either a deck of cards or with a remote "dumb" terminal and wait for the computer center to return the results. There was a queue for computer time. This condition held whether the computer was owned by commercial business, government, or educational institution. Even though a senior professor at a leading research institution would have a higher priority in the queue than a graduate student, the point is that both were subject to the queue, whose objective was to keep the *computer* busy for as close to 100 percent of the time as could physically be achieved. How the distinguished professor used his time while waiting was *his* problem. With the availability of low-cost computing time, computers' time became subordinated to that of their users. It would be difficult to exaggerate the historical importance of this reversal.

The acceleration of productivity growth that occurred in the 1990s reflected not only improvements in the goods-producing sector, but also reflected an even greater productivity improvement in the services sector, according to a recent study by Brookings Institution economists.[3] This finding is of special interest because there has been a past tendency to dismiss the possibility that the services sector was amenable to labor productivity improvement.[4] The services industries were heavy investors in computers and information technology in the 1980s and 1990s, and the evident improvements in these industries' productivity is consistent with the large-scale use of this technology.[5] These results were based partly on statistics from the US departments of commerce and labor, which are measures of output per worker hour. For this reason, these data could mislead because not all investment is made for the purpose of displacing people. For example, the cited study showed that productivity on railroads grew very slowly during the study period, 1995 to 2004. On the surface, this is surprising inasmuch as railroads achieved huge improvements in labor productivity in the early postwar era through such measures as dieselization, centralized traffic control systems, and other measures whose tendency was to displace labor. Since 1995, however, much railroad investment has been in the form of earthmoving—for the purpose of straightening and leveling rail lines and improving terminal efficiency. These efforts are aimed at speeding service and sometimes have little immediate impact on the number of people employed, the payoff being in the form of reduced maintenance, more rapid capital turnover, and a service level more attractive to shippers.[6] In another small example, cited in Chapter 9, the use of mini-solar cell and battery systems for yard switches does not reduce the number of people needed to operate the yard but pays off in terms of reduced worker compensation claims.

A Comparison

The Structure of Productivity Growth

One interesting difference between the productivity picture of 1900-1930 and that of 1970-2000 emerges from a comparison between two major new industries of the two periods: automobiles and computers. The early automobile industry grew because of a number of prototype inventions, such as the Otto-Cycle Engine, the hydraulic press, and others. Although there were some auto sales to businesses, the expansion of the industry depended largely on an expanding consumer market. This was aided by falling costs of producing autos that resulted from mass-production methods and the development of the consumer finance industry in the 1920s. The auto industry's impact on productivity in the overall economy resulted from the growth of itself and all the industries that supplied it. When consumer market growth matured in the 1920s, there was no industrial market available to take up the slack.

Like autos, personal computers benefited from a mass consumer market,[7] and this underwrote a dramatic decline in the cost of computing power. However, increasingly cheap computing power also created the growth of business demand that was widely dispersed throughout the economy and was the source of active demand for computers when the consumer market showed signs of maturing. The computer industry itself was not the productivity growth engine that autos had been in the 1910s and 1920s; it simply was not as big relative to the total economy as autos had been before 1930. The link between computers and the increase in productivity of the 1990s rested on the proliferation of computing technology applied to a variety of business problems. Figure 11.7 illustrates this comparison. A certain amount of general productivity growth took place as a result of the economic activity surrounding the growth of the two industries themselves, and this is referred to as "primary" impact in the figure. In the matter of primary impact on general productivity growth, the automobile industry's impact would have been proportionately greater for the reason that automobiles and support industries in the 1910s and 1920s were larger relative to the total economy than computers in the 1980s and 1990s.[8]

Figure 11.7
Impact on General Productivity

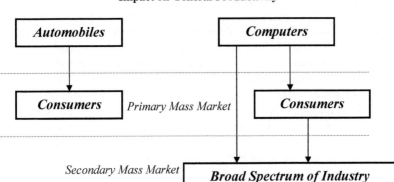

Of course, Figure 11.7 presents a highly simplified view of industry in the two time periods. There obviously was more industrial development taking place in the early century than automobiles, but it does not go too far to say that a good share of this development was in industries that in one way or another derived their prosperity from the expansion of the market for automobiles. This would include such industries as electrical systems, hydraulic control systems, tires, and steel. As for the more recent experience, the computer, a capital good that had a consumer market, had its impact on overall productivity via its application in general industry, as did several others of the period's major innovation streams—those based on the laser and fiber optic technology.

Inventions and Investment

The differences between the 1900-1930 period and the 1980-2000 period are easy to cite. If nothing else, the US economy in the later period was larger than in the earlier by an order of magnitude and far more diversified. However, there are some similarities. In both periods, the economy was confronted with a plethora of investment opportunities based on prototype inventions and their following innovation streams. In Chapter 4, it was noted that once the supply of investment opportunities had been largely exploited and the resulting industries attained a state of relative maturity, there was little in the way of innovation that could replace the kind of investment opportunities that had characterized the economy in the three decades preceding 1930. This suggested that the weakness of the technology push to the overall economy in the late 1920s and after 1930 was an aggravator of the depression conditions of the 1930s. The interesting question is: Was there a pattern in the way technological change progresses that might be repeating now?

Several developments with respect to technology in the late twentieth century appear roughly similar to developments prior to 1930. First, a handful of prototype inventions, most notably the integrated circuit, the laser, and optical fiber, have generated a number of different innovation streams with substantial induced investment. The general economic effects of the ensuing innovation streams became pronounced in the last two decades of the century, especially the 1990s. Prior to that, several prototype inventions from the 1930s generated innovation streams that contributed to earlier general prosperities, such as the 1960s. Examples include the jet engine, consumer electronics, and radar. Second, numerous startup companies were organized and financed to carry out the stage-two innovation and related investments spawned by the newly-invented technologies. This was especially pronounced in the 1980s and 1990s. Some of these companies have grown well beyond the small company condition in a short time span; for example, Microsoft, Oracle, and, the pioneer of them all, Hewlett-Packard.

Third, in both episodes, there was a noticeable withdrawal from inventing and stage-one innovation on the part of some of the established corporate research and development departments in favor of more intense development of and improvement in products that had already attained some level of commercial success. These operations had included such storied organizations as Bell Laboratories and GE Research. This is consistent with the idea that there is a massive diversion of inventive and productive resources from original inventing in order to digest the opportunities presented by prototype inventions.

Fourth, there is a similarity in the type of capital formed in the two periods, as between homogeneous and heterogeneous.[9] In the earlier period, the homogeneous capital example was that related to electrification with motors and lighting. In the later period, the outstanding example was computer-related applications. Prior to 1930, the capital that was committed to the early mass-production lines was specialized to those lines (heterogeneous). In the later period, the outstanding example of capital specialized to an industry was the capital used in such activities as manufacturing integrated circuits. Thus in both periods, substantial capitalization of both types was readily in evidence.

Important Differences: Early to Late 1900s

First, the timing of the maturing of a technology-inspired investment stream is important. During the 1920s, consumer market growth that spawned the automotive and electric power investment streams matured, and lack of new growth fomenters in the economy aggravated the 1930s depression. There is no direct parallel in the 1990s, for as one investment stream faltered, others showed no tendency to falter in lockstep. The telecommunications failure of 2000 was not accompanied by failure in other major investment streams, as is suggested by Figure 11.8, which shows that investment in computers and peripherals continued to increase, albeit more slowly, through the recession years. It is as though the investment opportunities of the late twentieth century have come in staged fashion such that the faltering in one stream is offset by strength in others. This reflects the high industrial diversity of the economy as compared with that of the 1920s.

Second, the hand of the US Department of Defense (DOD) has been very evident in the postwar technology picture and remains so. There was virtually nothing equivalent to the DOD role in technology development before 1930. DARPA remains highly active and funds university and other research, some of which is far more cutting edge than much of the work taking place in the corporate research and development shops. Hints of the nature of some of this research come to public view from time to time. For example, a bomb launched from an aircraft in the direction of a force of enemy tanks splits up into a number of smaller bombs that have the capability of targeting tanks individually and destroying them. Now, admittedly, it is not easy to say how such a device will benefit the civilian economy—except to observe that this capability indicates incredible speed in information processing. Who is to say that the civil economy will not eventually benefit from that? The history of the federal government's involvement with technology research and development reveals more than a few examples wherein technology that was developed for a specific and narrow purpose had a general economic impact not contemplated at the time of invention.

Third, the sheer size and diversity of the industrial economy argues that the danger of recurrence of the massive investment collapse in the late 1920s is unlikely. In the earlier time, much of the investment expansion was based on industries that were established after or just before the turn of the century. By 1925, this group of new industries was collectively much larger in relation to the total economy than is the case with a grouping of all the technology-based industries of recent times. Therefore, failure of investment growth in these industries at about the same time had a much larger impact on the greater economy than would be the case in recent and present times when the innovation streams are more diversified in their timing.

Mass Markets

The comparison between automobiles of the early twentieth century and computers in the 1990s in regards of their impact on overall productivity in the economy highlights the role of the mass market. A large consumer market can justify the cost of organizing production to realize economies of large-scale production. If the result of low costs is that a large spectrum of industry can profitably employ the new technology, then the impact on the overall productivity measures is likely to be more evident than when there is no follow-on industrial market for the technology. Two developments of the first years of the twenty-first century seem reminiscent of the early days of the personal computer when the mass consumer market was being promoted by pushing computers as devices for keeping recipe files. They are the ability to download and share music and picture files. Both of these kinds of files require large storage space, which is available and cheap today. Both have proved popular with consumers. Thus, the consumer mass market continues to underwrite falling costs of computing, thereby encouraging the expansion of business and defense-related computer applications.

The US consumer electronics industry, a mass market, grew rapidly from the late 1940s until late in the 1950s because of television sets and television broadcasting. This growth no doubt had some effect on the more-or-less general prosperity of the era, but inasmuch as there was no huge secondary industrial/commercial impact of television sets, the impact on the economy was probably confined to that period. During the 1960s, the US consumer electronics industry declined due to inability to compete with foreign producers, such as Hitachi and Sony, on a combination of quality and price. In the 1980s, US manufacturers became dominant in the development of semiconductor technology and production. By that time, solid-state electronic devices had become essential components of a variety of consumer electronic items, including television sets. While US producers retain a substantial market share in semiconductor products (approximately 40 percent of the world market in 1998[10]), the assembly of these products into finished consumer electronics items continues to be largely foreign.

Cellular telephony, a comparatively recent market development, may have a large impact. The primary effect may not be large in the US, in particular, for the direct impact of the cell phone industry is international. The assembly of headsets is largely non-US, but US manufacturers are very much in the business of design and manufacture of the integrated circuits that are used in headsets. The possible secondary impact arises from a considerable cell phone usage in commerce, industry, and the military, especially in the field of wireless data transmission. Here, there is an interesting mass-market development in the form of game playing on devices that are also telephones. The downloading of games is undergoing a consumer boom at the time of this writing, and its popularity is acting to sustain the demand for cellular phone handsets, which without the game market was showing signs of maturity. It is interesting to speculate how this technology will develop as a business tool, for the downloading of games can be viewed as a consumer application of wireless data transmission. Cell phones are also being integrated with digital cameras, and this combination is proving popular. It remains to be seen whether an industrial secondary market for this will develop.

Another example of a product that found a mass market is nylon. Developed initially as a fiber and applied to women's hosiery, nylon became popular quickly. The development of its market was interrupted by war and the rulings of the

War Production Board. In addition to being marketed as a fiber, nylon found numerous industrial markets as a plastic with desirable qualities.

Mass markets have traditionally been considered essential in order that an industry engaged in stage-two innovation of a new technology attain large size and profitability. Even after a prototype invention has been brought through stage-one innovation, producing the resulting product can involve at least some handwork, making the product itself relatively expensive. New companies working with technology in stage-two innovation may at some point face the problem of how to create the desirable mass market that would justify a large investment in production facilities that are able to achieve low-cost production. The response to this problem has taken the form of a gamble: price the product *assuming* mass-production costs and absorb what all hope will be temporary losses while the market grows. Ideally, such pricing would be based on engineering estimates of costs in mass production. One example of this has already been cited, that of Texas Instruments' pricing of transistor sets for the TR1 radio at $10.00 during the 1950s. This same tactic was emulated by Fairchild Semiconductor for the integrated circuit in the 1960s. This decision was credited to Robert Noyce, one of the company's founders and inventor of the integrated circuit, and was later said by Gordon Moore to be as important as the invention itself.[11] This pricing strategy has become widely followed in Silicon Valley and in similar businesses elsewhere.

Does Low-Cost Production Still Depend on a Mass Market?

At its beginnings in the early twentieth century, mass production was the means by which largely identical items (commodities) could be produced profitably at prices affordable by large markets. Since that time, mass-production techniques have been modified towards achieving product variability without sacrificing the economies of large-scale production. The proliferation of computer-based information technology (IT) in the closing decades of the twentieth century has hastened this process. The availability of computerized controls over manufacturing, distribution, and services has enhanced the possibilities for product differentiation in a physical sense. It has especially made possible the quick introduction of some products that really are different from the commonplace.

There has always been a powerful economic incentive for producers of commodity-like products to convince potential buyers that their output is preferable to that of the competition. This impetus has, among other things, supported the rise of an entire industry (advertising) whose purpose is to convince consumers that complex products such as automobiles are *not* basically all alike, and the effort to create product differentiation via advertising has run well ahead of any technology that enables product differentiation in physical reality. Indeed, the economic impetus toward product differentiation has been a primary driver behind development of the technology necessary to turn perceived product differentiation into physical reality.

Computer Control in Manufacturing

Computer-aided manufacturing (CAM) refers to the control of manufacturing processes by computers. The process is enabled by the computer's ability to process, store, and display large quantities of data. CAM systems grew in partnership with systems of computer-aided design (CAD), in which computers are used to create, modify, analyze, and optimize a design. CAD software replaces the traditional drawing board and allows the designer to test the object in various ways without having to construct pilot models, such as for heat transfer characteristics, as the optimum design is approached. The optimum design becomes the central data input to the CAM software, which regulates computer controls of the machining centers that produce the designed object. The two computer-aided processes have gone under the symbol CAD/CAM and began to appear in the 1970s. The impact of CAD/CAM technology has begun to show up in such traditional activities as automobile model changeovers with the effect of shortening the process. This technology has enabled producers to remain competitive in an era of shortened product cycles.

The sophistication and capabilities CAD/CAM systems have increased as computing power has increased. Their scope has expanded to include not only the automation of design and production, but also automation and coordination of the business functions of a firm, such as customer billing, cost accounting, production scheduling, product monitoring for quality control, and others. As a result of this wider use of computers, CAD/CAM has effectively become computer-integrated manufacturing (CIM). The principles of CIM are far from confined to manufacturing, as they have been successfully applied in distribution, transportation, and inventory control.

Production Flexibility and Product Cycles

The impact of CIM has been to diminish dependence on mass production of identical items to achieve low costs and accelerated productivity growth. In the past, serving mass markets has typically required specialized tooling, some of which became obsolete when the production facility of which it is part was adapted to produce a new product—i.e., as in a model changeover. The costs associated with this kind of changeover tended to be high, especially if the changeover process took significant time. Changeover costs are now being reduced by generalized machine tool centers that are computer controlled. Going to another product requires reprogramming this machinery rather than replacing it, at a significant savings of time and money. Thus, computerization of operations has created the ability to achieve production runs of limited size as cheaply as would previously have been possible only with large production runs with capital typical of the 1920s or even the 1950s. The production flexibility that this implies has enabled a heretofore-unattainable ability to customize products according to the customer's needs. Model T Fords were pretty much all alike, even in color. A customer ordering a product manufactured in a modern flexible manufacturing system can specify a variety of requirements and have these met at a cost that is little if any above what he or she would pay for a completely standardized product.

It is not necessary to rely on automotive examples. In the past, printing of large-volume products has been dominated by offset printing. This is a highly inflexible process, and customizing parts of the print run has been sufficiently costly as not to have happened very much. Flexibility within a print run can be had with digital printing, but that process has up until now been far too slow to be a contender for jobs such as printing of a widely circulated magazine or an edition of a major daily newspaper. This appears likely to change with the prospective introduction of inkjet printing technology that has been speeded up sufficiently that it can compete with offset's volume and permit customization within the print run in order to reach specific reader groups. This catches printing up to some of the marketing potentials growing out of the assemblages of large and detailed databases on consumer buying patterns.

If computer technology has enabled manufacturers to change products more cheaply than in the past and on a smaller scale, competition has resulted in a much foreshortened product cycle. It used to be that a major product could be sold, with the aid of minor incremental improvements, for a much longer span of years than has been the norm recently. With a long product life, investment in specialized capital whose value was tied to the value of a product that could be produced in huge quantity but in limited variation made economic sense. Historically, the shortest product cycles occurred in lines of products with a fashion dimension, such as apparel. Apparel was an example wherein changes in the product could be achieved without significant retooling of the production facilities. By the 1990s, however, the short product cycle had spread to many other classes of products, including some that previously had been produced by means of highly specialized production lines, such as automobiles. A foreshortened product cycle put a premium on speed of a product's development, for to be laggard in getting the product to market was to lose or greatly diminish the profit potential in the product. One of the side effects of shortening of product cycles is that hardware producers have an incentive to outsource manufacturing in order to reduce the risk that arises from investing in and maintaining specialized capital plant. Contract manufacturing is frequently done offshore in countries characterized by low wage labor.

A recent example illustrates the importance of being attuned to the shortness of product cycles. Laser printers at prices sufficiently modest to be affordable by individuals and small businesses appeared in the second half of the 1980s. The laser printer differed from earlier printer designs in several respects, among which was the very important one of greater memory capacity. This memory allowed the user to send sizeable files to the printer for printing and freed the main computer for other work while the print job was being executed. However, memory capacity in the printer and in the attached computer was still limited to the extent that the choice of typestyle was limited. To get around this problem, laser printers of the late eighties were provided with accessory slots into which could be inserted font hardware devices. These were, in effect, special-purpose memory units and there appeared a number of supplemental font products, both from laser printer manufacturers and from third party producers. By 1996, memory capacity in personal computers was sufficient to be able to store a large selection of type fonts. The producers of supplemental font hardware, thus, had to make their profits in the short period between the first appearance of the low-cost laser printer and appearance of the Pentium chip in the mid-1990s.

Coordinated Manufacturing and Distribution

Among the more important impacts of the modern computer has been in the field of inventory control. The "just in time" pattern has been widely applied in manufacturing with significant cost savings as the result. Perhaps an even more interesting application of computers to inventory control has occurred in distribution industries, especially retailing. The computerized electronic cash register is an innovation invention at the heart of retail inventory control. This device carries out multiple functions, including the recording of transactions, instant credit checks, and inventory control entries. Its effects are not limited to the operations of one retail outlet, for it is coordinated with the computer system of the supplying warehouse and, increasingly, with the manufacturer of the product in question. A transaction at retail will create information that is used not only to record the impact on in-store stock, but also to initiate a resupply from the warehouse and, in some cases, a resupply of the warehouse from the manufacturer. The mechanical registers that the computerized registers replaced did little more than record transactions. Electronic registers enabled material cost reductions through increased precision in inventory control.[12]

Not only did precise inventory control enable cost reductions at existing retail establishments, it also was a major element in an increase in the economic *scale* of the retailing establishment.[13] Precise inventory control has been a large element in the emergence of what has been termed "big box" retailing, in which the economic scale of the single retail establishment has grown rapidly. This pattern has been successful in such fields as hardware, grocery, and general merchandise. To take an example, the largest stores recently opened by Home Depot have had 140,000 sq. ft. of floor space, as compared with stores of the same company fifteen years previously of around 80,000 sq. ft. The thing that makes the big box interesting from a technology standpoint is that, while its principle rests on a prototype invention well into its stage-two innovation, it itself has had the impact of a highly disruptive prototype invention. To see this, visit a small or mid-sized city near to an interstate highway. The new big box, such as a Wal-mart or a Home Depot, will probably have located near the interstate. A visit to the traditional business district reveals a number of empty storefronts. For contrast, visit a smaller city that is remote from an interstate highway. Without the highway to attract the big box, there is at least a chance that the traditional business district will still be thriving.[14] Schumpeter used the term "new combinations" in his concept of engines of creative destruction.[15] The big box retailer certainly qualifies as a new combination.[16]

One of the more obvious consequences of the rise and expansion of information technology has been the production of huge amounts of data on many subjects. Indeed, the burgeoning of the overall availability of data has run well ahead of the ability to utilize the possibilities that arise from the expanded data potential. The basic problem concerns how data are to be made available to all who would use it. Dealing with this problem has been the driving force behind several lines of technology development that are now either familiar or becoming familiar. These lines of development include search engines by which the information accessible on the Internet can be found systematically. The Internet and its World Wide Web are still relatively new, but they have nevertheless undergone considerable evolution towards becoming systematically useful already. In its early days of popularity, there developed the term "surfing," meaning sitting in front of a computer and wandering haphazardly from one web site to another with the hope of discovering something interesting. The modern search engine is rapidly making "surfing" an obsolete term as the ability to investigate a specific topic through the use of a search engine improves. Much work in this direction remains to be done, as it is still common to type in a keyword and find that the search engine has yielded references in the ten thousands.

Once large amounts of information become systematically available, there remains the problem of how to use it. This is a very large challenge, but an example suffices to illustrate it. The field of marketing has produced large amounts of information about individual consumers and has offered the possibility of an increasingly fine subdivision of consumers by tastes, incomes, and other criteria. How does a mail-order business or the publisher of a widely circulated magazine take advantage of this capability? The problem up to now has been in printing technology. The desirable strategy would be to target each market subgroup with a publication tailored to the characteristics of that group. Here, however, the limitations of printing technology become a barrier. Traditional offset printing allows large numbers of publications to be printed at low cost per unit. The process has high speed but almost no flexibility. To cut one large-scale printing into numerous printings tailored to submarkets would be to incur massively increased costs, as the process calls for costly setup and its economies at large scale are lost with relatively small production runs. Digital printing has the required flexibility, but is much slower, so much so as to be prohibitively costly even for large runs. Here is the impetus toward increasing the speed of digital printing. In this example, the massive expansion of data availability has determined a direction for hardware development. This is a further stage-two development from the original prototype inventions: transistors and integrated circuits.

Outsourcing

Outsourcing is a term that is much bandied about in the media and for some it represents a valuable tool for cost control, and for others, it is an epithet. It refers to the decision to rely on the market to supply a part, an assembly, or a service that had formerly been produced internally by a company. In some instances, an outsourcing decision has resulted in shifting the activity out of the United States. The profitability of international outsourcing rests on two developments. First, there has been a diminution of barriers to international trade in goods and services ever since the end of the Second World War. Second, the possibility of taking profitable advantage of easing barriers to international movements of goods, services, and capital arises from the development of technology-based ability to coordinate operations over large geographical expanses. Thus, the development of such technology lies is the enabler of the revolution in world industrial configuration that has occurred, largely since 1980.

The outsourcing of the late twentieth century reverses many earlier decisions to integrate vertically. The outsourcing decision is typically based on considerations such as cost analysis (it is cheaper to do the job outside than inside) and as a way to divest operations not deemed to be satisfactory performers from the standpoint of the company's growth ambitions ("not a part of the core business"). Another side effect of outsourcing has been the geographic migration of economic activities towards the lowest-cost labor that can carry them out, which has been a factor in the trend toward disinflation that has occurred in the developed world since 1980. A relatively stable price atmosphere has materially contributed to generally prosperous times since the early 1980s and made it easier for monetary policy to maintain relative price stability. There is a limit to what amounts to a spreading of economic production towards the lowest-cost geographic areas. This is because there are but a finite number of places offering the possibility of producing with less cost than in all other regions. One has to speculate that the termination of this process will remove an important influence toward disinflation. Increased inflationary pressures in the future will increase the difficulty of maintaining a stable price environment with monetary policy.

A variety of activities has been outsourced. Historically, the process initially affected activities characterized by large production runs producing commodity-like goods. Certain labor-intensive activities in manufacturing have been outsourced to Mexico for decades and, more recently, to China and India. The required workers in these activities are characterized by low levels of formal education, and the work requires limited training. This should evoke recollections of the type of workers in the newly established factories of the nineteenth-century United States that employed people that previously worked on farms. Mexico, India, and China have large pools of such workers. As it has evolved, work of a more intellectually demanding nature has been outsourced to both China and India. This trend has been abetted in both China and India by the significant modifying of socialistic governance, largely since 1990, including the opening to international trade. India has long produced many highly educated people, especially in technical areas and for decades following formation of the country in the mid-1940s had a serious problem employing many of these people up to their potential. The result was export of talent to other countries, including the US. Today, demands of India's software industry for talent have employed many more natives and to some extent has tempted expatiate talent to return to India.

Outsourcing has created at least a temporary unemployment problem. Workers affected include textile workers and engineers formerly engaged in computer coding. From its beginnings, outsourcing has grown to include functions that call for levels of skill well above that of call centers and production line workers. These include the outsourcing of computer coding tasks that require sophisticated coding and mathematics talents. Some large firms such as IBM and Microsoft have established research and development facilities in India, and the effect is to displace American workers who have education and experience well above the minimum level. At least some of the opposition to reduced tariffs and other tools for the protection of workers in various countries grows out of the worker displacements that freer trade may imply. The publicized opposition at conclaves attempting to work out problems of freer world trade (the GATT Doha round) is composed of a number of disparate groups among which the common threads have not always been easy to detect. However, the fact of these demonstrations is evidence of the adverse reaction to some of the effects of globalization expressed by political action. This kind of resistance to the freeing of international trade added to resistance on the part of various governments' attempts to protect politically influential domestic industries may make further substantial progress toward the economic promise of globalization difficult. Temporary unemployment due to outsourcing is often ultimately traceable to changes enabled by technological change and is one manifestation of the social downside of technology-based changed.

Education and Technology

The principal thrust of this chapter is that economic growth is ultimately traceable to technological advance and that episodes of unusually rapid growth derive from especially fecund periods of original invention. The high technology flux that occurred in the latter twentieth century coincided with the development of substantial income disparities. The most prominent reason for this is that applying new technology tends to require the more highly educated and skilled among the population. The important technologies of the late twentieth century called upon skills, such as electrical engineering, to a great extent. Educated and skilled people command higher incomes than did those lacking these advantages. This became especially pronounced in the late twentieth century in contrast with the experience of applying the nineteenth century's technology legacy, for the earlier period was dominated by the emergence of mass-production manufacturing, which utilized workers of relatively low skills minimally trained.

Looking into the future has produced considerable concern regarding public education's ability to provide for the needs of a workplace increasingly dependent on high skills in the US and in the industrialized world at large. This concern is in light of demographic prospects of a shrinking active labor force and a relatively growing population of retirees. The problem in Europe looks particularly sinister because of fertility rates well below population replacement rates. The same holds true for China due to that country's historical draconian birth control regulation. In the US, fertility rates are near replacement rate presently, but the problem still exists due to accumulated effects of past demographic events. There are substantial political forces opposing any effort to adapt the US public school system to the future needs of the economy, and consequently the necessary reforms may be protracted.

In the meantime, the situation is not all bleak, for capitalist economies are nothing if not adaptive. There are at least three ways in which the technology-based economy is managing to find the human talent it needs. The first is a well-developed adult education system. This includes a variety of institutions, such as community colleges, technical schools, independent law schools, and many others. In spite of the criticisms leveled at the public school system, it remains possible for a motivated public school student to get an education adequate to prepare him or her for further training, at least in most places. Unfortunately, the necessary personal motivation to do this is not common among teenagers. Many will mature, some well into adulthood, to the point of identifying a career path. When this happens, there is an adult educational network to which they can turn to achieve their goals. This network is a strength of the US educational system that does not get much media attention.

A second factor is immigration. This has served to fill many jobs that would be difficult to fill without immigrants. The jobs these people take are often not high-skill jobs but some are. Immigration has engendered opposition among some elements of the native population, but the political action has so far failed to take decisive adverse action relative to immigration. This reflects the power of the political forces that recognize the need for immigration at various skill levels. Immigration is probably more useful in the US than in Europe, for the native populations of Europe tend to be even less immigrant-friendly than the US. A third adaptation to a shortage of labor at various skill levels is international outsourcing, discussed above. In some ways, outsourcing achieves the same objectives as immigration. The choice between international outsourcing and importing talent has come to depend of the comparative costs of the two options.

Prototype Inventions, Investment, and Growth

One of the more interesting ideas that was prominent during the immediate post-depression years was that the maintenance of investment was crucial to maintaining an economy's overall prosperity. This was an outgrowth of the depression, for which private investment failure was then recognized as a major cause. In the US and other countries, the 1930s saw faltering government attempts to fill the expenditure void left by the private investment failure. This kind of governmental reaction to slack periods in the economy has become a recurring pattern, exemplified by the less-than-successful attempts to stimulate the UK economy in the 1970s and the Japanese economy[17] in the 1990s with public expenditures on infrastructure.[18] The spotty record of investment as a public policy tool raises a fundamental question concerning the origin of growth itself: is capital investment a *primary* generator of economic growth, or is it only a transmitter of the effects of some other source of growth? The overall hypothesis of this book is toward the latter possibility. The first three decades of the twentieth century were like the final two in that both periods saw high rates of capitalization in private industry. The two periods were also similar in that the high rates of investment in both grew out of exploitation of business opportunities stemming from prototype inventions. In the earlier period, the prosperity of growth ended when the consumer markets served by the new industries matured and there was no industrial market to take up the slack. This experience points toward the conclusion that investment, in the absence of a fund of unexploited opportunities to replace existing capital with capital of substantially superior productivity, does not appear to have much strength to evoke growth. Capital of superior productivity can ultimately come only from technology improvement of radical dimension, exemplified by prototype invention.

The association between growth and prototype inventions appears consistent with the growth model formulated by Professor Robert M. Solow in the 1950s.[19] In essence, Solow undertook to determine the extent to which measured economic growth could be explained by measured accumulation of capital and the expansion of the labor force. His model was essentially an aggregate production function. The result was that 85 percent of the growth could *not* be explained by labor and capital growth; growth was highly correlated with the residual of the estimation. Solow hypothesized that the residual contained the effects of technological advance, which he treated as external to the economy (exogenous).

One of the problems with econometric estimation of the parameters of an aggregate production function as in the Solow growth model arises from the measure of capital stock used in the estimation. Statistical measures of the stock of capital in an economy always include a mixture of capital ranging from the newest and most efficient to old and obsolescent capital that has not yet been replaced by technology that is more efficient.[20] The potential for distortion resulting from using such data in the estimation of an aggregate production function is greatest in a time of high technology flux, when the efficiency difference between the newest and most efficient capital plant and the oldest plant is greatest. The aggregate production function is a static concept. Its statistical usefulness presupposes that there are time intervals in which the stock of capital has adjusted fully to a changed standard of capital efficiency. It is doubtful that such states occur in a dynamic economy, especially in a period of technological flux, for at such a time the economy is in a continuous state of replacing obsolete capital. The economy can be viewed as continually reinvesting its financial depreciation flows in the most productive capital that is available.[21] Any effort to estimate an aggregate production function will suffer if its measure of capital stock does not incorporate a cyclically variable measure of obsolescence as well as a measure of depreciation.[22]

The Solow model has been controversial for reasons in addition to the above-cited problem with the very concept of the aggregate production function. Another is Solow's dismissal of the possibility that technological change might be generated inside the economy (is endogenous). This aspect of the controversy induced a number of economists in the 1980s and 1990s to try to "endogenize" technological change as a part of mathematical economic models. The need for these attempts is evident. Consider a situation that is completely at the heart of widely accepted economic theory: that of a producer in a competitive industry. The theory blithely assumes that any producer in this situation will use the most efficient production function (the best technology) to remain cost competitive. The way this theory is often presented leaves the impression that this "best" technology is well known by all competitors in the industry.

This might be a realistic assumption in a world in which technology has been static for sufficiently long that all producers in an industry have had ample time to learn of and adopt the lowest-cost technology that is available. What happens, however, in a technologically dynamic world wherein technology is subject to incremental improvement most of the time and occasionally to dramatic improvement? A simple thought experiment regarding how individual producers react to technology change reveals several problems. The first is that a technological advance does *not* become known to all producers simultaneously. If the improvement were an isolated event in time, then all producers might adopt it in a reasonably short period of time. However, technological dynamism suggests a second problem: there can be many improvements, each building on the previous one over a number of years. At what point does a producer adopt? To wait until all improvements are available could expose a producer to increasing cost disadvantage relative to competitors who have already adopted the new technology at some point in its development.

It has been argued in these pages that investment that foments economic growth stems from prototype invention. Economic growth comes during the stage-two innovation of a cluster of prototype inventions. The question of whether technology is endogenous or exogenous makes most sense when applied to prototype inventions; some of these appear in ways having no obvious relation to the normal workings of contemporary national, local, or world economies. At least some prototypes, and maybe most, are exogenous. However, most other inventive activity supports innovation along directions influenced by prototypes. This is certainly true of second-stage innovation, the proximate fomenter of growth. Second-stage innovation works with ideas that *already are within the economic universe*, namely, prototype inventions. It is endogenous.

Conditions of technological flux create the possibility for one producer to think of a less costly way to produce a product *before* the competition has the idea, even for a purely a commodity product. Ford's production line revolutionized much of value-added industry. The production line was not a simple application of a single thread of technology development. It was a complex combination of a number of threads emanating from prototype inventions of the late nineteenth century; it is this very complexity that prevents all producers from instantly perceiving the wider possibilities growing from multiple threads of technology developments stemming from even a limited number of prototype inventions. This example has been repeated in the development of the tools of information technology based on the "big four" prototype inventions of the postwar period.

Notes

1. The reader will note that Figure 4.3 in Chapter 4 is in terms of year-on-year comparisons of productivity, and not in terms of the period averages presented here. The reason is the averaging process eliminates the 1911-1915 years, which are of interest in the presentation of the earlier chapter.
2. When the diesel-electric first emerged as a serious competitor as the freight prime movers of railroads in the late 1930s, the most obviously threatened class of workers was locomotive firemen, whose role was unnecessary on a diesel. While the firemen's union managed to stave off this threat to its members until about 1980, there was an immediate displacement of the labor force that had been necessary to keep steam locomotives in daily operation, which was much larger than that for the diesel.
3. Bosworth, Barry, and Triplett, *Is the 21st Century Productivity Expansion Still in Services? And What Should Be Done about It?* (New York: National Bureau of Economic Research, July 2006).
4. This notion has even acquired a name: "Baumol's Disease," after William J. Baumol.
5. During the late 1980s and very early 1990s, there developed the idea that as information technology was introduced in business worker productivity went *down,* not up. This idea gained the name "productivity paradox" and is attributed to Professor

David Solow. It seems to have grown from an observation that while computers had become very widespread in business, their effect had not shown up in the productivity statistics. With the aid of hindsight, this idea seems to have been premature. There was considerable reaction to it at the time among economists, and the one that came closest to prescience was that of Paul David of Stanford, who suggested that there can be a large time span while a new technology's possibilities come to be realized. David used the electric motor as an example.

6.   While these investments may enable higher traffic levels, which would imply a labor productivity increase, much of this work was of the nature of "catch up," to accommodate previous traffic increases; effects on labor productivity would appear only after a time lapse.

7.   The mass market appears to have been the first goal of the nascent personal computer industry as it emerged in the late 1970s and early 1980s. At that time, advertising for home computers emphasized such benefits as computers as a way to build and keep recipe files. The pioneer spreadsheet software of the 1980s hastened the progress of the personal computer into a serious business tool.

8.   The automobile and related supply industries at this time were the largest industry group in the country.

9.   The term "homogeneous" applied to capital means capital that can be retrofitted widely to existing and ongoing activities. "Heterogeneous" refers to capital that is specific to an industry and which will be rendered worthless if that specific industry declines.

10.   Columbia Encyclopedia, Article about electronics industry.

11.   Lowenstein, Roger, "The Integrator," *Technology Review,* Vol. 108, No. 5, (October 2005), p. 80.

12.   The use of the laser to read bar codes on items in transactions is an important marriage of laser and computer technology.

13.   Other factors, such as large-scale purchasing, certainly enter into the economics of retailing, but this advantage of bigness has a long history and does not explain the comparatively recent emergence of the big box.

14.   The author's favorite example of this case is Mount Pleasant, Iowa, which is a small service center in a farming region. Mount Pleasant has a thriving traditional downtown.

15.   Schumpeter, Joseph A., *The Theory of Economic Development,* (New Brunswick, NJ: Transaction Publishers, 1934), pp. 223-230.

16.   The development of the "big box" store in retailing provides a link between the developing field of information technology and the acceleration of the overall national measure of labor productivity in the 1990s. The replacement of less-efficient retailing units with more efficient big box establishments was suggested as the explanation of an acceleration of the labor productivity in retailing found in a recent study. See Foster, Lucia, John Haltiwanger, and C.J. Krizan, "The Link Between Aggregate and Micro Productivity Growth: Evidence from Retail Trade," (Working Paper 9120, Cambridge, MA: National Bureau of Economic Research), as cited in Triplett, Jack E. and Barry P. Bosworth, *Productivity in the U.S. Services Sector,* (Washington, DC: Brookings Institution Press, 2004), pp. 237-238.

17.   In both of these examples, public infrastructure investment was not effective at stimulating the overall economy because the infrastructure that resulted amounted, at best, to a minor improvement in the pre-existing infrastructure. In Keynesian terms, the marginal efficiency of the resulting capital was low. By contrast, the building of the farm-to-market road system in Texas and other states during the 1930s had significant local economic impact because there was no previous network

of good roads to serve rural needs. This capital had high marginal efficiency but the capitalization was on a small scale.

18. The use of public sector investment as a countercyclical policy tool has lived well beyond events that have cast doubt on the efficacy of this tool. The reason is that the idea has been embraced enthusiastically by politicians, a group of people among whom the basic logic of spending public money is never questioned very closely.

19. Solow, Robert M, "Technical Change and the Aggregate Production Function," *Review of Economics and Statistics,* 39, (1957).

20. Capital stock estimates tend to be based on an inventory concept and some assumption covering depreciation in use. The approach treats all capital as equal and makes no allowance for differences in efficiency between new and extant capital.

21. The problems of aggregate production function estimation are discussed succinctly by Professor Hicks. See Hicks, John, *Capital and Growth,* (New York: Oxford University Press, 1965), pp. 293-305.

22. During the 1950s and 1960s, there were attempts to incorporate an obsolescence factor into to formal models of the investment process. The obsolescence factors were almost always in terms of average rates whose effect was to diminish the life of capital equipment below that assumed in the depreciation rate, also an average across cycles. Inasmuch as the rate of technological obsolescence is highly variable over the long technology cycle, treating it as a constant factor (parameter) operating on a capital stock measure failed to shed much light on explaining why private investment is highly volatile over time.

12

Technologies that Develop Slowly

The Problem of the Lack of Enabling Technology

There are technologies, some of which have been around for many years, which for various reasons have not developed rapidly or have not achieved what some deem to be their full potential. These can be of interest today because their development, or lack thereof, can affect achieving the potential of some other technology going into the future. The threat goes to the very roots of economic growth. This is nothing more than saying that lack of enabling technology that holds up progress is still a problem as it was one hundred years, two hundred years, and longer ago. A list of examples of this would include solar power and wind power. Solar power's development has been hampered by the cost of a critical material, silicon. Also, both solar and wind power are *intermittent* sources, and their usefulness as base-load power sources has been compromised by the lack of an economical way to store large quantities of energy in the form of electric power or heat even for relatively short time periods such as intraday. Another example of delayed development of a technology is in the field of production of alcohols by fermentation. This ancient field has become of recent interest because of a seeming promise of a source of motor fuels alternative to petroleum-based fuels. The recent science of recombinant DNA appears to offer possibilities for creating bacteria that are more efficient in fermentation than any bacteria previously known.

In many cases, the disuse or underuse of a technology reflects that there has been a lower-cost way of achieving the same objective. Solar power, as of this writing, is simply more costly than power produced in conventional thermal-electric power plants. During the twentieth century, the chemical industry, starting in the US, converted its feedstock basis from coal to petroleum products for reasons of cost. In the opening years of the twenty-first century, this dependence on petroleum by the chemical industries of North America and Europe has become perceived by some as a competitive weakness in comparison with an expanding petrochemicals industry in the Middle East. However, the

259

solution is far more complex than a simple reversion to coal-based technology. For starters, the old coal-based technology clashes with modern clean-air regulation. It is not surprising that an abandoned technology should not have been kept up with the latest technological possibilities. While clean use of coal has been developed under some unusual circumstances, it still needs work. This chapter outlines some of the recent progress toward developing several historically slow-moving technologies and highlights these as bottlenecks in the development of other technologies.

Much of the current interest in technologies that, in some cases, have been around for many years draw impetus from historically high costs of hydrocarbon-based energy that have occurred in the opening years of the twenty-first century. There was a prelude to present conditions in the 1970s. It will be recalled that this too was an era of experimentation in ways to provide and distribute energy and that many of the experimental essays of that period were abandoned when the price of oil and gas declined in the 1980s. While there is an open question today as to whether oil and gas prices will again decline substantially, there are solid arguments for why they will not, at least in any foreseeable future. That leaves open a question of whether high hydrocarbon prices will bring about radical changes in the ways by which economic growth can be maintained. There are ample precedents for massive sea changes in how basic things are done. One that readily comes to mind is the conversion of the iron and steel industry from a charcoal basis to a coal-based coke basis that took place in England in the sixteenth century. The pressure of high energy prices not only induces redoubled research and development efforts on technologies that have been around for some time, but it also induces experiments with a number of combinations of new and existing technology in the search for economic energy.

Energy Storage

Full realization of the potential of solar and wind power depends on some sort of economical means for storing energy for at least a short period of time: say up to twenty-four hours. There are two principal ways for doing this that have been proposed. One is storage of electric power and the other is storing energy as heat. Electric power storage suggests batteries of some sort, but an alternative is to use solar energy to heat some fluid and use the heat energy to heat water to steam for the purpose of generating electric power via conventional turbines.

Large-Scale Power Storage

As noted in Chapter 6, the Clean Air Act caused a refocus of interest in several old technologies for producing electric power, solar and wind. Apart from the well-known problems of cost of power produced by these technologies, both suffer from being inherently *intermittent* sources of power, the former productive only when the sun is shining and the latter when the wind is blowing. The problem of intermittency will persist until a technology for economical intraday power storage becomes available in the case of solar power. As for wind power, the wind does not blow with perfect consistency even in places prone to wind, and economical storage technology for longer than intraday is needed in order to enable the use of wind power as a reliable base-load power source. Present large-scale solar installations are essentially daytime peaking facilities. For example, solar power farms in the Mojave Desert provide power during daylight surges in demand from air conditioners and other daytime loads. Neither solar nor wind will be useful as part of the base-load system as long as there is no general, low-cost means of storing electric energy at a meaningful scale in relation to demands on the power grid.

The potential relevance of electric power storage technology is not confined to solar and wind, for there is a huge and potentially very lucrative market in storing power produced by conventional thermal power plants. A technology that enabled storage power produced in low-demand parts of the day for use in high demand periods would not only mean that less power generating capacity is needed, it would also mean that the existing distribution grid would be subject to a more uniformly level load. In cases, such as exist in many parts of the country, wherein the present distribution grid is questionably adequate at peak demand times, there would be fewer brownouts and/or blackouts due to overloaded transmission lines. The environmental benefits of storage are considerable, for by directly substituting for power production capacity, it will reduce the need for peak load capacity. Less power would need to be transmitted, as less power would be lost in the form of heat from overloaded transmission lines.

Past solutions to the storage problem have included such measures as pump-storage facilities, in which power in excess of demand in low-demand times, such as at night, is used to pump water to a higher-level reservoir. This water can be drained during the daytime to a lower level, producing hydropower on the way down. Unfortunately, this kind of storage is very topography-specific, and not every place has a convenient mountaintop on which to place the upper reservoir. Recent attention toward this problem has focused on battery technology, where the main present-day problem is cost. As of the present writing, there is no battery technology that is economically attractive in relation to the cost of building enough capacity in power production and distribution facilities to cover peak demands. As of now, there has not been a period of rapid technological flux in battery technology that has produced cost declines comparable to what has happened in the field of, say, computers.

Flow Batteries

For storage of power from intermittent power sources, such as wind and solar, at the scale of the power grid, flow batteries offer some promise for the

Figure 12.1
Example Flow Batttery

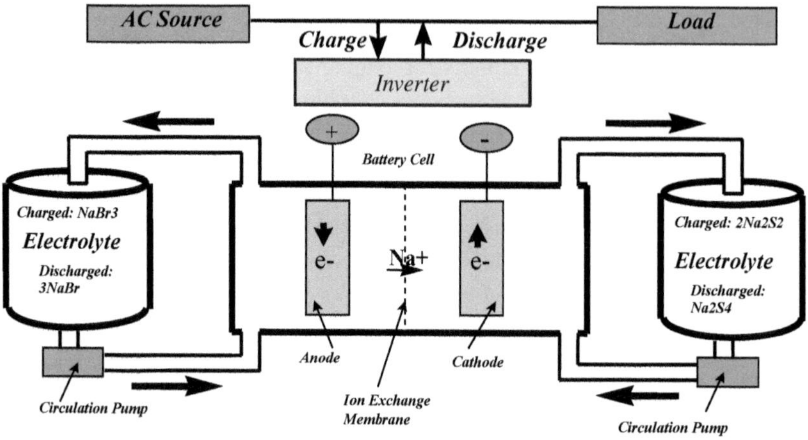

future. These have already been introduced in Chapter 6 in relation to present regulatory problems seen to be inhibiting the use of this technology. It is to be hoped that regulatory problems are of short-term concern, which becomes more likely as flow battery technology becomes increasingly attractive. According to a source within the American Electric Power Organization, flow batteries cost about $2000 per kilowatt.[1] The cost needs to be about half that in order for the technology to be attractive to the utility companies. The possible economic future of flow batteries is sufficiently interesting to warrant a brief discussion of how they work. Please refer to Figure 12.1, which depicts one of several flow battery systems that have been developed to date.[2]

The electrolyte consists of charged sodium salts of bromine and sulfur. The battery cell consists of a polymer membrane that passes positively charged sodium ions but is an insulator for electrons (similar to a fuel cell). The pumps circulate the electrolyte into contact with the membrane, where positively-charged sodium ions become dissociated and pass through the membrane. Electrons have to exit externally into the circuit that constitutes the load. Because the battery output current is direct current, it passes through an inverter which converts it to alternating current compatible with the power grid. Recharging reverses this process. The battery cell is composed of many units, each of which typically gets about 1.5 volts across the electrodes. The units are arranged in parallels and series in order to achieve the desired total voltage and current output.

The flow battery has the advantage that it can be built to any size needed, including very large size. The heart of it is similar in operation to the fuel cell, which is a battery-like technology that is being improved slowly at present. Another feature of the flow battery is that because the electrolyte is stored away from contact with the cells, and the electrodes and other circuit parts are made from inert materials, it is capable of many thousands of charge/recharge cycles without significant internal deterioration, which is a problem with lead-acid batteries wherein the electrolyte is stored in contact with the electrodes. It is not clear when fuel cells and/or flow batteries will become parts of truly economically competitive power options. At least one opinion holds that flow battery storage projects will become increasingly common in a five-to-seven-year horizon, but this kind of forecast should be taken lightly. The flow battery in its various versions has obvious applicability in which the large battery is teamed with intermittent power sources, such as solar and wind. It also has potential for leveling intraday power demand.

Small Batteries

Much of the discussion of the future of solar energy as a utility-scale source of electric power revolves around the question of improving the cost efficiency of the solar collectors themselves. In the usual small terrestrial solar power application, the load draws power from a battery whose charge is maintained by current from a solar collector. This arrangement has proved highly useful and reliable in certain *niche* applications that generally are small, especially where connection to a commercial power grid would be costly, or when a level of reliability in excess of what the grid offers is considered necessary. The system is designed such that the battery has enough charge to support its expected daily demand load, allowing for a cushion of safety. The economic attractiveness of this sort of system usually is to be found in the particular application. One interesting example is the use of solar collectors and battery systems to power track switches in railway yards. These applications have power from the grid readily available, but the solar-battery systems are more reliable than the grid power. The decisive reason for using them, however, is that they replace manually operated switch points that require a switchman to lift heavy machinery from a stooped position—a situation ready-made to produce back injuries. Thus, the electric switch point mechanisms based on solar power offer significant savings in terms of avoided workman compensation claims. Other applications abound.

Today's lineup of batteries for relatively small applications includes the Nickel-metal-hydride (Ni-MH) battery, which has up to twice the energy density of lead-acid and very favorable characteristics of speedy acceptance of charge. This last characteristic and its relative light weight (compared with lead-acid) made it the battery used in early models of hybrid-drive automobiles. An alternative modern battery is the lithium-ion battery (Li-ion), which has up to four times the energy density of lead-acid. Li-ion has become the battery of choice for laptop computers. It was not chosen as the battery basis for the early hybrid cars because Ni-MH was capable of faster charging. This may change, as the charging speed of the Li-ion battery has been dramatically improved with the aid of some nanotechnology, and Li-ion may have a future in hybrid automobiles. The latest Li-ion batteries with nanomaterial enhancement are said to be able to produce five times as much power as existing versions of the Li-ion battery. It is said to be the cheapest of the modern battery systems. A recent improvement in the traditional lead-acid battery has resulted from the replacement of some of the lead with graphite foam that is more porous and, therefore, offers greatly expanded surface area. This reduced weight and extended battery life. The improved lead-acid batteries are said to be on a performance par with nickel-metal hydride batteries used in hybrid drive cars.

The automotive market is far from the only source of demand for increased energy density portable batteries. The demand for power from small handheld devices such as cell phones is increasing more rapidly than is the capacity of batteries that power them. This reflects the addition of features that have proved popular with consumers, such as combining digital cameras with cellular telephone handsets. Small versions of the lithium-ion battery are the most usual batteries for small devices now. This type of battery produces power in the range of 110-160 watt-hours per kilogram (wh/kg) and can serve as a comparative standard for battery varieties still in research. One type under development is the thin-film solid polymer battery, which has an estimated wh/kg rating of 300. A second, and possibly more distant in time, battery possibility is the micro fuel cell, with an estimated wh/kg rating of 1500. The examples of this in prototype have the obvious appeal of much improved energy density, but its methanol fuel is flammable and creates a possible safety problem. A third possibility that is even more exotic is the nuclear battery, with wh/kg rating of up to 4500. There are some unanswered issues regarding shielding with this possibility. It also suffers the drawback of yielding very small current, for in spite of high energy density, it pays out its power very slowly as the radioactive elements decay. It may also face some suspicion because it is nuclear.

Fuel Cells

A fuel cell is an electrochemical device that produces electricity from an external fuel supply; it is a nineteenth-century technology (see Chapter 2 above). Typical reactants used in a fuel cell are hydrogen and oxygen, with hydrogen on the anode side and oxygen on the cathode side. In fuel cells, reactants flow in and reaction products, electric power, and water flow out; continuous long-term operation can take place as long as the flow of reagents is maintained. Because of their lack of any mechanical parts, fuel cells are efficient when compared to any power source that utilizes a mechanical step between fuel and electric power, such as a conventional thermal electric plant burning coal. A fuel cell has certain features similar to those of batteries. For example, the battery cell of a flow battery is essentially a fuel cell. The obvious difference is in the fundamental method of storage: in the flow battery, the stored energy is in the charged particles in the electrolyte; and in the fuel cell, the stored energy is in the form of hydrogen fuel. In this broad respect, the fuel cell is similar to any other system that converts the energy potential in a fuel to a different form of energy.

The cell part of a fuel cell is a proton-conducting polymer membrane (as in the flow battery) that separates to anode and cathode sides. On the anode side, hydrogen diffuses to the anode catalyst (a part of the membrane) where it dissociates into protons (hydrogen ions) and electrons. The protons pass through the membrane to the cathode, but the electrons exit externally into the circuit to the load, and the cathode side is the sink to which they return and combine with oxygen to form water, the only waste product. This is diagrammed in Figure 12.2. It is the harmless waste that is one of the fuel cell's most attractive features.

Much of the recent focus in research regarding use of hydrogen as a large-scale source of energy has focused on the automotive market. In this market, the internal combustion engine competitively benefits from all the advantages that over one hundred years of improvement plus mass-production economies can give it, as well as a ubiquitous and efficient fuel distribution and repair infrastructure. In spite of the massive disadvantage that this throws in the path of any technology that would encroach on the market of the conventional internal combustion engine, research on fuel cells has attracted some interest on the part of automakers. Ballard Power Systems is a producer of fuel cells and

Figure 12.2
Working Principle of a Fuel Cell

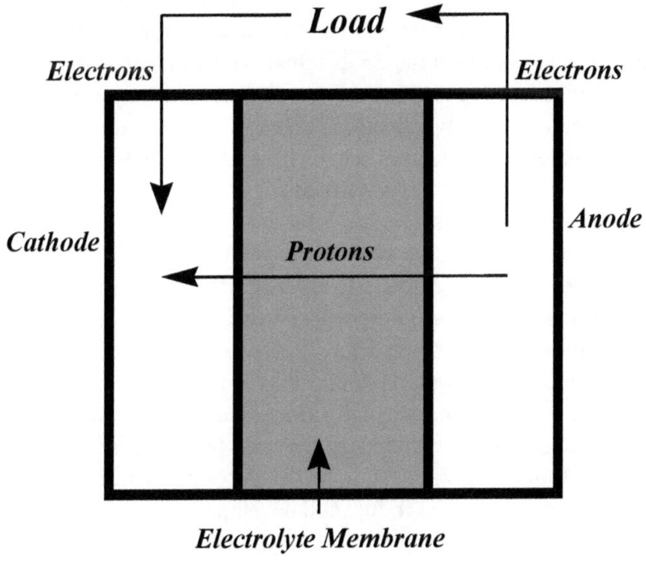

Electrolyte Membrane

claims to be the world's leader in automotive fuel cell technology. Both Ford and Daimler-Chrysler have taken equity positions in Ballard. General Motors and Toyota have elected to pursue their own fuel cell research programs, as have Nissan and Honda. There have been bus demonstration projects in Chicago and several other cities, and fuel cell-powered buses went into service in nine cities in the European Union in 2004, using power plants supplied by Ballard.

Fuel cell producers have recognized the difficulty of overthrowing the entrenched internal combustion technology, even given some dramatic reductions in the cost of the fuel cells themselves, and are turning their attention to the market for stationary power sources where there already exists a small market. This is the market for backup power to cover interruptions in power from the grid. In this market, fuel cell power is only twice as costly as alternatives, such as gas turbine-powered generators, a much smaller disadvantage than that faced in the automotive market. This disadvantage shrinks when the user is able to make use of the fuel cell's waste heat, which can be captured easily, for utilities such as building heating and cooling. Moreover, the fuel cell is quiet, thus saving the cost of muffling the noise of a diesel or gas turbine. Fuel cell producers anticipate reducing the cost of their products still further.[3] Fuel cells have also found a market as power sources in remote locations, such as in spacecraft, where the water by-product can be a source of drinking water for crewmembers.

As a storage device for the power grid, the fuel cell may someday be a part of an arrangement in which conventionally-produced power in the grid is used to make hydrogen and oxygen during off-peak times, via electrolysis of water, and the resulting hydrogen and oxygen is used to run the fuel cell to produce peak-demand power. Such schemes have received some attention in recent years. The cost of constituent materials remains a large part of the capital cost of any power scheme using fuel cells. For example, the catalyst that is a part of the cell membrane is usually a precious metal of the platinum group, and the need for inert materials as electrodes calls for such materials as gold. The fuel cell and the flow battery share this problem.

In addition to advantages of cleanness and quiet, the prospect that cost-efficient fuel cells may become available in relatively small units appears to open possibilities for electric power from geographically diffused sources. In some ways, this is arguably as interesting a prospect as the fuel cell offers, for the historical trend has been toward geographic *centralization* of power. In the context of decentralized power, the fuel cell's competition is solar power more than central station power. In such competition, the fuel cell's ability to operate night and day renders it independent of the need for low-cost storage. People who cherish the romantic notion of complete independence from the power grid should become very interested in fuel cell development.

Direct Heat Storage

Solar power advocates have proposed an alternative to using a photoelectric cell to produce electric power directly from sunlight. In this, solar reflectors concentrate solar energy to heat a fluid (ethylene glycol, for example) in a closed circuit. The heat in the circulating fluid is imparted to water in a heat exchanger to generate steam that is then used to operate a steam turbine and produce power. In order to deal with the intermittency problem common to all solar power arrangements, heat in excess of immediate generating needs would be used to heat a reservoir of molten salt, which efficiently stores heat.[4] The stored heat can then be tapped to generate power at night and at other times when the solar collectors operate at less than full efficiency. This arrangement's efficiency would be limited by the efficiency of the heat exchangers, another technology whose improvement over the years has been slow. A variant of this scheme that promises to be somewhat more efficient thermally would concentrate solar energy to heat salt directly. The molten salt would then be used to generate steam in a heat exchanger.[5]

Technologies Dependent on Oxygen

The modern industrial technology for producing oxygen goes back to Carl von Linde in the late nineteenth century (see Chapter 2) and is the basis for a number of processes that use precisely controlled large quantities of oxygen. These include the basic oxygen steel making process, the manufacture of ammonia fertilizer, and the manufacture of methanol, from which a number of industrial and consumer products are derived. Oxygen is produced industrially by lowering the temperature of air sufficiently to liquefy it and then separating the oxygen, nitrogen, and other gases by fractional distillation. This separation is conceptually similar to other fractional distillation processes except that it takes place at temperatures low enough that the components of air are in liquid state. Facilities that do this are known as cryogenic air separation plants, and their operating costs are sensitive to the cost of electric power used in liquefying air.

Now it appears that there will be an increased demand for industrial oxygen, for technologies for clean coal power plants all call for an industrial-sized supply of oxygen. While the price of oxygen has been declining over the past thirty years, it is still an expensive part of processes that have not yet attained costs competitive with the cost of power from conventional thermal plants.[6] The integrated gasifier combined cycle (IGCC) power plants that have been announced to date are on a very large scale, and the scale may help achieve lower oxygen costs. If the technology for producing oxygen cannot be improved substantially in the next several years, the new clean-coal plants may even so become competitive with conventional coal-fired thermal-electric plants as the costs of dealing with their combustion products rise with tightening environmental regulations.

Lower-cost oxygen will also promote other technologies dependent on hydrogen. As of today, the most common industrial processes for producing hydrogen are steam reformation or partial oxidation of methane (natural gas). The latter process takes place in the presence of a nickel catalyst, and can use heavier feeds such as napthas and gas oils when light feeds such as methane are not available. Both processes take place under conditions of severe temperature and pressure. Partial oxidation calls for precisely controlled inputs of oxygen. Both of these processes produce some carbon monoxide and carbon dioxide. The former is a valuable industrial fuel, and the latter can be reacted with steam and more methane to yield more carbon monoxide and hydrogen. One other possible future source of hydrogen is electrolysis of water, which is considered to be uneconomical at industrial scale except in cases where the cost of electric power is very low. An example of such a case is power produced by power plants during off-peak hours, at which time it is economical to sell the power for its marginal cost of production plus a small premium to contribute to capital and overhead costs.

Gasification of Hydrocarbons

Gasification is a process for converting any hydrocarbon substance into a mixture of hydrogen and carbon monoxide, or the most basic constituents of organic substances that can exist in gaseous form. Gasification goes back to 1800 and before, as crude forms of it, often based on coal feed, were used to produce town gas for lighting, cooking, and space heat. Eventually, natural gas took over many of these markets, but improved gasification has been the basis for the production of ammonia fertilizer and methanol, a chemical building block for a number of products. It was in these markets that the process was developed in the twentieth century. Also, the process was utilized for production of motor fuels by Nazi Germany in World War II and by South Africa during the apartheid period and was technically improved during both these episodes. Today, the gasification processes offer the advantages of being able to use any hydrocarbon substance and producing a slate of outputs almost totally free of polluting wastes. Figure 12.3 diagrams this.

The possibilities of gasifying coal have naturally attracted much attention in recent times. Behind this interest are several elements. First, there is the notion that the world is running out of oil and gas reserves (not probably so, but a reality if believed). Second, North America is endowed with huge reserves of coal. Straightforward use of this energy source, as in burning it under boilers, collides with the aims of the Clean Air Act and similar legislation around the world, but gasification technology offers the potential for almost perfectly clean use of coal. Production of electric power is only one application for gasification. It will be remembered (see Chapter 2) that the modern chemical industry originated with processes based on raw materials derived from the destructive distillation of coal tar. During the twentieth century, the world chemical industry shifted its raw material basis from coal to petroleum, mostly because of the cost attractiveness of the latter. Today there is a large element of uncertainty regarding the necessity for returning the chemical industry to a coal basis. Should the market ever deem such conversion to be necessary, gasification technology would play a large role. The Eastman Chemical Corporation has operated a coal gasifier at its Kingsport, TN, plant to provide feedstocks to several plant reactors, and these operations account for about 30 percent of the value of the plant's output in recent years. This gasification operation has been in place and under development for almost twenty years, and is an obvious source of technical experience should gasification be more widely adopted in the chemical industry.

Coal gasification offers a prospect even larger than that of a chemical feedstock: transportation fuels. Gasification's main product is synthesis gas (syngas), a mixture of hydrogen and carbon monoxide. From these two building blocks, diesel and other transportation fuels can be synthesized using Fischer-Tropsch processes, originally developed in Germany in the 1920s.

Gasification could play a significant role in future electric power production in the form of the integrated gasifier combined cycle (IGCC) plant. In an IGCC arrangement, a hydrocarbon fuel, usually coal, is gasified to produce synthesis gas as well as other gaseous products such as hydrogen sulfide (H_2S). All gasification products, including CO_2, are recovered in the form of marketable products, and the syngas is used to fuel a gas turbine that drives an electric generator. The waste heat from the gas turbine is captured and used to make steam, which is used in a steam turbine to produce more power. The thermal efficiency of the IGCC arrangement approaches 60 percent, compared with 35 percent attainable by a conventional fossil fuel plant. The only waste product that has to be dealt with is the CO_2 from the gas turbine, but this amount can be held to a very small amount by a process called water gas shift, which results in a hydrogen-rich fuel for the gas turbine.[7]

The economics of coal gasification are sensitive to a number of factors, of which two major ones are cost of coal and the salability of recovered CO_2. In a recent study, Ken K. Robinson and David E. Tatterson of Mega-Carbon Co. of St. Charles, IL, developed and compared several cases which highlight these sensitivities. In one case, Montana Rosebud coal (sub-bituminous) at $12.00 per ton was gasified and the resulting syngas used to make both transportation fuels and electric power. This case yielded a return-on-investment (ROI) of 15 percent. When the same case was run using Illinois Basin coal at $45.00 per ton, the ROI dropped to 9 percent. In addition to the differing coal costs, the difference in ROI reflects the salability of the captured CO_2 into tertiary oil recovery projects in the Montana case. It was not clear that there was such a market in the Illinois case. In both cases, all CO_2 was captured.[8] One source of uncertainty that appears to be holding up development of IGCC for power production is the recent trend toward deregulation of power production. The proposers of new IGCC plants have had difficulty in persuading state regulatory commissions to enable them to recover the capital costs of such plants from ratepayers.

To date, coal gasification has not been widely adopted either for power production or transportation fuel manufacture. The seemingly attractive economics in the Robinson-Tatterson study depend on the assumption of crude oil at $60.00 per barrel, and disappeared when crude falls to the $0 to $40 or $45 range. This amount of commodity price risk has discouraged investment in gasification to make transportation fuels. As for power production, the economic attractiveness of gasification in the absence of significant costs for CO_2 release into the atmosphere is low. The capital costs of an IGCC power plant are higher than those of a conventional thermal electric plant.[9] The differential total cost per kilowatt-hour for the two technologies can be reduced in favor of the IGCC plant by (1) the avoidance of having to dispose of the products of combustion from the thermal-electric plant and (2) using very low-cost feeds to the IGCC plant. The gasification process is very flexible as to its hydrocarbon inputs, and this opens the opportunity to use feeds that would otherwise be regarded as waste products that are costly to dispose of, such as certain industrial wastes. Such feeds effectively have negative costs. Wide use of gasification to produce power could result from regulatory designation of CO_2 as a pollutant.

Use of waste-product feeds was the basis for gasification's finding a commercial *niche* in the form of power production in petroleum refineries during the early 1990s, where the hydrocarbon input was high-sulfur petroleum coke, high-sulfur vacuum flasher bottoms, and various refinery waste products that would otherwise be costly disposal problems (See Chapter 7). Sulfur occurs in virtually all the natural fractions of crude oil, and the heavier the fraction, the greater is the sulfur concentration, and the more costly is its removal. Refiners have long been able to deal with sulfur in light and gaseous fractions of the crude oil, for in these the sulfur exists in the form of hydrogen sulfide, which is relatively easy to strip out and convert to environmentally benign elemental sulfur. In heavier products, such as gasoline blending stocks and diesel fuel, the sulfur exists in larger, more complex molecules, and has to be removed by more complex processes, such as hydrotreating and hydrorefining. However, the sulfur content in petroleum coke, the residual product of the delayed coking units, is not removable by any economical means, and its presence has turned it from a product with a market to a waste product as sulfur regulation has tightened. Some refiners' response has been to gasify the coke and use the resulting synthesis gas to produce electric power in an IGCC plant. In one of the first US refinery gasifier installations the effect was to enable the refinery to avoid an annual bill for purchased power of approximately $12 to 14 million and to replace it with a standby facility at the local utility costing roughly $2.0 million. Moreover, the plant avoided nitrogen costs of $1 million per year and cut its bill for process steam in half.[10] At about the time this plant (in Kansas)

Figure 12.3
Hydrocarbon Gasification and Products

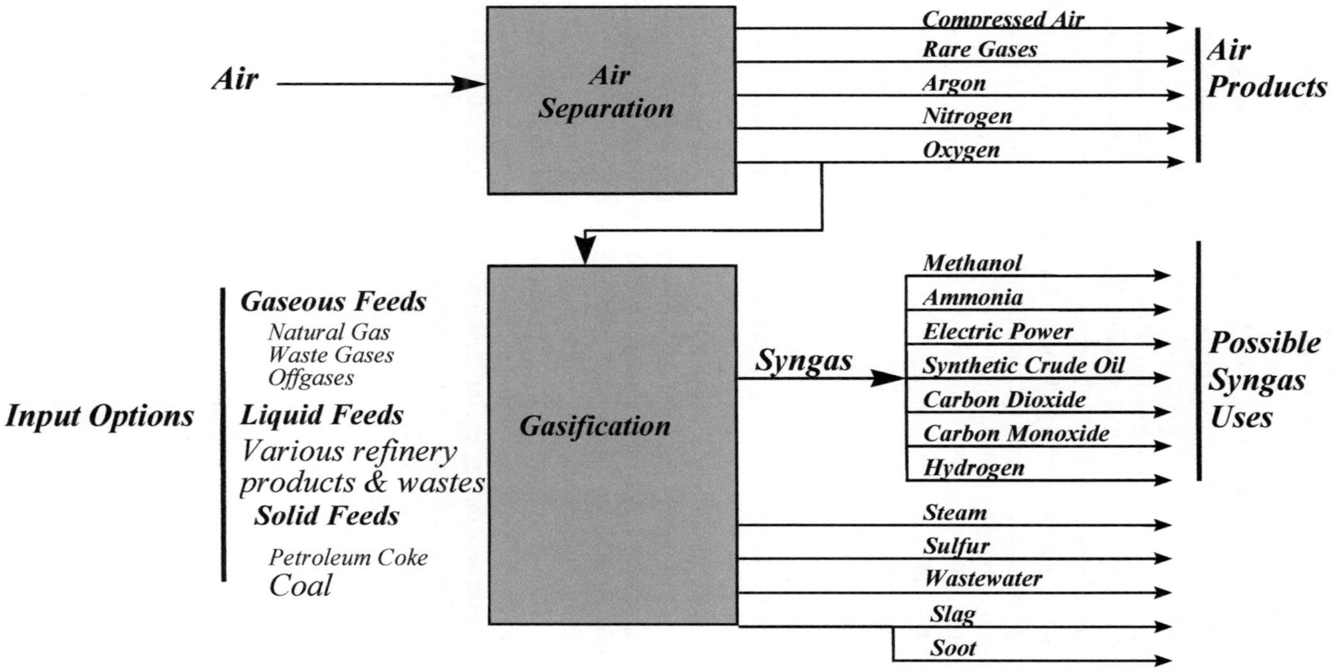

was started up, several refineries in Italy also started gasification plants using visbreaker bottoms as feed.[11] Since the mid-1990s, a number of gasification facilities in refineries have been installed in the US.

There is a very interesting IGCC example in Italy on the Island of Sardinia at the Sarlux refinery complex. This is a very large unit, comprising three gasifiers, two gas turbine/heat recovery units, and several steam turbines. The unit was designed not only to supply the refinery power demand, but also the demand from the entire Sardinian population. The input is high-sulfur residual fuel oil (visbreaker bottoms) from the refinery. This fuel, based on Algerian crude oil, has long been the typical fuel for the Italian power industry, and the resulting smog emission was one of the main factors driving interest in IGCC technology. Instead of burning the residual fuel oil directly, the IGCC unit converts it to a number of marketable products, especially synthesis gas. The unit was ordered in 1996 and went on stream in 2000.[12]

Examples of IGCC plants that consume coal for the production of commercial electric power in the US have up to the present writing been confined to several projects that have been subsidized by the Department of Energy.[13] Both projects date from the mid-1990s and have been technical successes. Both are full-scale, and the Polk plant is used by its operator, Tampa Electric, as base-load capacity. Since these plants went on stream, several economic events have worked to improve the attractiveness of IGCC technology for power production. First, the price of natural gas, the basis of most peaking capacity built since 1995, has approximately tripled since then. Second, even though the US was not a signer of the Kyoto Treaty as of 2005, the prospect Kyoto-like restrictions on carbon dioxide emissions is widely viewed as real to the point where they will be imposed at some future date, and the IGCC technology is better suited for controlling CO_2 than conventional coal-electric technology. Third, the economic performance of the two extant demonstration IGCC plants has been better than originally projected. At least three large coal-dependent utilities have announced the intent to build large IGCC power plants in the US.[14]

In order to find what is arguably the world's most intense interest in coal gasification, one has to look to China. The Chinese are building a huge coal-to-liquids plant in Inner Mongolia that will utilize a huge nearby coal reserve for feed. The Shenhua (the builder of this plant) plant will use an alternative to Fischer-Tropsch technology known as direct liquefaction. With Fischer-Tropsch, syngas is the basis for building molecules that can be refined into motor fuels. Direct liquefaction cuts out the syngas step. In it, most of the coal is pulverized and blended with some of the plant's synthetic oil. The blend is then treated with hydrogen and heated to 450 degrees Celsius in the presence of an iron catalyst, with the result of breaking long chain hydrocarbons of the coal into shorter chains that can be refined into motor fuels.

The primary motivation for Shenhua's project is apparently a desire to reduce dependency on imported oil rather than air pollution control, inasmuch as the product will be transportation fuel. This highlights that there are two general motivations for interest in coal gasification: relief from dependency on imported oil and air composition/quality. Chinese interest in coal gasification reflects the former. As for air quality and composition, it should be remembered that most of China's power production is from coal-burning conventional plants. This capacity is expanding rapidly in response to the country's growing power needs, and, on average, one such plant opens somewhere in the country each week. The result is some of the dirtiest air to be found on the planet. China is cognizant of the problem and will undertake to deal with it, but it is uncertain as to when this will happen. As in the US, the incentive for utilities to upgrade power production by widely adopting coal-gasification technology is lacking. The importance of what is happening in China is that improvements in gasification technology are being tested on a large scale.

In addition to the problem of greenhouse gas emissions, the hydrocarbon gasification technologies are enormous consumers of water, an increasingly scarce resource in many places on the globe. Unlike conventional processes such as thermal-electric power plants, which use water for cooling but returns it to the biosphere, gasification of hydrocarbons actually destroys water molecules. Gasification of a hydrocarbon feed produces hydrogen plus carbon monoxide or dioxide. Some of the produced hydrogen comes from dissociating it from water molecules in the feed, and this is especially important with feeds having very high carbon composition, such as coal. The probable scale of these facilities suggests that their impact on area water resources could be serious. This problem is certainly in prospect with the large Chinese project cited above. In global terms, the water problem could be less that it first appears to be because when the hydrogen is burned, water molecules are formed. However, this might be of small comfort to the near neighborhood of a gasification-based facility.

Solar and Wind Power

Solar Power

The solar idea is from the nineteenth-century legacy. The solar cell was discovered by Charles Fritts in 1883. From that time until the present, the idea of unlimited power directly from the sun has had a strong appeal, an almost romantic aura, which is largely responsible for solar power's receiving research and development attention over the years. In 1953, solar research achieved a key breakthrough when Bell Labs scientists discovered that silicon is a far more efficient semiconductor than selenium, the basis for Fritts' discovery. Even then, however, solar has failed to live up to early visions of it as a truly abundant source of power. To the present time, it has always been considerably cheaper to produce power in central power stations by burning fossil fuels: on a dollar-per-watt basis of comparison, solar power was three to four times as costly as conventional power from fossil-fuel burning central power stations as of this writing.

In spite of its limitations as a base-load supplier to the power grid, its use in markets that it has attained has grown very rapidly in the opening years of the twenty-first century. Annual growth in the world solar power industry in the six years up to 2005 averaged over 30 percent. This growth has placed a strain on the world's capacity for producing the highly pure polysilicon used in most solar panels, the result of which has been a spike in the price of this material. Inasmuch as demand has been interrupted by demand setbacks from time to time, the industry has been reluctant to add capacity without the protection of consumers' commitments to buy the product. However, new capacity is coming on, and the shortage is presumably of a temporary nature. A permanently elevated price would in time curtail the rapid growth.[15]

As of the end of the twentieth century, solar power accounted for one-hundredth of 1 percent of the world's power consumption.[16] The future of this technology is not clear as of this writing. The emerging science of nanotechnology has produced some materials with interesting photoelectric properties in recent years, and there have been some other developments having favorable implications for photoelectric materials. Another new material that may have interesting implications for solar power cost is electrically conducting plastic. Such development could enable a wider market for solar power in the future by lowering the cost of the materials going into solar cells.

Apart from the need for relatively efficient solar cells made from less expensive materials than in the past, the future of solar power as a base-load provider for power consumers depends on progress in several enabling technologies. First, as already noted, much improved storage batteries are needed. Battery technology has improved over the twentieth century, but only slowly in comparison with any number of other technologies that could be cited. Second, inasmuch as the most favorable conditions for generating solar power tend to be found in such places as the deserts of the southwest, there is a need for more efficient technology of power distribution than is available now. Such improvement may be in the form of carbon nanotubes. Development of this technology would certainly support the development of base-load solar power, but it would also support any other form of power generation as well.

Wind Power

Wind power has developed almost to the point at which it can be a competitive source of power under favorable conditions. During the twenty years preceding 2004, the cost of electricity from wind power dropped from 80 cents per Kwh. to 4.5 cents (current dollars), which looks competitive with power from conventional fossil-fueled power plants in some parts of the US. However, this comparison depends on highly favorable wind conditions (steady 18-20 mph) and utility-scale production.[17] Technical development presently takes the form of larger generators with increasingly efficient gear trains, taller windmills, and the capacity for generating power with lower-velocity wind conditions. Wind power is more advanced in Europe than in the US. Offshore wind projects either presently announced or in construction imply a capacity of up to 30 gigawatts worldwide.[18] Germany, possibly the most advanced country in the application of wind power, has set a goal of 20 percent of its power needs to come from this source by 2020.

Power Transmission

The reader will remember the conflict from the earliest days of commercial electric power between proponents of direct current and those of alternating current distribution. That conflict was resolved in favor of alternating current on the basis of its much lower line loss in transmission from power plant to consumer. At the end of the twentieth century, the line loss problem, once considered negligible, is again in the spotlight, for in the US almost 25 percent of the electric power generated is lost as pure heat in the process of its delivery to consumers. Part of this loss is due to inherent limitations of the power transmission technology, but another large part can be attributed to the age of the system. There are many instances in which the capacity of the distribution grid has not been kept apace with the growth of peak demand placed on the grid, with the result of overloaded and overheated wires. Naturally, these conditions have promoted thinking on how to reduce or eliminate the loss. The payoff would be double, for in a world where there is widespread concern over warming of the atmosphere, the elimination of this toaster effect would combine with the avoided necessity for producing the power that has historically been needed to cover line loss, and the problems of dealing with any greenhouse gases implied. It is difficult to exaggerate the importance of low- or no-resistance transmission, for its lack will condemn solar and wind power to their present *niche* markets. At such time as very low-loss transmission becomes possible, not only will base-load solar and wind power be enabled, but the efficiency of *any* power technology will be enhanced. The benefits do not stop there. For example, it would eventually become unnecessary to expend energy in moving huge quantities of heavy fuels, such as coal, from source to market.

One obvious solution would be to rebuild the transmission grid to greater capacity using metallic conductors. Such rebuilding would be costly, even when a tradeoff between enhanced transmission capacity and reduced need for generation capacity is taken into account. One alternative possible technological basis for eliminating line loss lies in the science of superconductivity or the property of some materials under certain conditions of conductivity without electrical resistance. The reason why superconductivity is not an immediate solution of line loss is that so far, all materials that demonstrate the property do so only at very low temperatures, such as liquid nitrogen. Superconductivity was discovered in 1911 by the Dutch physicist Heike Kammerlingh Onnes, and the discovery was directly enabled by improvements in mechanical refrigeration. Onnes dedicated his career to low-temperature physics and was awarded the Nobel Prize in physics in 1913. The first superconductors performed at 4 degrees on the Kelvin scale,[19] the temperature of liquid helium. Research in low-temperature physics over the course of the twentieth century has resulted in the discovery of materials that exhibit superconductivity at increasingly high temperatures, and, at the present time, this quest has yielded superconductors that perform at above 130 degress Kelvin, or halfway to zero centigrade, but still below the range of ordinary temperatures.

A second technology for reducing transmission line loss is high-voltage direct current (HVDC). On the basis of recent research and experience in the operation of about five hundred miles of HVDC lines in the US, this technology offers significant reduction in line loss compared with alternating current over equivalent distances. High reliance on solar power implies a radical change in the geography of power production inasmuch as it would tend to be concentrated in areas of high sunlight, such as the southwestern desert country of the US. A very large problem with such a concentration of power production is that of moving the power to markets. A practical solution requiring little further technological improvement would be to build HVDC trunklines from producing areas to points convenient to major markets, where the voltage would be reduced and the power converted to AC for distribution regionally in the existing grid. HVDC does not mean zero line loss, but it does offer an efficient long-distance transmission means that does not require waiting for some breakthrough such as would be necessary for a transmission strategy based on nanotubes.

One promising avenue of research at the present time is nanotubes, which under some conditions exhibit near-superconducting properties. The present speculation seems to be that any practical application of these materials to power transmission is many years in the future. At present, one key problem is that not all nanotubes have this property. Moreover, there is so far no systematic method for creating nanotubes with the desired properties in the laboratory much less for producing them industrially.

Fermentation

Fermentation may be as old a technology that humankind has. It has become of high interest in the current and recent spate of high prices of petroleum-based fuels. Interest in the production of fuel ethanol has come to the point where significant investment is going into the production of ethanol from corn in the US. As a relief from the high cost of motor fuel and dependence on questionably stable foreign sources of petroleum, ethanol from corn appears to be a long-term nonstarter, inasmuch as the product yields relatively little energy in excess of the energy expended in the total corn-ethanol process. Ethanol would be a far better prospect as a solution to present-day problems of the supply of motor fuels if it could be made cheaply from agricultural wastes and other biomass. The promise derives from the ability to use feedstocks that grow freely in nature and are free except for costs of harvesting and movement to ethanol production facilities. Examples of such feeds include switchgrass and other weeds. These offer the added benefit that their growth benefits soil quality rather than depletes it. Moreover, they grow on much more land than is cultivated, and their use has little or no implication for the supply of food, unlike corn.

The rub is that the fermentation of weeds into ethanol is presently much more costly than fermentation of corn or sugar. The requirement here is for more efficient bacteria in the fermentation process—so that the costs involved in processing it into alcohol would more than offset the low costs of the feed material. An episode of high oil prices induces research efforts to use the new science of bioengineering to produce the needed bacteria. Historically, improvement of fermentation bacteria has not occurred rapidly, and today the improvement of bacteria to realize the promise of switchgrass appears to be a truly slowly developing technology. What *is* new is biotechnology itself, and the economic impetus toward using it toward the goal of improved fermentation. Success would enable a number of interesting possibilities in the next several decades.

Notes

1. This cost is said to be falling and may be expected to do so as the number of applications increases.
2. The battery depicted is the Regenesys Polysulfide Bromine System, one of which is being built by the US Air Force at its Columbus, Mississippi base as a load leveling and standby power source. Another example is the so-called redox system.
3. Freedman, David. H., "Fuel Cells vs. the Grid," *Technology Review,* vol. 105, No. 1, (January/February 2002), pp. 40-47.
4. Molten salt loses only about 1 percent of its heat during a day, making it possible to generate steam for more than long enough to maintain power production for a full day.
5. The United Technologies Corporation and the US Renewables Group have formed a joint venture to commercialize this very technology.
6. According to Raymond F. Drnevich of Praxair, as of October 2000, the real cost of oxygen had historically dropped about 3 percent per year prior to that date. See Olliver, Richard A., "Applications of BGL Gasification of Solid Hydrocarbons for IGCC Power Generation," paper presented at the 2000 Gasification Technologies Conference in San Francisco, CA, Oct. 8-11, 2000.
7. A small amount of CO_2 in the turbine fuel is necessary to maintain flame stability.
8. Robinson, Ken K. and Tatterson, David E., "Fischer-Tropsch Oil-From-Coal Promising as Transport Fuel," *Oil & Gas Journal* Vol. 105.8, (Feb. 26, 2007), pp. 20-31.
9. It has been estimated as of 2005 that a 1000-megawatt IGCC plant can be built for $2 billion. The equivalent thermal-electric plant using coal fuel can be built for about $1.5 billion. On a life-cycle basis, the cost comparison is relatively in favor of the IGCC plant, but IGCC is still more expensive.

10.   Anne K. Rhodes, "Kansas Refinery Starts Up Coke Gasification Unit," *Oil & Gas Journal,* (Aug. 5, 1996).
11.   Visbreakers are common in European refineries whereas delayed coking units are the rule in the US. The difference is that the European refineries are not as committed to total maximization of gasoline production as are US refineries and are configured accordingly differently.
12.   Please refer to http://www.power-technology.com/projects/sarlux/
13.   These are Tampa Electric's Polk Station and Cinergy's Wabash River Station in Indiana. See Makansi, Jason, "From IGCC Emerges Holistic Approach to Coal-based Plant," *Power,* (May-June 1997); and Makansi, Jason, "Wabash Proves Out Next-century Powerplant," *Power,* (September-October 1996). The US energy bill, which became law in 2005, provides for tax incentives specifically to encourage this technology for producing power in the US.
14.   Public Service of Indiana (PSI), an operating arm of Cinergy Corporation, has announced interest in building an IGCC power plant to be operating by 2011. American Electric Power has announced intent to construct an IGCC plant of over 600-megawatt capacity. Both of these companies have let contracts for design.
15.   Recent rapid growth for silicon has been promoted by substantial government subsidies in a number of countries. For example, Germany has guaranteed producers of solar power up to 55¢ per kwh. Concerns about the future availability of fossil fuels have inspired these subsidies.
16.   John Perlin, "Solar Power: the Slow Revolution," *Invention and Technology,* (Summer, 2002), p. 25.
17.   US Department of Energy, National Renewable Energy Laboratory.
18.   General Electric Co., *Annual Report,* 2004, p. 27.
19.   Zero on the Kelvin scale corresponds to the complete cessation of molecular motion, or the total absence of heat. Zero Kelvin corresponds to -273 degrees on the Centigrade scale.

13

On Invention and Innovation

What Makes an Inventor?

Traits

An inventor is someone who has demonstrated skill at identifying and solving new problems. The problems identified can take a myriad of forms but typically relates to some activity that is familiar. Cyrus McCormick sought to find a way to harvest grain crops that was less costly in terms of labor. Rudolph Diesel sought to use the new science of thermodynamics to improve on the thermal efficiency of stationary engines.

The "solving" part of the definition of an inventor is far from a trivial problem, inasmuch as if the solution were obvious, it would not have been a problem in the first place. A command of scientific principle is not absolutely necessary, but it can help. The independent inventors of the nineteenth century, such as Edison, typically were not well educated in a formal sense but had a sense of what might work. This approach was fallible and did not prevent investigation of many blind alleys. Invention and discovery by persons formally educated in the best science of their time developed in Germany, France, and the United Kingdom before the pattern became common in the US. It was supposed that a good scientific and engineering educational grounding would make the inventing process more efficient than the trial and error approach typical of the early independent inventors. However, given that an invention, especially a prototype invention, takes technology beyond the known, there is no reason to suppose that even a good education in the technical known will be an efficient guide beyond the known. The essential point is that to invent successfully, one has to think *beyond* the received knowledge of the technology with which he has chosen to work. It also helps to have enough innate skepticism to challenge widely held assumptions and to be able to visualize results of the inventive effort. There can be no better example of an individual with these traits than Theodore Maiman with his development of the ruby laser (see Chapter 9 above).

Where the formal education really shines is in helping recognition of the principles underlying a prototype invention and its possibilities after it is a reality, something the inventor may or may not be able to do. An example illustrates this. As noted in Chapter 3, the triode tube had been invented by Lee de Forest who thought of his invention as a receiver of radio signals. Frank Jewett, a University of Chicago Ph.D. working for AT&T, learned of the triode and suspected that it had wider amplifier possibilities. With a greater understanding of amplifier circuits than that of de Forest, the AT&T scientists, over a period of three years, developed the triode into a powerful and reliable amplifier, and this enabled coast-to-coast telephony for the first time (1915). The scientists could not claim the original prototype invention, but once it was reality, they made a dramatic adaptation of it. Their contribution was not in inventing, but in advancing the second stage of innovation.

An inventor, in pursuing an idea, can run into major resistance. This can take any of a number of forms, including technical blind alleys and personal ridicule. William Kelley conceived the idea that steel could be made cheaply by blowing air through molten iron. His family, already established in the iron business, indulged him by letting him pursue his experiments at a furnace deep in the woods that was as out of sight as it could be, and did not offer much in the way of encouragement otherwise. He succeeded in discovering what the world came to know as the Bessemer process for steelmaking, which he invented simultaneously with and independently of Henry Bessemer and for which he was awarded the US patent. After Chester Carlson had proved in principle what became known as Xerography and obtained a patent on his invention, he shopped it to almost twenty corporations that logic told him should be interested, but none were. For a while electrophotography looked like the invention that no one wanted.

Motivation

Time spent in inventing is probably time spent away from other activities, such as making a living. There are some cases in which inventors have pursued their ideas at considerable material sacrifice to themselves and their immediate families, as well as their financial backers. Charles Goodyear went through several sets of investors before he discovered how to keep rubber from being sticky when warm. Overcoming resistance calls on the trait of determination. This raises the question of where inventors get their motivation to invent. There has to be a large element of sheer intellectual curiosity involved. Another element that is probably present is the will to be proven right. Money and wealth, which sometimes accrue to successful inventors, might also enter the equation, although the experience of the twentieth century says that this is not a universal motivating factor. For example, Stephanie Kwolek, discoverer of Kevlar, was a DuPont employee on salary and eventually retired on her employer's pension plan. Chester Carlson had a day job with which he supported himself while working on photocopying on off hours.[1] This apparently left him with little time to maintain his marriage, which he lost. Cases such as this point to the possibility that inventors simply *like* what they do with sufficient passion that they do not want to do anything else.

Support

One problem that all inventors and stage-one innovators have had to face is in obtaining an income, given that inventing and stage-one innovating normally do not produce income until the results are brought to successful commercial conclusion. Table 13.1 shows a selection of inventors, many of whose results have been mentioned in this book, and how they generally supported themselves. Not surprisingly, a number of the earlier inventors/discoverers started out essentially on their own and received support only after their achievements were recognized. This group includes Goodyear and Kelly. Faraday benefited from sponsorship of Sir Humphrey Davy, but was on his own resources as a worker in a bookbinding company until he contracted his relation to Davy. Chester Carlson fit this pattern and reminds us that the independent inventor had not disappeared even well into the twentieth century.[2] Edison was also of this group, although by the time he got seriously into his work on the incandescent light, he was an established inventor who benefited from the resource of the laboratory that he had founded that was supported by investors. Another group held faculty positions in universities, such as Foucault and Haber. This group reflects the comparatively early interest taken in technology among the continental universities.

A major change that the twentieth century brought was the emergence of the corporate research and development laboratory, starting with the facility organized by the General Electric Company, which was widely emulated. As is pointed out below, these corporate invention shops were often successful, at least at inventing. The characteristics of a successful inventor or discoverer, while not widely distributed in the population at large, can be recognized by personnel recruiters working for corporate research shops. While the success of the corporate strategies that led to the founding of the invention shops was mixed, as discussed below, on balance these shops can be regarded as providing a means of support for people who possess an interest and knack for inventive activity. They thereby opened this career possibility for some individuals who might not have otherwise dared the challenges facing the independent inventor dependent on own resources. Also, they could provide laboratory and other support facilities needed, especially in fields such as biology and chemistry. This benefit came to be of increasing importance as twentieth-century technology developed.

Table 13.1
Selected Inventors: Means of Support

Inventor/Discoverer	Invention	Date	Source of Support
Pascal, Blaise	Hydraulic Press	1646	Family wealth
Faraday, Michael	Electric motor/dynamo	1832	Own resources, sponsorship from Humphrey Davy
Goodyear, Charles	Vulcanized Rubber	1839	Own resources, occasional jobs, intermittent investors
Kelly, William	Kellly process for steelmaking	1850	Subsidized by family business (iron smelting)
Foucault, Leon Jean Bernard	Gyroscope	1852	University faculty
Otto, Nicklaus August	Otto cycle engine	1867	Professional engineer
Linde, Carl von	Mechanical refrigeration	1876	University faculty
Edison, Thomas Alva own laboratory	Incandescent lamp	1878	Established inventor: headed
Fritts, Charles	Solar Cell	1883	
Tesla, Nicola Edison	Induction (AC) motor	1883	Employee of Continental
Hall, Charles Martin	Hall process for producing aluminum from alumina	1886	University postgraduate student
Roentgen, Wilhelm Conrad	X-Rays	1895	University faculty
Marconi, Guglielmo	Radio	1896	Professional Engineer
Wright, Orville and Wilbur	Airplane	1903	Own bicycle business
Haber, Fritz	Haber process for fixing nitrogen from the atmosphere	1908	University faculty
Baekeland, Leo Hendrik	Bakelite	1908	Established inventor; wealthy from sale of successful business based on earlier invention
Carlson, Chester	Xerography	1939	Own resources, unrelated day job
Noyce, Robert	Integrated Circuit (co-invented independently)	1958	Corporate R & D Employee
Kilby, Jack	Integrated Circuit (co-invented independently)	1958	Corporate R & D Employee (Texas Instruments)
Maiman, Theodore H.	Laser	1960	Corporate R & D Employee (Hughes Aircraft)
Kwolek, Stephanie	Kevlar	1964	Corporate R & D Employee (DuPont)

Grant Money Support

Government and private foundations in the US have played a large role in providing financial support to the kind of original research needed to produce prototype inventions and to support stage-one innovation based on such inventions. There has been recent concern among scientists and entrepreneurs over reductions in various government budgets for such support. Is this a problem that can be solved simply by restoring and expanding the budgets of research-supporting government agencies? If it were that simple a matter, the level of worry now in evidence seems exaggerated. Unfortunately, the problem goes deeper.

Grant money is like any commodity in that if it is priced below its market value, the quantity of it demanded will exceed any amount supplied. Grants are not entirely free, for they require a formal written proposal that represents a cost of time for the would-be grantee; but this "cost" in no way performs a rationing function comparable to the interest or equivalent that has to be paid on financing from a capital market. In the absence of a price mechanism as rationer, an administrative rationing scheme has to be substituted. Granting agencies typically employ professionals of recognized competence in the disciplines to which they grant, and committees of these people serve as peer referees in the process of judging the merits of the applications that they receive. This is similar to the system commonly used to judge papers that have been submitted to scholarly journals for publication. Regardless of how well this rationing system should work, it is still an administrative process operated by humans. Its effectiveness can be compromised by the same budget costs that are causing worry among would-be grantees, and this means that the refereeing process can potentially be swamped, especially in a time of technological flux.

On the other side of the table, there is a vigorous competition among would-be grantees for the rationed funds. This has given rise to an education sub-industry entirely devoted to the subject of how to write grant proposals. The apparent threat here is that skill at composing proposals will or has transcended the actual potential of the proposed research as the decisive element is securing grant approvals.

Another problem is that in the actual practice of awarding grants, past research success is a very real plus for a grant applier. This probably gives the process a conservative cast. As already noted, an original invention is a novel rearrangement of existing elements that may step outside the bounds of generally received knowledge. Moreover, someone inventing anew probably has had one or more false starts, or no inventing record at all, Even if he or she is on the track of something that later turns out to be a breakthrough, the would-be inventor faces a good probability of funding cutoff should the referees lose patience or perceive a better use for the funding.

Finally, any rationing system that is dependent on human judgment for its operation is always in danger of becoming politicized, corrupt, or subject to emotions of the referees, such as jealousy, and because the system is human, no one has any real right to surprise when one of these events happens. This potential may be greatest when there is more than one vision of the future direction that technology ought to take. For example, in the opening years of the twenty-first century, there are a number of technology-based proposals as to how this country should provide for its future electric power needs. One group seems dedicated to a future commitment to solar power. Another has some sort of commitment to the "hydrogen economy." Another is committed to wind, or nuclear, and so on. The Bush Administration has expressed interest in the hydrogen option, and to some, this gives the appearance of "playing favorites" among technology alternatives by politicians. Whether or not politicization has actually played a part in determining how research grants have been approved so far is an open question, but there is considerable talk that it has been, especially from disappointed would-be grantees.

The present system of rationing grant money among applicants also creates a potential conflict between the grant applicant's own interest and the public interest. This is because a successful grant beneficiary has a massive personal commitment to a line of innovation that has heretofore been successful at attracting funding and depends on the flow of grant money for his or her own support while continuing the line of innovation. In a circumstance in which there are multiple technical options for achieving the general goals of the research, the system offers no incentive for someone pursuing one line to consider a different, but possibly superior, line pursued by someone else. Each would-be grantee has an agenda that might or might not yield an eventual best solution to the problem that a number of people are working on. The technical problems involved in a general endeavor, such as achieving a future means of clean and safe electric power generation, can be so great that no individual or small group can be presumed to have the best solution when so much of the technology is highly underdeveloped. The would-be grantees become evangelists promoting their particular agenda, using whatever opportunities they have to promote their agenda before professional peers and the general public.[3]

Is there any superior alternative to the present system of administrative rationing of funds to support university inventors and innovators? Probably throwing university researchers on the mercies of conventional lending sources is not the answer, for if the peer-review system has a conservative bias, conventional private lenders would have even more of such a bias. Recent history suggests, however, that the private sector is not unwilling to take gambles in technology development, and with private funding, comes judgment regarding the commercial merits of the project. Judgments on the part of market-oriented funders may be no more privy to the future than foundation and government grantors but have the virtue of being yet another view of a project proposal. Many startup companies based on patents and few other assets and have zero sales have found private-source means of support and have succeeded without government support. Indeed, university researchers have diversified their funding options by forming startup companies. The relation between Stanford University's research and Silicon Valley startups is legendary. Stanford early set up an office whose purpose is to assist faculty researchers in the legal aspects of setting up companies. This model has been emulated by other universities.

Can Large Companies Invent?

The answer to the question, "Can large companies invent?" has to be unequivocally "yes." Name one of the prototype inventions of the later twentieth century and the chances are good that it will have originated in the research and development shop of some household-name corporation. The transistor is from Bell Laboratories; Kevlar is from DuPont Laboratories; the integrated circuit originated in the research arms of Texas Instruments and Fairchild Semiconductor; and the personal computer and laser printer originated from Xerox PARC. The list could be longer. This level of success is not hard to explain. First, the companies succeeded in staffing their research arms with people having inventive skills. This reflects, as noted above, that inventive talent can be described and recognized. Second, in many cases the corporate research and development shops offered very good conditions for inventively talented people to work, both as to laboratory facility availability and, often, insulation from senior management pressure to focus work on near-term development of salable products.

One of the major criticisms of large-company inventiveness in recent years has been over the issue of using the research and development arm merely to perfect and improve existing products rather than invent new products. No matter how successful a research and development manager is at defending the creative atmosphere of his division, a senior management desirous of focusing the research effort on the short-term will probably have its way eventually. It has been estimated that in the 1990s, the research and development arms of major US corporations spent 80 percent of their effort on existing product lines and only 20 percent on truly new lines of development. US public corporations, under pressure to produce favorable financial results in the short term, have been suspected of steering their research efforts in a short-run direction, or toward stage-two innovation. Some have seen this development as a threat to the future supply of prototype inventions.

Pressure stemming from the company's need to achieve near-term financial results may be part of any tendency to steer research and development in the direction of innovation related to existing products, but this is not the entire story. Suppose a company acquires a prototype invention. Creation of a business from such an asset often requires significant inventive effort. Making a business out of the telephone required many innovation-related inventions, such as switchgear. Exploitation of the prototype invention can require such a levy on inventive resources, both inside and outside the company, as to divert such resources from prototype inventing. Three examples show that truly basic research has not been totally given up by large corporations. As of this writing, Lucent Corp's Bell Laboratories is working on quantum computing. IBM is conducting research on the fundamental workings of the human brain with the ultimate object of using the knowledge gained to design a supercomputer that emulates the human brain. Intel is working on a project in which Raman spectroscopy is being used to detect single molecules. None of these projects has much near-term profit implication for the companies engaged in the work.

A Cycle of Prototype Inventions?

Against a recent history in which a supply of prototype inventions has initiated innovation streams with associated investment opportunities that have underpinned an era of relative prosperity, what can be said about the future of original technology? A future supply of new inventions does not seem to be a problem in general. For now, there is some concern that federal budget cuts in funding to granting agencies will impair inventive activity in universities. There is an additional problem growing out of business corporations' perceived shifting of their research efforts away from original invention work, but the market seems to be responding to this in the form of the independent invention laboratory. The real problem seems to be what it historically has been: financial support for inventors and stage-one innovators.

What does the economy do when presented with a host of opportunities such as resulted from the internal combustion engine, electric power, all the other new technologies that supported the automobile industry, the electrification of industry and homes, not to mention all the business and industrial possibilities of the computer and other devices based on the integrated circuit? It turns its attention and *resources*, real and financial, toward taking advantage of its opportunities. The outward signs are the appearance of numerous new companies that carry out the investing necessary to realize the potential of the new technology, or stage-two innovation. The first two decades of the twentieth century offered many examples of new firms, as did the last two decades. Schumpeter was very taken with the idea that entrepreneurial activity tends at times to appear *en masse,* as, for example, in the 1910s and 1990s, and evidenced by the organization of new firms.[4] This process is accompanied with a general loss of interest in prototype inventing and stage-one innovating. These are risky and compete for resources with innovation inventing, which works with markets and applications already in sight. Moreover, an invention that is perceived as threatening to an established market could be vigorously resisted when there remain plenty of attractive investment opportunities stemming from earlier prototype inventions. In short, the economy goes through a period in which it does not need prototype inventions. Interest in the results of original inventing returns only upon exhaustion of the investment opportunities that grew out of a previous slate of prototype inventions. Under some conditions, a substantial time interval elapses between the exhaustion of the investment opportunities from a slate of prototype inventions and the start of a new technology-based upswing based on new prototype invention, and, as the experience of the 1930s demonstrates, such a delay can cause economic trouble.

This argument points to *timing* as a necessary element in the likelihood of an invention's being developed to the point of commercial usefulness. What happens to inventions, including those of potentially prototypical magnitude, that occur with bad timing in the sense that no one is interested due to being fully occupied with innovating other prototype inventions? The probable answer is that stage-one innovation requires some one or some entity with enough interest to undergo the risk and expense inherent in this phase of innovation. Stage-one innovation can be viewed as a screen through which an invention or discovery must pass in order to become commercially relevant. Sometimes an invention languishes for years, decades, or lifetimes before a confluence of interest and enabling technology encourages its stage-one innovation.

The purely statistical evidence of such a cycle is not new, for it harks back to the long cycle theory that grew among the Austrian economists. The existence of long cycles was demonstrated by N.D. Kondratief, who described a long cycle of approximately fifty years' duration. This was developed by A. Spiethoff, who identified shorter cycles that are in effect superimposed on the long Kondratief cycle and account for the fact that the long cycle upswing is typically interrupted by setbacks.[5] Theoretical underpinnings of this idea have also appeared in the literature. J.R. Hicks recognized that the appearance of a major invention leads to the effective devaluation of a part of the existing stock of capital (obsolescence). The adjustment to the invention consists of the replacement of the obsolete capital with capital compatible with the new technology.[6]

The whole purpose of the concept of prototype invention is to separate out from the field of all inventions those that are of the highest order of magnitude as to their ability to induce fundamental change in the way people live. One way to look at potential to change the way people live is to relate it to the potential for monetary and other return that flows from the possibilities that an invention creates and to ask who captures these. In some, but not all, cases the capturer of some of the return will be the inventing entity, whether individual or corporate. The entrepreneurial effort by which the invention is brought to commercial success will share in the return. For people and businesses to want the new technology presupposes that there is a share of the return for them, also. For a business or industrial consumer, the return comes in the form of increased efficiency. For an individual consumer, the return comes in the form of more convenient living; indeed, a share in the total return potential is a necessary condition for commercial success. The return potential that accrues to the general public above and beyond the purely monetary return to commercial interests is what economists have called positive externalities (see Chapter 1). A prototype invention can be viewed as one yielding the largest and most widely dispersed positive externalities.

The postwar period has not so far produced its great depression, but its richness of investment opportunities based on technology definitely has produced some resistance to prototype inventing. The evidence of this is in the de-emphasis of this kind of inventing by the corporate research and development shops, the difficulty faced by the likes of Chester Carlson in shopping his invention, and the various examples of companies neglecting their internally produced inventions. The general prosperity of the era encouraged the formation of numerous new companies which were interested in and able to develop the possibilities of the handful of really big inventions that did appear. How long can the prosperity go on? Moore's law has described the dramatic increase in computing power and much of the original research that is taking place has implications for even more computing power in the future. Unless something interrupts this process, the promise for future investment opportunities based on technological development seems bright. No upswing lasts forever, however. Perhaps the best indicator of the impending end of a period of prosperity based a technology surge is widespread deterioration of demand for products that are lineal descendents of the prototype inventions that triggered the boom in the first place.

Revival of the Independent Invention Shop

The apparent lack of interest in inventing new things and exploiting the results on the part of large corporations in recent decades has, in the view of a handful of individuals, created an opportunity for the revival of the independent invention laboratory. This is not a new idea, for it envisions organizations resembling the laboratories founded by Thomas Edison, Nicola Tesla, Charles Kettering, and others before 1920. The idea is to assemble a staff of talented inventors and create the kind of cross-disciplinary atmosphere and other conditions to encourage creativity. Such an organization would concentrate on invention independently of any consideration of what to do with the results, or who was to do it. There would be no senior management to discourage investigations that do not relate to existing product lines or that threaten existing product lines. Such an organization would not obviously be in position to undertake stage-one innovation generally, but would presumably live on license fee income.[7] A stream of inventions from independent laboratories would fit well with the strategy of not inventing but improving the inventions of others, which GE has historically followed. It would get around the problem that is said to have inspired the establishment of the company research and development arms in the first place—the risk of not having new technology as a basis for replacing products which are on the declining end of their commercial cycle.[8]

The idea of taking inventing in-house was very much of a part with the overall corporate strategic thinking that was common early in the twentieth century: that extensive vertical integration could reduce overall risk of doing business by keeping the basic production facilities, for example, the auto assembly plants, working at full capacity, for these facilities had high fixed costs and required operation at full or near-full capacity to be profitable. This is no better exemplified than with Ford and General Motors, which came to dominate the automobile industry, and U.S. Steel (USS). At one time, Ford had divisions doing everything from producing iron ore to assembling finished automobiles, including at least one railroad. The same was true of USS, which not only had divisions producing the raw materials of steelmaking, but also operated a number of railroads and Great Lakes freighters. These railroads were operated as common carriers, but each had some role in the haulage of raw materials and the delivery of them to USS plants. GM accumulated divisions making many of the components of automobiles, and one of GM's early technology acquisitions (in 1919) was the Delco Corporation, the invention laboratory established by Charles F. Kettering in 1909.

Belief in the risk-attenuating benefits of vertical integration lasted well into the post-World War II era, but has been sharply questioned by corporate managements in recent decades. Breaks in the belief in industrial integration began to appear in the 1980s, somewhat under the influence of the Japanese model, which at the time appeared to be highly successful. In the type of arrangement that emerged, an automaker and, for example, a window glass supplier, would take equity positions in each other, the idea being that each would share in the prosperity of the other, and each would have a stake in the supplier-supplied relationship that existed between them. This sort of arrangement had disadvantages and gave way to outright outsourcing in the 1990s. The idea that then arose and now is widespread is that the market can be relied on to supply inputs from lines of business that are not of the large corporation's "core business." Outsourcing is often seen as a way of getting underperforming assets off the large corporation's balance sheet. It also eliminates problems that might arise from the supplying firm having different economies of scale from that of the supplied firm.[9] In these overall circumstances, it would not be surprising to see the emergence of a substantial independent inventing industry as an encouragement to following a strategy of buying new technology rather than inventing it.

Getting an Invention to Usability: Stage-One Innovation

Once an invention has been born, at least two elements are needed to bring it to commercial usefulness. They are entrepreneurship and capital, which together are necessary (but not sufficient) to transform the invention to a commercially viable product. The trait of entrepreneurship and ready access to capital can be combined in one individual, but that is not a common combination. In some cases, the pattern of an invention brought to commercial viability within a comparatively short span of years does not fit reality. There are cases where stage-one innovation can take multiple human lifetimes. The outstanding example is the solar cell, whose invention dates from the 1880s, but which became more than a laboratory curiosity only in the 1950s when silicon replaced selenium as the semiconductor.[10] What kept this stage-one process alive for so long? Possibly it was that the original invention inspired visions of unlimited and abundant power, and that this vision was sufficient to win the solar cell a place, albeit a tiny one, in the budget of one or more corporate research and development divisions from time to time. Indeed, the silicon breakthrough was by Bell Laboratories scientists.

The Entrepreneur

The term "entrepreneur" became introduced into economic theory in the late nineteenth and early twentieth centuries by several different economic writers. The sense of that term to be used here is that of Professor Schumpeter: "The carrying out of *new* combinations we call 'enterprise'; the individuals whose function it is to carry them out we call 'entrepreneurs.'"[11] The emphasis on the word "new" is added. Other writers did not place this restriction.[12] It has not been the purpose of this book to elaborate all possible fields of entrepreneurial activity, but the two brought into focus here are, in this author's opinion, particularly important from the standpoint of economic development. The entrepreneurial role is omnipresent in the history of bringing invention through the critical and risky stage-one innovation. The bearing of this risk is one of the characteristic roles of the entrepreneur, which suggests the trait of having a vision of the finished product and what could be done with it. This is something that the entrepreneur shares with the inventor and, in some cases but not all, the traits of inventor and entrepreneur turn out to be embodied in the same person.

Professor Schumpeter hypothesized that the primary (or long-term) business cycles are started by and reflect unusually intense periods of entrepreneurial activity. Entrepreneurship requires a combination of qualities of leadership, persistence, and ingenuity that makes them very much a minority in a population. Entrepreneurial activity, to be successful, must overcome various forms of resistance, as the advancing of a new combination often will be opposed by persons and organizations with strong vested interests in existing combinations. However, once some entrepreneur overcomes all the resistance and is successful, he or she will be emulated by others who will encounter less resistance than the pioneer. The reason is that the pioneer's work will have called attention to the advantages of new combinations such that they will receive increasing support that will offset the opposition. In other words, the pioneer entrepreneur generates a follow-on effect that accounts for the essential point that entrepreneurial activity comes in surges and does not occur uniformly through time.[13] It is easy to see how someone observing and writing in the mid-twentieth century might be impressed with this "bunching" phenomenon. Bunching of entrepreneurial activity is crucial to Schumpeter's view that the business cycle is a normal mechanism of economic growth in a capitalist system and not a contradiction of capitalism as Karl Marx held.

Schumpeter is famous for his making the entrepreneur the leading actor in the process of economic development, and no one has really developed a better explanation as to the human force behind the long business cycle. However, it is not absolutely clear to this writer what brings on the pioneer entrepreneur at irregular intervals of time. One can speculate. It may be that the pioneer is the first to overcome the resistance after he or she or others have attempted to introduce the new combination without success. This study offers an alternative hypothesis on this question. While entrepreneurs deal with a variety of problems that includes the exploitation of new technology, new technology is a very important starting point for entrepreneurial activity. As has been shown in earlier chapters, the appearance of inventions, especially prototype inventions, has a tendency to cluster in time. The nineteenth-century legacy is a superb example. The powerful new technologies that were available in 1900 represented combinations of recent inventions, and some important inventions of the more distant past that had been waiting for sufficient enabling technology to be developed and exploited. An entrepreneur requires something with which to work; and the entrepreneurial surge grows out of the technology surge. The answer to the question of what starts the long cycle thus lies in the answer to the question of why prototype inventions sometimes emerge from stage-one innovation in bunches.[14] It should be recognized that there are many fields of entrepreneurial activity of which developing new technology and reacting to massively changed regulatory climate are but two. However, the essence of Schumpeter's idea is that something foments a surge in the level of entrepreneurial activity, and for this "something," there are few better candidates than the emergence of new technology, or a basic change in the rules of doing business.

While it is not the primary purpose of this book to undertake extensive international comparisons, one such comparison fits well here. In order to capture the growth-producing potential of active innovation of a slate of powerful prototype inventions, it is necessary to have a favorable climate for the formation of new companies, for these are often the most important innovators. Ease of starting new companies is a necessary (but not sufficient) condition favoring entrepreneurship. Other important conditions include a court system that will enforce contracts, and ease of access to capital. It is difficult to exaggerate the importance to a country's growth prospects of conditions that encourage the entrepreneurial energies of its population.

There is a huge variability among nations as to how well these conditions are achieved. This comes out vividly in the World Bank's recent report on its annual "Doing Business" survey.[15] This survey produces an overall index of the general business climate based on eleven criteria, including "Ease of Starting a Business." The results are presented as comparative rankings, in which increasing scores correspond to increasing difficulties of conducting business.

In general, the OECD countries offer comparatively favorable conditions for business. In overall ease of doing business, the US ranks third, after two much smaller economies, Singapore and New Zealand. The worst conditions for entrepreneurship are heavily concentrated in Latin America. In the category of "starting a business," the United States ranks fourth most favorable, and ranks first among the largest economies. For contrast, the so-called "Bolivarian Revolution" countries of Latin America, Venezuela, Bolivia, and Argentina rank, respectively, 172, 140, and 114.

What characteristics make an entrepreneur? On the basis of data on people who have started businesses from thirty-four countries compiled by the Global Entrepreneurship Monitor research consortium, entrepreneurs tend to be relatively young: under thirty-five years of age and otherwise employed. The traits found in these people include the vision to imagine beyond what seems possible, the ability to combine resources in unconventional ways, and the confidence to pursue their ideas to fruition. Entrepreneurs are also distinctive in their ability to cope with the psychic and financial consequences of uncertainty. The group includes many who would assert greater control over their work circumstances than can be had as an employee, especially of a large organization.[16]

The ranks of independent inventors include examples of inventors who have undertaken stage-one innovation. An independent inventor does have the freedom to innovate provided he or she can find an income while innovating. Edison had the resources of his own invention lab, an organization supported by investors. Charles Goodyear and Chester Carlson, both working essentially alone, struggled at times. Goodyear, especially, was able to raise investor capital several times before his eventual success with the vulcanization process, but prior to his success, rubber products he sold all failed commercially because they became sticky when warm, and each failure tended to dry up his sources of capital.

The rise of the corporate research and development division had the effect of shifting the focus of inventive talent to improving existing products and away from *de novo*, unconstrained inventive activity. This was a shift in emphasis rather than elimination of prototype inventing, for the four most important prototype inventions of the postwar period come from corporate research. A question for now is whether inventors who work for corporate research and development divisions are cut off from engaging in entrepreneurial activity by constraints imposed by their employers. Stated slightly differently, what does an inventor, who is also a company employee, do with an invention in whose future he believes strongly, but in which his employer takes no interest? The question is especially cogent if the employer perceives the invention to be a threat to its existing interests.

The answer comes from the experience of history. If the employer is not interested, the employee has the option of resigning and forming his own company. The problem becomes one of financing the new enterprise, or the same problem faced by the independent inventor. Indeed, the overall record does not provide conclusive evidence that inventors who are company employees are, as a group, not interested in entrepreneurial activities. Some individuals are interested and some are not. For those who are not, there is the additional option of changing jobs, assuming the employer does not attempt to suppress the invention through patent rights. If the employee can find a company interested in the invention in question, then no one is better than an inventor at the task of figuring how to invent around hostile property rights.

This leaves a question of under what conditions can an inventive company employee with an interest in entrepreneurship make the transition to independent inventor. The financial constraint suggests that one condition be a period of history in which there is high general prosperity. A second condition is a period characterized by a high rate of technological flux. The former condition suggests availability of financial capital and the latter suggests investors willing to make bet on a new technology. The 1980s and 1990s meet these conditions, but there are good examples from earlier in the postwar period. Witness the founding of Intel Corporation by Robert Noyce and Gordon Moore in 1968. Both had been employees of Fairchild Semiconductor, itself a postwar startup. The first two decades of the twentieth century met the conditions. Charles Kettering had been an employee of National Cash Register when he co-founded the Delco Corporation in 1909 based on his invention of the automotive electric starter.

Large Companies and Entrepreneurship

There are numerous examples in which companies have failed to profit from their own successful in-house inventions. To have missed profiting from an invention that subsequently turns out to have been a prototype invention suggests that somewhere along the line, no one in a position of authority in the inventing company had the vision to see the future of the new thing. It is not realistic to say that *no* one at all saw the future, for the fact that some other organization was interested in undertaking stage-one innovation, as Texas Instruments did with the transistor, suggests that someone saw some possibilities. It may be that the Xerox case offers some insight. The fact that Xerox missed out on its own inventions has been well documented. Moreover, this kind of failure was far from confined to Xerox. Perhaps the thing that justifies the attention that this case has received is that Xerox's weak exploitation of its PARC inventions came less than fifteen years after the culmination of the company's own successful entrepreneurial experience with the stage one development of Chester Carlson's copier technology. Having so recently had a successful entrepreneurial experience, one might think that the company that did it would be good at it, but the record says otherwise. One is entitled to ask "why?"

There may be no definitive answer to this question, but the circumstances were these. First, Xerox had a change at the top in 1968 when Joe Wilson, who had been the entrepreneurial spirit behind the development of the plain-paper copier, was succeeded in the CEO position by Peter McCollough, who was widely regarded as a bold manager in his own right on the basis of his work in selling the early copier models to business. The other change that took place about the same time was that the company, after extremely rapid growth based on the market success of its copier products, found itself without the tools of internal control needed to operate a large business. The result was a chaotic quality in the operation of the company. The response to this was the organization of staff functions and the hiring of people to fill them. In short, Xerox was forced to adapt quickly to its own newly attained size. Many of these changes were in place by 1970, the year PARC was organized.

One may speculate that the transition into a large corporation with staffs, vice presidents, and other trappings of bigness effectively kills the entrepreneurial capacity of a company.[17] Where before, crucial decisions had been made by individuals, they became the province of committees in the expanded organization. It is a curious thing that people, who as individuals are bold and imaginative, can compose committees that are nothing if not cautious and risk averse. A quotation sums up this idea: "You may take the most gallant sailor, the most intrepid airman, or the most audacious soldier, put them at a table together—what do you get? *The sum of their fears.*"[18] In view of how much of the modern corporate organization has been borrowed from the military, it is not much of a stretch to place this observation into the context of a corporate bureaucracy. The very evolution of the organization may have suppressed its entrepreneurial capacity even when there were individuals of this spirit still present in the company. Apparently, no one involved in these decisions foresaw the future of the personal computer, and the laser printer may at first have appeared to offer no real advantage over the existing copier technology. It remained for others to dominate these markets.[19]

The Capitalist

The problem of financing the stage-one innovation follows the problem of supporting the inventor while inventing. This problem is perhaps the most daunting in the case of the independent inventor who undertakes to carry his own discovery through its stage-one innovation. The example that comes to mind here is Chester Carlson. Carlson started the work that ultimately led to xerography in about 1933, working in his kitchen and supporting himself with his day job with a patent attorney's firm. At the time he gained his first patent, in 1937, he had moved his operation to small rented quarters (at the behest of his wife), and went on in this vein until just after World War II, when he was able to attract corporate interest and support from the Haloid Corporation of Rochester, then a producer and distributor of photographic supplies, that was looking for a basis for expanding its business. Carlson's was a clear case of the inventor and the entrepreneur being the same person.

When the invention comes from a corporate research and development organization, the problem of stage-one innovation financing is, in theory at least, one of drawing support from the stream of corporate cash flow. The realization of this source of financing of stage-one innovation presupposes that the corporation is interested in developing the invention, which is not always the case, as has been noted. It often happens that even a good and promising invention ends up being stage one innovated by an organization other than the one that invented it. Examples of corporate assumption of the stage-one-innovation burden from internal-source inventions certainly include DuPont's successful bringing of such discoveries as Nylon, Orlon, and Kevlar to market. A contrasting case is that of AT&T, which sold the rights to the transistor for a price that almost guaranteed that another organization (in this case, TI) would develop this invention.

A new small company engaged in stage-one innovation on the basis of some invention or discovery sometimes can be financed by venture capitalists (VC). These are people or groups that manage pools of capital raised from investors. They take equity positions in startup firms that they judge promising and make their money by selling stock in the startup to the public in an initial public offering (IPO). In order to make any money this way, the VC companies' managements must first make a judgment as to which technology-based startups are sufficiently promising. They spread risk by taking positions in a number of technology startups that pass their quality screens. The hope is to sell their equity interests at a considerably higher price than they paid. The VC companies, thus, shoulder part of the stage-one risk. The drawback to VC financing from the standpoint of the technology startups is that the financing is of limited duration, as the VC companies can grow impatient to realize their returns by an IPO. VC financing, thus, places new companies under pressure to complete their stage-one innovation in order to make the company salable to the public, which can be a problem. For this reason, entrepreneurs often will resort to any other financial source within their reach in order to avoid VC financing.

There has been a variety of other startup financing sources. At very early stages of a startup's development, capital can come from such sources as second mortgages on homes and even credit card debt. A number of private financing sources frequently get used, such as family and friends. There are a number of firms specializing in the business of attracting investors to an enterprise. The shares of these startups are typically unlisted and require sufficient interest on the part of the investor to persuade him or her to own an illiquid asset. Another possibility is research contracts from organizations intrigued with the innovation in progress. Rudolph Diesel was supported by corporate investors interested in the possibilities of the technology set out in his patent of 1892. One of these was Krupp, which was interested in a thermally efficient engine that could use blast furnace gas as fuel.

The Role of Government

As noted above in Chapter 6, the federal government has assumed a major role in the financing of stage-one innovation of some inventions that later proved to have significant commercial potential. It is the common pattern that the government supports stage-one development of technology that serves some government purpose seemingly unrelated to commerce, often but not always related to national defense. From this activity a number of very important results for the general economy emerged, including the Internet and much basic development of the components of modern computers. The importance of this type of government activity is that it supports stage-one financial support of projects whose apparent lack of connection to the interests of the private sector probably mean that private support would never have happened.

Stage-Two Innovation

The Problem

Inasmuch as a new technology can have the potential for application in markets in addition to the markets that were in the minds of inventors and stage-one innovators, stage-two innovation opens opportunities for any entrepreneurs, individual or corporate, who might see a business opportunity in an application not necessarily contemplated by the original inventor. This can be a corporate research and development shop that invented in the first place, or bought technology invented by someone else, but frequently it is an entrepreneurial enterprise working on a commercial idea for the technology in question.

A new invention or discovery emerges from stage-one innovation as a viable commercial product in some market. This first market tends to be an existing market wherein the new device can substitute advantageously for some previous device that performed the same function. A leading example is the transistor, whose earliest applications included use in a radio amplifier circuits as a direct replacement for vacuum tubes. While preexistence of a market may offer a clue as to the first use of an invention after stage one, it is probably more accurate to expect the device to go to the market that is most willing to pay for it. The development of the modern automobile offers numerous instances of this, inasmuch as first markets tend to be at the high end in the finished product market. Such inventions as electric starting systems went first to the Cadillac, a luxury model even in 1909. Applications down the luxury scale took place as design improvements and rising production volume made the device cheaper to produce. Thus is suggested a rule determining the path of an invention's application during the stage-two innovation process: at any point in time, the next application will be that wherein the cost of the new technology will have the least impact, proportionately, on the total cost of the assembly to which it is applied.

Silicon is a present-day example of a material whose markets have so far expanded according to this rule. Pure silicon is costly to produce even though it is among the largest-volume elements in the Earth's constitution. Pure silicon has to be won from sand, the oxide of silicon, at considerable cost. The largest market for pure silicon until recently was as substrate for integrated circuits. In this, silicon is used in the form of a chip, generally less than several inches in diameter. As a component of larger devices, such as computers, the cost of the silicon is a relatively minor component in percentage terms. The coming big market for silicon is in solar panels, wherein the silicon is a major cost element. In this market, the cost of silicon has so far proved an inhibition to the growth of the solar energy market. The demand for solar panels as sources of energy for residences and commercial buildings has so far been confined to very high-end structures. At the high end, the cost of the silicon sheets going into solar energy collectors has proportionately less of an impact on the total cost of the structure than it would have on the total costs of less costly structures. Expansion in the use of solar energy will take place in response to either or both of two kinds of development: a decline in the cost of producing silicon or the substitution of a cheaper substrate for silicon in solar collectors.

One way to look at the transition from stage-one to stage-two innovation is as a point at which the emergence of the invention as a viable commercial product greatly reduces the risk associated with further innovation. As riskiness diminishes, the field of possible financial backers can expand beyond those individuals and organizations whose initial faith in the value of the invention led them to shoulder the risks of stage-one innovation. This is the point at which large corporate research and development organizations might be expected to take an interest. The corporate parents of these organizations are equipped with marketing and distribution arms that can realize the commercial potential of the product. It is not uncommon for established companies to rely on buying newly developed technology rather than inventing it in-house.

Established Companies and Stage-Two Innovation

A number of observers have raised the question of whether a large organization can profitably see its internal inventions through both stages one and two of innovation. The answer is not clearly "yes," for the record is spotty at best. The idea that animated the establishment of the corporate research and development arm at the turn of the twentieth century was that it was too risky to depend on independent inventors to provide inventions that could be developed into commercial products, and that taking the inventive function in-house was another form of vertical integration that would reduce that risk. The subsequent success of some of the research organizations at producing prototype inventions attests to the wisdom of taking the inventors in-house.[20] However, the anticipated stream of highly profitable products resulting from the inventions produced, generally speaking, failed to materialize. Examples are easy to cite. AT&T was not the principal beneficiary of Bell Laboratories' inventions of the laser and the transistor. Xerox did not become a major player in the personal computer and related industries that grew out of its PARC staff's efforts. Eastman Kodak was very much a latecomer to digital imaging. Examples are not confined to recent decades. Charles Kettering was an employee of National Cash Register Company at the time he and several others invented the automobile self-starter.[21] While the starter's being an off-hours project means that it was not really an NCR internal invention, the company did not show any real interest in the invention and apparently raised no objection when Kettering and others left to found Delco.

There may have been at least one success story in the overall picture of companies' not profiting from in-house inventions or discoveries: DuPont, which successfully exploited discoveries of its laboratories such as nylon, Teflon, Orlon, Kevlar, and others. One factor may have encouraged success in this case. The people who staffed DuPont's laboratories, generally speaking, were trained in the discipline of polymer chemistry. This skill set would have been highly compatible with the company's general goals and the abilities of DuPont's established sales and marketing organizations.

As for the poor overall record of exploiting inventions originating in-house, one is entitled to ask why. There are several possible reasons.

- Cash flow is likely to be deemed insufficient to advance on all the fronts presented to senior management, whether the opportunities come from the reasearch and development department in the form of new inventions or through proposed acquisition. In some cases, the inventors can simply overwhelm the parent company's ability to undertake stage-one innovation. Many companies had (and have) a "rule of thumb," which said for every dollar spent on researching a new technology, ten dollars are needed for what this book calls stage-one innovation, and a hundred dollars is needed to bring the product to the customer. Whether this is or is not realistic, the perception of it creates a real barrier to the stage-one development of at least some inventions.

- If cash flow has to be rationed, then a choice has to be made regarding which opportunities to fund. Standard analytical techniques for budgeting capital are less than useful, for they require future projections of cash flows from an investment, and if the line of investment involves a new product, the analysis may suffer from inability to foresee the revenue potential of the product. The risk attending a new invention is multiplied if its stage-one innovation requires borrowing money (by way of extending internal cash flow). Risk and uncertainty are things that managements are paid to deal with, and managers will often react to them by avoidance when that seems like an option. There is often a bias against the pursuit of something new. Given its preference, the typical corporate senior management will opt for stability and shy away from anything that appears the least bit disruptive of stability.

- A newly invented product may be perceived as a competitive threat to an existing line of products. One of the classic first reactions of an established organization to a perceived threat to its settled order is denial. A prototype invention may not support the improvement of existing technology and can be threatening. Kodak does not appear to have been interested in pursuing digital imaging after having pioneered the technology, and one can suppose that its senior management underestimated digital technology's threat to its existing business that was based on silver halide chemistry. It is not possible to know how important the threat factor from new inventions has been without intense investigation of the record of individual cases of failure to profit from in-house invention. The logic of trying to hold a threat off the market seems suspect at any rate, for if the new technology offers a serious threat to an existing market position, then someone sometime is going to develop it, and the threat will become fully realized. This is especially true if the inventing company yields to the temptation to gain a revenue stream through licensing its newly invented technology to others.

- In spite of the aforementioned reasons why large companies appear to have trouble exploiting their in-house inventions, there have been many people in these large companies who have been disturbed at the idea of "letting a big one get away." One idea for coping with the big company's dilemma as to what to do with a given invention is to go into the venture capital business. In this, the big company forms a division whose activity is to take equity positions in startup companies founded by former employees with promising inventions. The intent of this kind of arrangement is to free the inventor to seek capital backing on the outside. In this way, the large company retains an interest in what happens, but the risk inherent in stage-one innovation is shared with outside investors. There were a number of such arrangements made in the late 1990s, but the ensuing technology recession appears to have dampened interest in such arrangements on the part of some companies.

- One of the problems that has been encountered by large companies whose research and development divisions have produced prototype inventions comes from the timing of these inventions: there is no apparent law that says that the new invention will appear with optimal timing in relation to the product cycles for existing products. The original idea behind taking inventing in-house was to increase the chances that something new would be in the wings when needed, but what was to be done with new things that did not conform to this timing? In so many instances, the answer to this question came in the form of settling for the possibility of a stream of royalty income resulting from the licensing of the new idea to others. The question that has been asked is, "Can large companies invent?" Perhaps an equally interesting question would be, "Do they really want to invent?" The GE history (see below) seems to suggest that a strategy of buying new technology can be a profitable alternative to inventing.

Buying versus Inventing Technology

Large companies that are well established often have the option of buying into a line of business based on new technology. This can be in the form of an outright acquisition by the established company of the firm having the new technology or some other arrangement. One example of entering a new line of business short of outright acquisition is a partnership in which the established company undertakes to market and distribute the product of the smaller technology company. This kind of arrangement can be advantageous for both companies. The innovating company may have put all of its financial resources into the technical problems of stage-one innovation and later finds itself with a viable product and no capability to undertake and finance stage-two innovation. The established company can offer a working marketing organization, which can fill in the innovating company's deficiency. This would describe the relation between Immunex, the developer of Enbrel, a bioengineered drug found to be effective in the treatment of rheumatoid arthritis, and Wyeth (formerly American Home Products, AHP), the established pharmaceutical company that markets Enbrel. AHP acquired an equity position in Immunex and at one point had a majority interest, but this has since been reduced. When a company, such as General Electric, buys another company in order to possess itself of a desired technology, there is always the question of whether this was less costly than what would have been spent on stage-one innovation by GE had it invented the technology. There does not seem to be an obvious answer.

Another variation on acquiring, rather than inventing, technology is the practice of acquiring license rights to a technology of interest. General Electric's formal research program dates from the turn of the twentieth century, and was the model for several other large companies' organization of their research and development divisions; yet, it is not easy to think of an example of a truly prototype invention from GE. The company has prospered reasonably well by licensing technology from inventors and improving it. GE purchased Austrian technology for the tungsten filament for the incandescent lamp and improved it to the point where tungsten became the universally preferred filament material. This pattern was repeated for a number of the technologies in which GE enjoys a leading position today: X-rays, CT imaging, and jet engines are examples. This strategy regarding technology makes sense when there is confidence that the world will produce a supply of new technology on a sufficiently regular basis.

The Position of the Technology-Based Startup

If large companies have trouble exploiting their own inventions, is there any magic that helps small companies to be more successful at such exploitation? The short answer is "yes," for technology-based startups are inherently one-product firms who have to stage-two innovate for as long as they can survive.[22] The alternative to pushing stage two to a successful conclusion is corporate death. Big established companies have options, but frequently, technology-based startups do not.[23]

In bringing new technology to the marketplace, the role of new firms is very evident historically. As Professor Schumpeter noted, ". . . in general, it is not the owner of stage coaches who builds railway lines."[24] The diesel engine provides an apparent counter-example to Schumpeter's observation. Rudolph Diesel's stage-one innovation lasted approximately from 1892, the year of his patent, to 1897, the year of his first viable engine. In this period, Diesel received substantial support from several firms that were established builders of stationary steam engines. Apparently, these companies saw no problem with serving their existing markets with a more efficient power plant even though doing so implied obsolescence of their existing line of products. Does the diesel example really constitute a major exception to the generality that it is new small firms that bring new technology to its commercial potential? Perhaps insight can be had by reflecting on who produces diesel engines now. The list includes General Motors, Cummins Engine, and Caterpillar in the US and Maybach in Germany, Sulzer in Switzerland, and others. These were small companies once, and most were creatures of the twentieth century. All but one was and is associated with the diesel engine as a land transportation prime mover. In other words, new firms appeared as the diesel's application moved beyond the initial concept of a more efficient stationary engine. On balance, it would seem that today's diesel evolved from the activities of new companies. Of the original backers of Diesel, one of the few survivors is Gebrüder Sulzer, the Swiss-based engineering firm associated with marine engineering and stationary engines.

In spite of the Kodak example, it may be that valid examples of exceptions to the small-firm new technology generality can be found in the camera industry; for Nikon and Canon, long established in conventional camera manufacture, became major players in the digital imaging business. Apart from this and several other examples, it appears that there is a strong bias in favor of development of truly new things by newly formed companies.

In view of the prominent role of startup companies in the stage-two innovation of technology, it is legitimate to ask what conditions are favorable to the organization and sometime success of these organizations. There is no mystery here. The list of necessary conditions includes, but is not necessarily limited to, a strong independent judiciary that will enforce contract and private property (especially intellectual property) law, a wide and deep capital market to which there is ease of access, a technically educated population, and a low-tax environment. In looking around the world, it is surprising how few countries actually meet even these conditions.

Diffusion of Technology

The huge variability in the time it takes for a prototype invention to find the full extent of its potential in an economy has already been detailed in these pages. The time duration of stage-one innovation is, as already stated, often dependent on the progress of other lines of technology. Indeed, much of the technology blossoming that took place before 1910 rested on the recently developed capability of the machine tool industry to achieve very close tolerances in large quantity. The time duration of stage-two innovation is also highly variable and, in some cases like that of the solar cell, realization of the potential market still depends on the progress of related technology. However, the problem of stage-two diffusion is more complex than that, especially where prototype inventions are concerned. Because of the prototype invention's creation of possibilities not previously foreseen, its results can lead to industrial developments not anticipated even by the inventor. Entire new industries can emerge in the course of stage-two innovation stemming from a prototype invention. The prototype invention thus affects the dynamics of industry itself, not just a particular industry.

The economics profession has given some attention to the technology diffusion problem. In general, recent research has dealt with the problem of diffusion of technology among a well-defined group of industries. Paths of diffusion tend to follow a sigmoid (or S-shaped) pattern. The rate of diffusion is generally found to depend on a number of variables, especially the profitability of the new system compared with that of what it would replace.[25] In general this work rests, at least implicitly, on the assumption that there exist well-defined industries that can potentially use the new technology to their advantage. If the industrial structure is well defined and stable in time, than the corollary assumption is that the number of firms and/or establishments that can use the new technology is known for each given price. Given these assumptions, it can be said that the work does not recognize the dynamic character of the very structure of industries, which is in play when the new technology stems from prototype inventions.

The writer who has fully recognized the threat to existing industries that new technology can bring is Professor Schumpeter. This threat is embodied in the concept of *creative destruction*, wherein a new combination competes with an established combination and eventually causes its obsolescence and abandonment.[26] The working of this idea can be seen readily in the demise of the line-shaft. Even in instances where the new technology results in the creation of human possibilities not envisioned before the new technology, there is likely to be found a trail of destruction of earlier combinations. For example, the introduction of the automobile and its infrastructure of good highways created a revolution in land transportation possibilities and eventually contributed massively to the virtual abandonment of railway passenger transportation in the US.[27] The personal computer has satisfied wants that did not exist before its introduction, and in many instances, it is not possible to identify any directly-caused destruction. This is an idea of "nondestructive creation," advanced by Amar Bhidé of Columbia University's Business College.[28] The creative destruction idea highlights an essential difference between inventing and innovating, especially in stage two. It is that stage-two innovation brings an invention to the point where its threat to an existing order (if any) comes into focus, and it meets resistance from those interests subject to being creatively destroyed.

The standard textbook teaches that technological innovation results in the shifting of one or more supply curves upward and to the right. The issue of how much time this shift takes assumes that the shifting process will be allowed to progress uninterrupted. During a period of high technological flux, there is no guarantee that this condition will be met. Any shifting of the supply curve is subject to being interrupted by the appearance of an even newer technological option. Again, there is the example of the line-shaft, wherein the possibility of replacing the entire factory design that it implied appeared even as its operation was being improved by the simple replacement of its prime mover. It would be interesting to do a study whose object is to discover whether there is a systematic increase in the probability of a line-shaft-like history when there are conditions of relatively high technological flux and a relatively long period of potentially applicable technological innovations.

Notes

1. Carlson worked for a law firm that specialized in patent work, and, for part of that time, he himself was qualified as a patent attorney. This vocation was not totally unrelated to an inventor's needs.
2. 1932 was the first year in which more patents were granted to corporations than to independent inventors in the US.
3. The traditional way for a new faculty member to gain tenure called for publication of papers in scholarly journals. So many were successful at this that academic departments devised a system of grading publications based on the number of citations a given published paper received. One effect of this system was to encourage young faculty to write only in currently "hot" subject areas, and this augmented the conservative cast of the peer review system.
4. Schumpeter, *The Theory of Economic Development,* p. 215.
5. Haberler, Gottfried, *Prosperity and Depression*, pp. 272-274.
6. Hicks, *Capital and Growth*, pp. 294-305.
7. One such recently founded enterprise is Invention Science, organized by Nathan Myhrvold and Edward Jung, former Microsoft executives. See Schwartz, Evan I., "Sparking the Fire of Invention," *Technology Review,* Vol. 107, No. 4, (May 2004), pp. 32-40. There are at least a half-dozen of these invention laboratory startups.
8. Xerox justified the establishment of PARC on the grounds that it was basically a one-product company and saw the digital world as a logical direction for its growth. A selection of products from independent invention shops might provide an alternative to setting up an in-house research entity for companies in this situation in the future.
9. For a discussion of the theoretical issues surrounding the "make or buy" decision, see Williamson, Oliver E., *Markets Hierarchies: Analysis and Antitrust Implications,* (New York: The Free Press, 1975), pp.82-105.
10. This is not totally true. Even the inefficient solar cells of the 1930s found a market in photographic exposure meters.
11. Schumpeter, Joseph Alois, *The Theory of Economic Development,* p. 74.

12. For example, Alfred Marshall treated "entrepreneurship" as synonymous with "management."

13. Schumpeter, *The Theory of Economic Development,* pp. 228-230.

14. "The specific problem of leadership arises and the leader type appears only where new possibilities present themselves." Schumpeter, p. 88.

15. International Bank for Reconstruction and Development (IBRD), *Doing Business* survey, (October 2007).

16. Gerena, pp. 17-18.

17. The loss of entrepreneurial spark in the course of company growth and success has occurred in many cases other than Xerox. For example, Eastman Kodak grew to dominate the market for silver halide-based film under the lead of George Eastman, an entrepreneur's entrepreneur. After Eastman's death in 1932, the company enjoyed four decades of high prosperity in its chosen market. During that time, it rejected Chester Carlson's plain paper copier invention and instant photography. It tried to enter both the copier market and the instant photography markets belatedly and half-heartedly, and gave up on both eventually. The company had attained such a state of comfort in its basic business that it fell into denial of the potential of digital photography in the 1990s. The company is now encountering difficulties because of its tentative approach to digital photography, which displaced silver halide technology very rapidly. In spite of its entrepreneurial beginnings, the company lacked the flexibility to benefit from any of these new technologies.

18. Churchill, Winston, "On the Chiefs of Staff System," Nov. 16, 1943. As quoted in MacMillan, Harold, *Blast of War,* (New York: Carroll & Graf Publishers, 1967, 68), p. 352.

19. Schumpeter makes the point that entrepreneurship ". . . is not a profession and as a rule not a lasting condition." Schumpeter, *The Theory of Economic Development,* p. 78.

20. Many of the successes can be found in the chemical industry.

21. Kettering had a strong inventive reputation within NCR. He became interested in automotive-related problems when he became involved in a neighbor's project of building a Packard from a kit in his barn; he had consulted Kettering on the electrical system. The eventual result was the power starting system, built with the help of several other NCR engineers working in off-duty spare time. NCR knew what he was doing. Apparently they took no serious interest and gave Kettering no difficulty when the latter resigned to found his own invention laboratory.

22. Even as its copier success unfolded in the 1960s, Xerox remained an essentially one-product company. In this respect, it was more like a technology startup than the corporation that it later became.

23. This line of reasoning tempts one to suppose that lack of choice is an important element in any list of the traits of an entrepreneur. Professor Schumpeter does not seem to have discussed this possibility.

24. Schumpeter, p. 67.

25. For a collected description of this work, see Waterson, Michael, *Economic Theory of the Industry,* (Cambridge: Cambridge University Press, 1984), pp. 152-164.

26. Schumpeter, Joseph A., *Capitalism, Socialism, and Democracy* (New York: Harper Perennial, 1950), pp. 81-86.

27. The New York Thruway was opened between New York City and Buffalo in the summer of 1958. Over the course of that summer, ridership on the New York Central Railway (serving the same two end points and intermediate points with what had previously been regarded with a quality service) fell by one-third.

28. Bhidé, Amar, lecture given in Nov. 2004, as cited in Gerena, Charles, "Nature vs. Nurture," *Region Focus,* (Fall 2005), p. 19.

14

On Choice among New Technologies

The Problem

If anticipating the precise direction of future technology was a treacherous business at the turn of the twentieth century, the same is at least as true today. Research into the question of how choices among competing new technologies are made has turned up a few principles, but so far, there are no definitive theoretical guides as to how the technological future will unfold. Apart from theory, the historical record is of great interest, not because it provides definitive clues as to how to view the future of technology from the standpoint of a given instant in history, for it does not; but because it illuminates the kind of influences that can affect the course of technology.

One of the more interesting lines of thought bearing on how technologies are chosen comes from the work of W. Brian Arthur and others. The problem that Arthur addressed was that of what influences the choice among two or more technology options that are developed to the point of being clearly superior to an established technology. The choices are market and non-market influences. As will be seen in the discussion below, this problem statement fits well with some historical cases, such as which power plant was best suited for the automobile in the early twentieth century. Today, the choice problem appears somewhat different. First, the impetus for change in some major problem areas such as how electric power is to be produced comes from political as much as strictly economic considerations. Any new technology in this field has to satisfy constraints imposed by the Clean Air Act and its amendments in the US and equivalent legislation in other countries. Second, there are few instances in the world of commerce in which the technology to replace an established technology is developed to the point where its commercial characteristics are clearly discernable. If public needs force consideration of a technology before its commercial characteristics are fully known, market considerations have little or no role in the decision process. In computing, for example, technologies with promise for dramatic improvement over today's computers, such as quantum computing and nanotube computing, are today little more than laboratory re-sults. The immediate choice problem is, therefore, how to allocate the research and development effort. The competition is not between or among alternative proven technologies, but for research and development funding.

The problem of choosing among multiple technologies that are still in stage-one innovation, such as futuristic means for producing and distributing power and for computing, is reminiscent of the problem of choosing among proposed designs for weapons systems faced by the Department of Defense. These can involve elements that have never been proven in practice and sometimes depend on technical breakthroughs of highly uncertain timing. Prior to the 1960s, the practice was to let each proposed design be built and then to test the results against each other as to performance characteristics. As weapons systems be-came increasingly sophisticated and costly to develop, this approach became less and less satisfactory. If two systems proved to be viable, the result often was that different armed services, e.g., Navy and Air Force, would often claim differing performance requirements to justify production of more than one system. Any desire for standard designs to be used by more than one service arm threw the choice problem back to that of choosing based on proposed designs.

The Department of Defense attempted to respond to this problem by vari-ous changes on contracting procedures. A number of such changes were made during the 1960s, and they then included a contract definition phase and in-centive contracts. The former was aimed at resolving whatever uncertainties could be resolved prior to commitment to specific items. Incentive contracts were designed to replace the earlier form of cost-plus contracts by rewarding a contractor for performance in excess of what it had promised and penalizing the contractor for costs in excess of bid estimates. Defense contracting procedures have been tinkered with extensively since the 1960s, but it is not clear that the basic problem of choice among technologies that are only partly formed has been solved. Ultimately, the problem arises for one reason: at a given instant in time, the future is uncertain.

Even if the government had been able to develop a choice system that would reliably pick the best among alternatives that have no reality beyond early-stage design proposals, applicability of such a system to the general problem of choos-ing among technologies in early stage-one innovation is doubtful. The reason is that when government agencies are faced with such choices, there is often an element of hurry that can influence the decision and result in a non-optimal pattern of path dependence as suggested by Dr. Arthur. Recall the choice of the light-water reactor as the US power standard by the AEC. In spite of an an-nounced policy of patiently amassing experience with the various alternatives, the AEC made the LWR standard in response to pressure from the Navy and from private sector firms that wanted to be able to develop a standard power plant package that could be marketed to domestic and foreign customers. One can conjecture that the more pressure there is to make a choice, as is often the case with defense systems, the more room there is for a sub-optimal choice. When there is little concentration of pressures to make a choice among early stage-one systems, the government can still exert influence by its allocation of research support funding. Any favoritism in funding that develops is itself a subtle form of pressure and may not be conducive to reaching the best solution to the problem.

What the Theoretical Literature Says

Theoretical work on the question of choice among competing technologies has been associated with the name of W. Brian Arthur and others.[1] In this literature, the problem is stated as follows. There are two or more new technologies that can be defended as being superior to an existing technology. An often-offered example is the Otto-cycle (gasoline) engine versus the steam engine or the electric motor as automotive power sources. The adoption of the new technologies takes place over a period of time as users of the old technology (horses in the auto example) retire their horses and buggies and choose among the possible replacements. At the time the new technologies appeared, none could be proved clearly superior to the others as to operating costs, repair costs, and other costs, and all are new enough that the future patterns of improvement are not clearly foreseeable.

How each individual technology adopter chooses among the replacement choices is influenced by two broad kinds of considerations. The first of these has to do with whether the new technologies can be adopted under conditions of increasing, constant, or decreasing costs. By increasing costs, it is meant that costs of adopting a technology rise for all adopters after some number of adoptions short of that technology's becoming the sole technology in use. By decreasing costs, it is meant that costs borne by all adopters will fall with increased numbers of adoptions of that technology to the point where that technology option has a monopoly of their entire market. The neutral case is when an additional adoption has no effect on cost borne by all preceding adopters. In the case of increasing costs, the two technologies will permanently coexist in the market, as neither will enjoy a cost advantage over the whole range of the market. If conditions of decreasing costs are present, there are several possible outcomes, of which one is that one or the other of the new technologies will take over the market in its entirety. In decreasing cost situations, the history of the process does make a difference, for with each incremental adoption, costs for *all* adopters decline. The pattern of adoption of a technology under these conditions is referred to as *path dependent*.

The second kind of influence over the choice of technology is random events, seemingly unrelated to the technical merits or economic characteristics of the technology under consideration. These events can take any of a number of forms, such as a chance advantage of marketing talent favoring one technology over the other, or some engineering consideration arising from limitations imposed by inadequate enabling technology that creates a preference for one of the competing technologies over the others. If conditions of decreasing costs are present, then these outside events can push adopters toward one choice over the other. Once one of the new technologies gains an advantage over its rivals in terms of "market share" among adopters, it tends not only to keep the advantage, but also to expand it.

There are several ways by which decreasing cost conditions can arise. One is that new technologies often have infrastructure requirements in order that they attain their economic potential. In such cases, the greater the population of one of the new technologies, the greater is the business opportunity for providing support services for that technology, such as inventories of replacement parts, fuel supply infrastructure, repair shops, and even the growth of entire industries that have complementary relationships to the technology in question. The more this infrastructure develops, the greater is the incentive for a prospective adopter to choose the option with the best-developed infrastructure. As this process develops, a losing technology becomes increasingly thinly supported by whatever infrastructure that it might require, and this acts further to discourage any choice but the leading technology. When this happens, the winning technology is said to be *locked in.*

The bottom line with regard to this choice problem is that (1) nonmarket influences can promote adoption of one technology to the ultimate exclusion of the others if there are decreasing cost conditions, and (2) two or more technologies will share the market under increasing cost conditions. Beyond that, the theory says nothing about what the ultimate market shares will be after the selection processes work themselves out. One of the conclusions from this theory is that the influence of what might be called market forces does not guarantee that the technology chosen is the superior one in the long term in the decreasing costs case. This conclusion is due to the influences of the random events that tip the choice one way or another during the competition. The following discussion describes some of the possible random events that can and have influenced the choice process in the past

What Historical Examples Say

Keyboards

The standard typewriter and computer keyboard is known as the QWERTY keyboard, after the left-to-right sequence of keys of the top rank. When the typewriter first appeared in the 1870s, early makers offered a variety of arrangements of letters on the keys. Some of these experienced jamming of striker bars when the typist typed too fast, and the QWERTY arrangement was found to slow typing sufficiently to minimize the jamming problem. It was a solution to a problem that would have yielded to greater precision in manufacture of the parts within a decade or two. As it was, the rights to the machine with the QWERTY arrangement were acquired by arms maker Remington in the 1880s. With the marketing effort of a major producer, QWERTY machines became widely used. As these machines proliferated in offices, there grew a large pool of people trained in the QWERTY keyboard. Inasmuch as competent typing involves high training of small motor skills, conversion to another keyboard pattern of a typist accustomed to QWERTY involved costly retraining. Offices, thus, increasingly had an incentive to use QWERTY machines, as these afforded the highest probability of finding already-trained typists. People with QWERTY skill had an incentive to hire out to an office using these machines to avoid having to retrain and to retain the ability to change jobs to another QWERTY office.

This was a near-perfect example of how a mechanical problem unrelated to the needs of business offices promoted the keyboard arrangement whose subsequent adoption benefited from a positive feedback effect, as in the Arthur model. Superior designs in the sense of enabling faster typing appeared from time to time, but none of these had any success in displacing the established keyboard design. It has been possible to build and demonstrate keyboards of superior design, so as to establish an experimentally solid case that the proposed replacement keyboard really is better. Another element of this example is that the design of the keyboard could be permanently fixed without interfering with possibilities for improving the typewriter itself, with such innovations as electric drive, daisy wheels, or type balls, or for its eventual replacement with personal computers with word-processing software coupled with laser printers. This means that the cost of maintaining an inferior keyboard standard is confined to that of the loss of some increment of typing speed.

Automotive Engines

In the very early twentieth century, the gasoline and steam engines were perceived as viable competitors for the automotive market. The Otto-cycle engine was a new development while steam was from a technology that had been around for almost two centuries. Each offered a slate of advantages and disadvantages. The gasoline engine was relatively easy to start from cold and had a comparatively high power density. Its disadvantage was that its efficient speed range did not correspond well with the vehicle's desirable range of speeds, and this problem could be corrected only with gearing; in other words, with a transmission and a clutch. The steam engine's speed characteristics were more compatible with the vehicle's speed requirements so that no gear transmission was needed. Its drawbacks revolved around its boiler, which was maintenance-intensive and, in some cases, dangerous. The competition between the two was reasonable even as long as the car was little more than a hobby, but when the market began to grow towards what became a mass market, the gasoline engine began to capture an increasing share of the automotive market.

In terms of the Arthur theory, one random event that pushed the decision in favor of the gasoline engine was the emergence of mass markets and the beginnings of mass production, which Henry Ford and others developed so well. There was no offsetting attempt made to foster a mass market for steam cars. Production of steam cars remained in the hands of a group of companies whose managements never, for the most part, showed any interest in developing a capability for large-volume, low-cost production that would require developing a mass market for their product. Thus, the initial impetus toward the Otto cycle came from the concentration of a combination of high engineering talent and business acumen in the gasoline engine camp. For this there seems to be no explanation in terms of the characteristics of the technologies themselves, and it ranks as a major "random event" that worked toward an early advantage for the gasoline engine. Decreasing operating costs were present with the choice of the Otto-cycle engine. As the population of these increased, there appeared a growing repair and fueling infrastructure, and development of fuels by the petroleum industry that decreased the costs of owning and operating these engines for all owners.

There is the opinion among engineers even today that the gasoline engine represented an inferior technology relative to steam. Arguments for the steam side take points from hypothetical improvements that could have improved the competitiveness of the steam engine had they been applied; but at the time, no one was interested in this kind of investment after the market showed a predisposition toward the gasoline engine.[2] This may seem like the ultimate Monday-morning quarterbacking exercise, but if it is taken seriously, the counter might run something like this: in the absence of long-range foresight, the market bet on the newer technology under the assumption that the newness of it assured a greater potential for improvement. In 1900, the Otto-cycle engine as a practical power source was about fifteen years old, and steam technology had been around for close to two hundred years. This example differs from the keyboard example because in the latter case, it was possible to build typewriters with alternative keyboard arrangements and test these with actual typists to establish any claimed superiority to the QWERTY arrangement. With the automotive engine question, latter-day arguments of steam's superiority rest mostly on speculation.

Nuclear Power Plant Reactor Designs

Following World War II, the Atomic Energy Act created the Atoms for Peace Program and charged the Atomic Energy Commission (AEC) with fostering the development of atomic energy for peaceful purposes of generating electric power. At the time, there were several designs for power reactors, including the light-water reactor and the gas-cooled reactor. The AEC initially took the position of letting construction and operating experience inform as to the best technology option, but this commitment to making the choice based on technical merit was soon overridden by events. As it turned out, the early choice between light-water (LWR) and gas-cooled (GCR) technology was decisively influenced by nonmarket events in accordance with the Arthur model.[3] One of these events was the Navy's adoption of the LWR for its submarine propulsion program. The decisive characteristic for the Navy was the compactness of the LWR relative to the alternatives. The other decisive event that influenced this choice was the desire of the US government for a standardized reactor design that could foster an export market for nuclear power plants. This responded to the interests of several manufacturers of nuclear power systems. At the time, the LWR, based on the Navy's experience, was the most tested of the options. As predicted by the Arthur model, the LWR technology took the lion's share of the US nuclear power market; it became locked in.

This case has been argued as one in which an inferior technological choice was made, at least in the long run. Part of the basis for this position comes from the Three-Mile-Island incident (TMI) of 1979 in which one unit of a light-water plant melted down. The causes of this were subsequently attributed to plant design flaws plus a large amount of human error. Supposedly, the GCR design was less vulnerable to the latter kind of failure. In the 1950s when the LWR/GCR choice was made, not only was the technology of the power reactor not well understood, but its promoters did not have a sure awareness of their own level of understanding. Such awareness grew, and there is no better evidence of this than the experience of the 1970s, in which nuclear power plants under construction were served with rapid-fire regulatory changes in details of design standards that had to be incorporated into the construction. While some of this kind of regulatory action may have been justified, it did succeed in running up the construction costs of the various plants to the point of destroying the economic appeal of nuclear energy for the utility industry. Three-Mile Island did not kill the nuclear power industry, for it was already dead due to regulation. The question of whether LWR or GCR was superior was moot, as after the incident, no nuclear plant proposal could be permitted regardless of design.

The point that the GCR design might have been less prone to human error than the LWR seems to argue that had it been the design of TMI, the accident would have been less likely. This might have been the case, but altogether it seems like a weak argument in light of the findings of the Kemeny Commission, appointed by the president to investigate the TMI incident. At least some of the blame for the human errors was found to stem from the culture of the utility companies operating TMI and other nuclear plants. In most of these utilities, virtually the entire fund of operating experience was from operation of conventional thermal electric plants, mostly based on coal fuel. In that culture, it had become a widespread practice that when problems emerged, to let them ride as long as the plant could be operated, and then deal with them only when it was necessary to shut down for massive repairs. With the fossil fuel plant, this approach was workable, for it posed no serious threat to the population surrounding the plant. The penalty for this approach with a nuclear plant is potentially far more severe, and as TMI demonstrated, a near-disaster can develop very quickly.[4] This is a case in which the choice of technology was decisively influenced by factors that were non-economic. The same can be said, however, for the virtual abandonment of nuclear power as a choice for new generating capacity in the US after the 1970s. Any future resort to nuclear power will have many aspects of starting over (with the benefit of a fund of experience).

In light of the managerial and operational problems that the TMI incident highlighted, it is of interest to reflect on how the Navy has been so successful in the management and operation of its fleet of nuclear-powered vessels, all of which were and are powered with LWR-design steam plants. The Navy faced two basic problems, of which one was the very complexity of the nuclear steam generators, and the other was a high turnover rate among operating and managerial personnel. The Navy's solution was to develop an institutional culture that possessed the technical operating knowledge and also had the ability to impart this knowledge as necessary to individuals new in the organization. The institutional memory was the same as that which allowed many kinds of operations requiring highly coordinated teamwork in the face of high personnel turnover. The operation of aircraft carriers requiring the highly coordinated launching, recovery, maintenance, fueling, and arming of aircraft is a prime example. The Navy's approach accepts that individuals new in the organization have to be taught, while private industry typically insists on hiring only people who are already trained. Private industry typically rejects any teaching role. Private industry's historic reluctance to incur training costs is probably a contributor to the durability and longevity of the QWERTY keyboard.

Replacement of the Line-Shaft

The Arthur problem, as already noted, was stated in the form of two or more viable technologies competing for the role of replacing an established but demonstrably inferior technology. This leaves open the question of how common is it that there are two or more replacement technologies sufficiently well developed to be plausible improvements over an established technology. Moreover, during a time period of high technological flux there is a real possibility that there will emerge a further combination of technologies that was not contemplated in the initial contest. For a shining example of this, one only has to remember the line-shaft and how it was replaced.

Before 1900, replacement of the stationary steam engine that powered the line- shaft might have seemed like the kind of situation to which the Arthur model can be applied. There were three plausible candidates at the beginning of the century: the Otto-cycle engine, the Diesel engine, and the electric motor. All of these were new but well enough developed to be recognizably superior to the established line-shaft power, at least in terms of thermal efficiency. Moreover, one can make the case that decreasing costs to adoption were present, for any of the three candidates required an infrastructure of fuel, repair parts, and other supply. In spite of these conditions, this contest was won by none of the above, as noted in earlier chapters. The line-shaft had many problems apart from the thermal efficiency of its prime mover, and eventually it disappeared to be replaced by machines individually powered by electric motors. The Diesel's factory role became that of powering a dynamo to provide plant power when commercial power was not available, and the Otto-cycle engine ended up with virtually no role as a stationary factory engine. If the options available to a factory designer in 1900 are compared with the slate of available options circa 1910, they will be found to be radically different.

In a period of technology flux, it is not always certain that the problem will remain as originally perceived. In the line-shaft example, the effects of whatever random events were working to influence the technology choice before and around 1900 were overwhelmed by some substantial technical developments and market forces that completely changed the nature of the problem from that originally perceived. This is the sort of thing that can happen when the apparent choices for replacement technology are based on prototype inventions, many of whose possibilities are still unfolding and not fully foreseeable at the time the choice must be made. The point comes home forcefully when one thinks of comparatively recent cases of competing technologies. During the 1970s, magnetic tape for recording music was packaged in cassette and 8-track formats. Both were plausible technologies for replacing reel-to-reel taping. Within only a short span of years, cassette overwhelmed 8-track, only to be overcome competitively by compact disks (CDs). These, in turn, may face total replacement by DVDs. It would seem that the probability for permanent commitment to a "locked-in" technology is low, especially in the field of consumer electronic goods, because of the shortness of product cycles that characterizes this class of goods.

One has to ask: From where did the technology that enabled complete line shaft replacement come from? Formal economic theory has a long history of regarding new technology of all kinds as coming from outside the economic system (that is, of being exogenous to economists' models of the market in question or to models of the entire economy for that matter). One of the healthier developments in the economics profession of recent decades has been the questioning of this idea. The key to line shaft replacement was improvement in the electric motor in the form of reducing its size and bulk (increasing its power density) to the point where it made economic sense to power individual machines with a dedicated motor of appropriate horsepower. This application of the electric motor can hardly be considered external to the economy. The improvement of the motor was the result of increased knowledge of how to accomplish tasks with reduced expenditure of the traditional factors of production. Business competition has always strongly incentivized this kind of thinking. It is an integral part of the economic system.

An Example from the World of Oil and Gas Exploration

Dr. Arthur has offered several examples in which standardization was more or less imposed on developing technology before the possibilities of the several options could be fully realized. In the case of the light-water nuclear reactor, the proximate imposer was a government agency, but several large companies contributed to this decision by the pressure they put on the AEC to enable them to market a standard nuclear power package in world commerce. The QWERTY keyboard was a standard effectively imposed by a large manufacturer. In the automotive engine case, there was no externally imposed standard that dictated the eventual dominance of the internal combustion engine as automotive power, but the steam solution was overwhelmed by massive competition from the gasoline engine makers that developed in a fairly short span of years. To see this, reflect that Stanley, the leading steam car manufacturer, did not get around to easing the boiler water supply problem with a condenser until 1914, the year *following* the opening of Ford's mass-production line. Abner Doble, who came closest to civilizing the steam car in the years while it was being produced and marketed, did not perfect his improvements until the 1920s, by which time the gasoline engine's competitive dominance was fully established.

If this experience suggests anything, it is that the longer the period of time that elapses before standardization is imposed for one reason or another, the more likely it is that the advance of technology will produce a result, like that of the line-shaft, in which the nature of the originally perceived technological choice is overturned by even better technological options. Oil and gas exploration in the US is an activity dating well back into the nineteenth century, and it has had significant periods of advancement due to improved technology. Many if not most of the estimates of ultimate recovery of oil and gas that have been published in the last fifty years have been based on a model developed by M. King Hubbert between 1955 and 1965. At a time like the present, when the world's potential supply of oil and gas has become a question of much interest in the political sphere, there has been considerable questioning of the Hubbert model.[5] A recent study by Richard Nehring of NBG Associates of Colorado Springs undertook a rigorous post-audit of the Hubbert method applied to two of the largest five oil provinces in the history of the US industry and found that the Hubbert-based estimates have been conservative to the point that the method is highly questionable.[6]

Reduced to its essence, the Hubbert model sees oil discovery following a symmetrical, bell-shaped path (see Figure 14.1) from first discovery to exhaustion of discovery potential. Discovery is followed, after an interval $(t_p - t_D)$ by a similar curve depicting the path of production. The area under either of these curves represents ultimate production. For the US, the $t_p - t_D$ interval has been estimated at ten to twelve years. At the peak of the discovery curve, it is assumed that approximately half of the ultimate recovery from the province

Figure 14.1
M. King Hubbert Model for Predicting Reserves

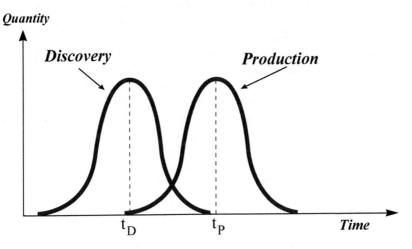

has been discovered. Ultimate production can therefore be predicted by noting the timing of the peak of discovery and assuming that decline in production will reproduce the pattern of the decline of discovery, which in turn is assumed to mirror the rise in discovery to its peak. Note that the assumption that both discovery and production follow symmetrical, smooth, bell-shaped paths is crucial to making the prediction of ultimate recovery. One of the early problems in applying the Hubbert model was that the annual statistics on additions to reserves are composed of two elements: new field discoveries and "revisions and extensions." In order to bring the latter component within the compass of the model, it was attempted to treat the revisions and extensions as an empirical correction to the model. Even with this correction, the model tended to predict on the low side.

Another difficulty in applying the Hubbert model is that the history of many, if not all, of the large oil producing basins does not follow the prescription of a smooth increase to a peak production. The two basins analyzed by Nehring had a number of peaks both in the discovery and production record. This causes a problem of recognizing the true peak of discovery, for the model can only be applied after this point has been passed and recognized. One of the basins Nehring studied, California's San Joaquin Valley, was initially discovered well back in the history of the industry but much of its reserve was not developed intensively for many years because of its physical characteristics. It was known to be a very large reserve, and was relatively shallow, but the crude was of such high viscosity that it did not flow through the formation to the production hole. It defied exploitation by what are referred to as conventional production means.[7] The other basin studied, the Permian Basin of West Texas and New Mexico (one of the world's "supergiants"), was initially discovered in 1920. Nehring applied the Hubbert model from the standpoint of two dates: 1982 and 2000. In both cases, the model, even with the correction described above, yielded predictions of ultimate recovery dramatically short of what actually transpired.[8]

Nehring's principal finding was that the method failed because of the inability to foresee ultimate recovery from older oilfields within the basins studied. The "correction" that was applied to the Hubbert method to take the extensions and revisions into the model's compass failed to recognize the massive impact of technology on the extensions and revisions, especially in the older fields. Nothing can illustrate the impact of secondary and tertiary recovery methods on the production profile of an older field as dramatically as an actual history (Figure 14.1). The Weyburn Unit was created in the mid-1950s.[9] A quick comparison with the hypothetical production decline pattern of Figure 14.2 suggests that Weyburn should have been depleted by 1980. As of 2005, the unit was estimated to hold eighteen years additional production.[10] The Weyburn profile illustrates the results found by Nehring with regard to San Joaquin and the Permian Basin.

To some extent, the impact of technology is driven by economic conditions. After 1970, the price of crude oil rose and made many technological projects economically attractive. The San Joaquin enjoyed a surge in production as a result of the steam flood technique, introduced about 1975. The Permian has benefited from a number of techniques developed over the years. These include secondary recovery methods (e.g., water flooding) and tertiary techniques, such as carbon dioxide floods.[11] The Hubbert approach implicitly assumed that the technology of oil production would be stable for the duration of its predictions. This is a weakness common to a number of prediction models based on history, including the Arthur model. Nehring found the Hubbert model yielded plausible results when applied in the 1960s, but from the late 1940s to after 1970, there was a comparatively quiet period in regards of application of new technology in oil production. That changed in the price environment that came into being after 1970.[12] Thus, the best that the Hubbert model could do was to predict ultimate recovery in an environment of static technology and conventional production techniques.

Technology Choice in the Face of Uncertainty

It is fairly unusual that one has the simple problem of choosing a replacement technology from among two or more established and proven technologies, all of which are economically superior to the technology currently in use. At the outset of the twenty-first century, it is more often the case that the candidate(s) for the role of replacement technology are themselves only newly developed and have an uncertain future of benefiting from further (innovation) invention. At present, there are four technologies for producing electric power with little or no damage to the environment that in a pure engineering sense are ready

Figure 14.2
Historical Oil Production Profile: Weyburn Unit

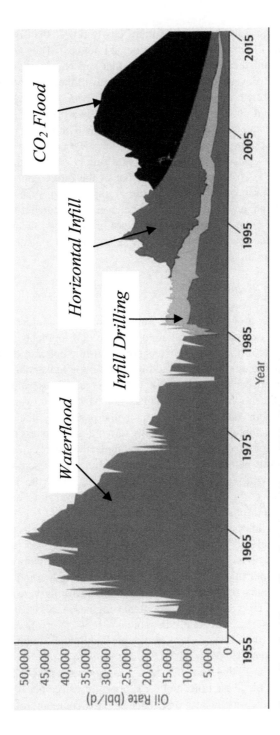

Source: Petrofund Energy Trust, Ltd., Annual Report 2005, with permission.

for application now but promise more costly power than is available from conventional thermal-electric plants fueled with coal. They are the integrated gasifier/combined cycle (IGCC) plants, solar panels, wind power, and nuclear power. Private development of power using any of these technologies faces the problem that any power produced from them is at a cost disadvantage relative to conventional power plants. As long as this is a problem, private development of these and other new sources of power will be severely limited. Private developers face two kinds of uncertainty that were not major problems at the outset of the twentieth century: that of how quickly these technologies can be improved to be cost-competitive and that of the future course of air quality regulation. This means, among other things, that projects such as a major IGCC plant in Indiana may be delayed, or may never be executed if the uncertainties of future technology and/or regulation are not resolved favorably. Any such project announcement today should be interpreted as a gamble on the future course of regulation.

Uncertainties Regarding the Future Course of Technology

Uncertainty grows out of the inability to foresee the future possibilities for development of a particular technology. Such developments include not only improvements in the particular technology itself, but also developments in enabling technologies. Thus, would-be automobile makers of the first decade of the twentieth century would not have been able to foresee the improvements in fuels for the gasoline engine that materialized with improvements in refining technology. This lack of foresight opens the possibility that a choice among competing technologies may eventually turn out to be the less desirable choice, as Arthur has pointed out.[13] In another example, the diesel cycle engine became practical shortly before 1900, but it did not find all of its present-day applications until well into the post-World War II years, during which the diesel became the dominant power for motor trucks, railway locomotives, towboats on the inland waterways, and for seagoing merchant ships.

There are other examples. In the early days of diesel-electric locomotives (the 1940s and 1950s), the superiority of diesel power over steam power was not always obvious. In one instance, a major railroad company undertook a systematic and controlled comparison of the two types of locomotive that resulted in roughly comparable operating cost figures. The results were not conclusively in favor of either type of power. The steam engines in this test were of the most modern design then available.[14] In another case, the Norfolk & Western, a major coal-hauling railroad, operated very profitably with steam power well past the time (the late 1940s) when most other railroad companies had committed to diesel power. In that case, the steam locomotives were of the most efficient design attainable at the time, and the support infrastructure for steam, such as repair facilities, was modern and efficient. This company finally adopted the diesel in the late 1950s in response to various costs that stemmed from being a lone steam holdout in an otherwise all-diesel industry.[15] In the 1950s, when the major steam-to-diesel transition took place, most US railroads had been operating with steam engines of 1920s design, whose replacement had been thwarted by depression conditions in the 1930s and by the War Production Board in the first half of the 1940s. As the transition was taking place, and for a number of years afterward, there was a body of engineering opinion which held that had the steam engine benefited industry-wide from the latest designs and from a modern and efficient infrastructure, as on the Norfolk & Western, that steam technology would not have been retired nearly as early as it was. This speculation had a brief vogue in the mid-1970s when the price of diesel fuel attained high levels. Interestingly, the diesel seems finally to have won this hypothetical war, as there seems to be no speculation on the potentialities of steam power in the present era of high-priced diesel fuel. One of the most important differences between conditions now and those of the 1970s is today's heightened concern with the environmental consequences of burning coal under steam boilers that arises from perception of the greenhouse gas problem.

Uncertainties of Regulation and the Technology Choice Problem

Environmental regulation has introduced many uncertainties at the outset of the twenty-first century that were not factors at the beginning of the twentieth. An example illustrates this. Changing World Technology (CWT), a West Hampstead, NY, company, owns technology that it calls thermal depolymerization (TDP). TDP processes animal wastes, as from a meat processing plant, into a combination of methane and oil that is a near-perfect substitute for conventional diesel fuel. The gas is used to fuel the process and the oil is available for sale. TDP is presently being tested in a full-scale plant in Missouri that is adjacent to and processes the waste products of a turkey-processing plant owned and operated by ConAgra Foods, Inc. CWT owns the TDP plant, and ConAgra owns a minority share of CWT.

The revenues to the TDP plant include proceeds from sale of the diesel fuel (up to five hundred barrels per day) plus credit for any avoided costs of alternative disposal of the turkey offal. If the offal were purely a waste product, then the cost of its disposal could be high, and the avoidance of this cost would be a strong positive credit to the revenues of the TDP plant. Unfortunately for profitability, there is a market for turkey offal: it goes into processed turkey food. As matters presently stand, the offal can produce a revenue stream that goes into the TDP profitably calculation as a negative. ConAgra has gambled that turkey offal will be outlawed as an input to turkey feed. Its reason for so supposing was based on a similar prohibition against the use of beef offal in processed cattle feed that resulted from precautions against mad cow disease.

Thus, ConAgra's investment in TDP is based on a gamble on the course of future environmental regulation.

This is an example of events regarded as probable but that have uncertain timing. A prominent example of this of the late twentieth and early twenty-first century comes from regulation of carbon dioxide emissions. As the US has not ratified the Kyoto Treaty, its restrictions on CO_2 emissions are not binding in this country. However, all industry must deal with the prospect that it will be subject to the treaty's rules at some time or place, or of a similar constraint. At the time of this writing, there are several kinds of reaction to this uncertainty. Companies that have substantial revenues originating in countries that have ratified the treaty, such as General Electric, have effectively embraced the treaty and are acting as though it affects all their operations. Other companies, such as electric utilities, that frequently do not have significant foreign earnings are not moving aggressively to embrace the treaty. However, at least some utilities are nervous regarding its future in the US and how future CO_2 regulation will affect their operations. Fortunately, CO_2 is like any other commodity in that if it can be obtained in sufficiently high concentration, it can be recovered at reasonable cost and sold in various industrial markets. Moreover, CO_2 can often be disposed of in an environmentally satisfactory manner by sequestering it in geologically suitable subsurface formations. If there is an oil field reasonably close to the source of CO_2, then it can be sold into what is known as tertiary recovery to enhance oil production. This is a potential as well as existing market for carbon dioxide, and there are instances in which it is transported by pipeline for a hundred or more miles in order to reach a tertiary oil production project.[16]

Integrated gasifier/combined cycle (IGCC) technology for producing electric power provides an excellent case in point as to how utilities and supporting industries are reacting to regulatory uncertainty. The capital cost of an IGCC plant is said to be approximately 15 to 25 percent greater than that of a conventional thermal electric plant. American Electric Power has announced plans to construct a large IGCC plant with construction to commence possibly in 2007. If the capital costs cited are realistic, then such a plant might experience competitive difficulties in an era of unregulated power production. One way to express the difference between an IGCC and a conventional thermal-electric plant is that the former deals with pollutants in the coal fuel before combustion while the latter deals with pollutants in the products of combustion. If this announced project moves into construction, it will possibly be because regulation has substantially increased the costs of dealing with pollutants from conventional plants.[17] Thus, the announced IGCC project can be viewed as a bet that costs of dealing with pollutants from conventional plants will increase substantially in the next several years.

IGCC technology is a response to regulatory uncertainty that would deal with pollutants of coal combustion by eliminating them. However, uncertainty regarding future regulation, especially of CO_2, has produced another bet that yields a strategy very much different from that which the power generators that have expressed interest in IGCC appear to be taking. A major Texas power generator announced plans to build a number of conventional coal-burning thermal-electric plants over the next several years. The bet here is that there will be a grandfather clause in any national or state legislation that designates CO_2 as a pollutant. If this building program had been carried out and if future CO_2 regulation does include such a grandfathering, then this power generator would have ended up with capacity sufficient for demand growth well into the future that would enjoy a cost advantage relative to power producers affected by future tightening regulation. Events since announcement of the coal strategy appear to have caused the its abandonment, but the episode still serves as an example of the kinds of business strategies that are growing out of regulatory uncertainty.

Networks and Adoption of Technology

The network problem can arise when the existing stock of capital has a number of owners but works as a network across the entire industry. In the marketplace, any single firm that attempts to invest in more efficient capital that is not compatible with the existing network can incur a competitive disadvantage *vis-à-vis* the competition. No one owner has an incentive to invest in this way if the other owners do not do the same simultaneously. The raw market incentives add up to a big game of "steal the bacon" in which each actor waits for the others to act. The reader will probably note that this problem resembles that faced by any would-be user of a superior technology in the face of an existing "lock-in" situation (in the Brian Arthur sense) whose effect is to preserve an older but inferior technology.

Railway coupling between freight cars provides a prime historical example. Until the 1870s, coupling was achieved by a device known as the "link and pin." In this system, a worker was required to stand between two freight cars as they moved together and to guide a link into a socket with one hand and to drop a pin into the socket housing with the other hand. This system demanded an order of hand-eye coordination that most humans do not have, and the result was a huge toll of smashed hands and fingers. It was commonly said that you could identify a railroad switchman at a distance by the lack of one or more fingers. Managements took the attitude that the link and pin system was cheap and therefore satisfactory. There were no workers compensation laws in those days, and no class action lawyers, so the human damage was of no cost to railway managements. There were numerous immigrants willing to take the places of those injured beyond the capacity for the work.

Then, as now, freight cars moved in interchange from one railroad to another (the network). This avoided the necessity of emptying and reloading cars at interchange points. Thus, there was no market incentive for one company management to incur the cost of applying safety couplers before the connecting carriers did likewise. A safety coupler had been invented by a dry-goods clerk named Eli H. Janney whose patent of 1882 reflected a number of design improvements following his original patent of 1873. While there were many designs for improved couplers in this period, the Janney coupler emerged as the most satisfactory. The network problem was solved by an act of Congress that imposed a standard of safety couplers and air brake equipment in 1893. Full compliance with this standard was achieved after the turn of the century.

The idea of the network as a barrier to technological progress points toward a need for imposition of a standard on an entire industry. The federal government often turns out to the logical source of such standards. State governments are often unsatisfactory sources because of limited geographical reach. In the railroad coupler case, a number of state legislatures passed legislation requiring safety couplers on freight cars, but these were largely ignored. Yet another source of industry engineering standards is the monopoly such as the early Bell Telephone system. This is not an aspect of the old Bell system that is often cited, and it had perhaps come to the end of its usefulness at the time the system was broken up by antitrust action. However, the early Bell system financed research and development that produced results cited elsewhere that might not have emerged had the early telephone system taken the form of a patchwork of local systems covering the country. Many standards have come from various engineering associations. Such organizations have been prominent in promulgating standards in various engineering and construction fields.[18]

In a relatively recent example of a network-imposed gridlock that arrests progress toward efficiency, gypsum produced in stack gas scrubbers at coal-burning power plants was disposed of in landfills for over twenty years while gypsum was imported from Canada for the purpose of making common wallboard. As mentioned in Chapter 6 above, the reason for this strange circumstance was that the utilities' cost of landfilling their gypsum sludge was included in their rate bases. They rejected the idea of selling their sludge to wallboard producers because to do so would have, in their expectations, removed the costs of sludge disposal from their rate bases; in other words, they had no profit incentive to work with the wallboard people. The deadlock was broken only with widespread deregulation of power production, which had the effect of inducing the power producers to seek ways of eliminating costs of sludge disposal by selling it. As in the railroad coupler case, the impasse was solved by a government intervention—deregulation that was not primarily intended to deal with the sludge disposal problem. Before deregulation, there was no ongoing carnage among workers to be ended. There was just the inefficiency of the practices, and deregulation did bring about the reduction of the costs of haulage of thousands of tons of wallboard to all parts of the country.

Yet another example of how a network effect relates to the adoption of new technology can be found in certain labor markets. In a highly simplified labor market, the firms in an industry compete for the services of labor factors in various skill groups. Generally speaking, it would not pay an individual firm to undertake the internal production of desired worker skills by means of an internal training program. Training is costly, and any single firm that undertakes a training program not only accepts costs that competitors do not have, but risks having newly minted trainees leave to work for competitors. That would place the firm in the unenviable position of bearing the training costs for the entire industry. In actual historical practice, business and industry have avoided this problem by massively outsourcing worker training—to universities, public schools, technical schools, union apprenticeship programs, the armed services, and others. Most of the time, this system has proved adequate to supply business with the skilled workers needed. As a rule, business firms do not ordinarily subject newly-hired employees to extensive training. Exceptions to this rule can occur when workers of particular skills become scarce. In such cases, a training program is one of the available strategies for creating a supply of workers with desired skills. There is a variety of such strategies, including importing foreign workers and outsourcing the work to offshore affiliates.[19]

A network-type problem can appear in spite of this overall satisfactory worker supply system when new technology is being placed into service. This was part of the problem that led to the Three-Mile-Island incident described above. The utility company in question attempted to apply long-established operating procedures that had been successful with thermal-electric power plants to its new nuclear facilities, but which turned out to be inappropriate for nuclear plants (see above, this chapter). The accident highlighted the need for special training for the operators of nuclear facilities that no one was providing. Under the assumption that all utility companies installing nuclear facilities were doing more-or-less the same thing, it did not pay any one firm to undertake the specialized training needed for the same reasons as were already cited in the case of the simple labor market above. Moreover, none of the normal outside sources of these skills, such as engineering colleges, was providing the requisite training. The technology was new in the 1950s and 1960s, and it took a potentially tragic accident to highlight this training deficiency.[20]

Another example of the network effect is the QWERTY keyboard. It is the network effect that gives the QWERTY keyboard its durability as a standard, for there is not a large potential reward to any firm or other entity that would create a desire to try to substitute a superior keyboard. The case demonstrates how a locking-in process can create a network that could be based on an inferior technology as well as be a barrier to improvement. What about other examples? The internal combustion automotive engine had a long run of improvement, but now its dominance is being at least nibbled at by the electric car in its modern form of the hybrid gasoline/electric. This technology has freed the electric drive from its tether to a stationary power source. The light-water power reactor can be said to have self-destructed, at least figuratively, because its early promised economic advantages evaporated during the 1970s due to regulation. The current talk of nuclear power revival will involve a careful examination of which power reactor design should be the future standard. The link-and-pin railway coupler technology dates back to railroad prehistory (the eighteenth-century mine trams) and may have come to be a standard through something resembling an Arthurian lock-in.

The deficiency in the training of operating personnel for the nuclear power plant could be regarded as growing out of the lock-in of the light-water technology, but this problem is really a manifestation of a potentially much larger problem that grows out of private industry's historic reluctance incur the costs of training people. Inasmuch as this kind of problem arises when an established industry deals with a new and unfamiliar technology, there is potential for more examples. For example, if one contemplates the many technical combinations by which the electric power industry may develop in the future, one realizes the possibility for unforeseen and possibly dangerous situations due to poor training. It is not clear that these will grow out of future lock-in processes.

Compatibility with Existing Capital

There are many examples wherein proposers of replacement technology have considered it necessary to put their proposals forward as compatible with an existing body of capital in place. Where there has been no network effect barrier to private sector adoption, compatibility has probably been a legitimate selling point. However, a compatibility requirement constitutes a restraint on what the new technology can accomplish and tends to lose importance when further technology developments provide additional improvements. The farm tractor is a good example. Initially, it was a straightforward adaptation of the internal-combustion engine as a prime mover to pull plows, threshers, harvesters, and other farm implements previously drawn by horse or mule teams or steam traction engines. These implements were powered by a "bull wheel," a heavy wheel that operated the implement's machinery as the implement was drawn forward. Bull wheels were notoriously inefficient, particularly in damp soils. The early tractors were viewed simply as mechanical horses, compatible with existing implements. The bull wheel itself was retired only after the invention of the power take-off on the tractor itself as a greatly improved drive for the implement's machinery.

In another example, the heating and cooling of railway passenger cars was based on steam at the time straight-electric and diesel-electric locomotives were beginning to appear in the 1930s and earlier. In order to make these new types of motive power compatible with existing passenger rolling stock, each locomotive unit was equipped with a steam generator. While the new motive power offered the possibility for electric heating and cooling, the changeover was regarded as being prohibitively expensive by the railroads. The changeover did not take place until well into the Amtrak era (after 1970) and was forced by, among other things, the virtual disappearance of the work force skilled in the maintenance of the old steam heating and cooling equipment through non-replacement of retirees. By that time, passenger service had become so diminished that the conversion had ceased to be financially daunting. Unlike in the tractor example, rail passenger service was regarded as a money-losing business, and the failure to maintain the work force reflects the declining nature of the business. There are many other examples of companies using compatibility to attract sales in the computer age. These include the font cartridge add-ons for the early laser printers.

The ability of a new technology to be sold as compatible with an existing body of equipment is probably an effective encouragement to its early adoption, although not an important element in its long-term development. Compatibility is certainly to be considered in assessing the future of a number of the nascent technologies of today. In particular, it would seem as though a working technology that can be presented as compatible with existing technology would have a greater chance for near-term adoption than one facing requirements for massive redesign of existing capital. Even less likely to be adopted near-term is a technology that depends on prospective breakthroughs in enabling technology that have uncertain timing.

Present and recent experience provides some conspicuous examples wherein promoters of this or that novel technology have simply not worried about compatibility with an existing body of technology. Ethanol as a motor fuel extender is not compatible with the existing pipeline distribution system for motor fuel because of its tendency to come out of phase with the gasoline with which it is mixed in the presence of water. This has meant that fuel ethanol can be distributed to consumers only by truck or rail, both more costly than pipeline distribution. The most glaring example of non-compatibility with existing distribution capital under discussion today is hydrogen, for which there exists only a limited-area pipeline distribution system serving petroleum refineries and other large industrial users.

Technology Choices of the Early twenty-first Century

Electric Power Production

While it is not the purpose here to attempt to predict the future, one forward-looking observation that seems unavoidable is that the world of the future will require a dramatic increase in the production of electric power, and achieving this will have to honor many environmental constraints. Generally speaking, the alternatives to present day power production technologies are more costly than technologies presently in use. All of these offer the prospect of power production with fewer negative externalities (cleaner air) than present fossil-fuel power plants. The proximate reason for considering replacement at all is government regulation in the absence of any market pricing of clean air. Inasmuch as the latter is not priced by a free market mechanism, government regulation is in effect pricing higher air quality, and the difference in the production cost per kilowatt-hour as between the replacement technology and existing technology implicitly attributes a value to air quality. Unfortunately, this is a measure that has definition only in terms of technology as it exists at present. The central issues governing how electric power will be produced in the future are as follows.

1. Can fossil fuels continue to be used to produce power in the long term?
2. Does nuclear power have a future?
3. Will it be possible to use a completely clean power source, such as solar or wind, as a base-load power source?

The corollary question relating to the long term future use of fossil fuels is whether to deal with the air quality issue by cleaning up the combustion products from conventional fossil fuel plants or to pre-process the fuel to recover all impurities prior to combustion. How this choice is made may turn on how regulation regarding CO_2 emissions unfolds. So far, the most promising technology for disposing of CO_2 appears to be geologic sequestration, in which the gas is pumped into a natural subsurface trap. This has been developed in the context of tertiary oil and gas recovery. Conventional fossil fuel power plants can be retrofitted with facilities for recovering CO_2, as can integrated gasifier/combined cycle (IGCC) plants. Present evidence suggests that recovering the CO_2 may be easier with IGCC, and this issue may dispose future power plant investment decisions in favor of IGCC. Thus, the big uncertainty facing investors in power plants is how CO_2 regulation will unfold. There is one huge advantage that IGCC technology enjoys: it makes use of coal, the far-and-away most abundant of energy resources.

Future prospects for nuclear power rest on two uncertainties: (1) can the public accept that nuclear is safe, and (2) is there a way to dispose of spent nuclear fuel such that radiation will not threaten populations in the future. The outlook on number 1 does not seem favorable, as opposition is built on the emotion of fear. A strong case that nuclear power is safe given competent management and operation can be made. There is a substantial fund of operating experience with power reactors that has been built up over the years. Three-Mile Island can be viewed as a part of this experience, as it highlighted problems of operation and management philosophy that can be avoided in the future. There are a number of reactor designs that appear to promise reduced sensitivity to human error than the LWR technology. For example, fast-neutron reactors would enable the extraction of more energy from recycled nuclear fuel, minimize the possibilities for weapons proliferation, and greatly reduce the time nuclear waste must be isolated. The payoff to investing in nuclear technology is that nuclear power plants are absolutely free of air pollution.

Against this, however, is the idea, common among environmentalists, that there should be an absolute guarantee against an accident. This, of course, is a requirement that is impossible to fulfill. It is difficult to envision real progress in the development of nuclear power investment in the absence of draconian federal overriding of state and local ability to delay and block power projects. Inasmuch as no such federal intrusion in the permitting process seems politically likely, the realization of the technical potential of nuclear power seems to be a thing of the distant future. In spite of these obstacles to a nuclear power revival, high prices of crude oil and its products and the environmental problems of burning coal have fomented an interest in nuclear power in government and among utilities. This interest has also been whetted by the provision of federal subsidies to nuclear power in the Energy Policy Act of 2005. As of the present writing, a number of power producers have expressed interest in building nuclear plants that range from formal study to application for permitting. If and how quickly a nuclear revival takes place, its technical configuration remains to be seen.

As for the problem of disposing of spent nuclear fuel, the present does not offer strong grounds for optimism. The Department of Energy's Yucca Mountain facility in Nevada is the center of intense controversy for a variety of reasons. One of these is that the facility lacks the capacity to accommodate expected volumes of nuclear wastes based on present technology. The solution may lie in some form of spent fuel reprocessing whose effect would be to reduce the absolute volume of highly radioactive wastes by utilizing a higher proportion of the energy potential of the original fuel to produce power.[21] One possibility is the fast neutron reactor technology, which would greatly reduce the amount of time that wastes have to be isolated—from 10,000 years to 150 years. If present governmental interest in reviving the nuclear power industry is to bear any fruit at all, the waste disposal problem will have to become the subject of a believable official policy.[22]

Both solar and wind power continue to progress as *niche* players in the grid power picture. Recent solar developments in California's Mojave Desert are aimed at providing daytime power sources, and this may represent the immediate future of solar power. These projects are designed to serve daytime peaks in load from such demands as air conditioning. The uncertainty surrounding solar and wind as full-fledged, base-load sources of power comes from the lack of low-cost power storage technologies. Thus, a large bet on solar and/or wind as a base-load power source is a bet on the speed with which the costs of flow batteries or some other storage technology can be brought low enough to be competitive with fossil fuel plants that embody clean coal technology. A further uncertainty grows out of the speed with which the costs of materials going into solar cells can be reduced with, say, conducting plastics or some other development. A further uncertainty comes from CO_2 regulation. Both solar and wind, with suitable storage technology, would avoid the CO_2 emission problem completely, and either of these technologies could become competitive with integrated gasifier/combined cycle technology under some foreseeable regulatory conditions affecting CO_2 emissions from conventional thermal plants.

According to the American Wind Energy Association (AWEA), in especially windy places, wind power sells at 4 to 5 cents per Kwh, roughly comparable with prices for power from conventional thermal-electric plants. The location issue—solar to sunny places, such as deserts, and wind to windy places—in many instances means that these power sources are located remote from population centers. This means that progress is ultimately constrained by the rate of progress in the development of low-loss transmission technology in addition to storage technology. In spite of these constraints to progress, wind power capacity in the US had grown to 6,740 Mw by 2004. US development is behind that of Germany (16,629 Mw) and Spain (8,263 Mw) in 2004.

The Clean Air amendments generate a regulatory climate that encourages a change of power technology. Of the technology options, nuclear and IGCC are the closest to being ready. The regulatory factor constitutes a large element outside of conventional economic considerations (such as costs, etc.) could be decisive in promoting serious experimentation with both of these options—and qualifies as a random outside factor of the sort whose decisive role Arthur has highlighted. While the power choice problem does not fully conform to the problem that Arthur stated, it might be one that is heavily dominated by non-economic considerations.

Power Transmission

To muddy the future vision further, there is the problem of power distribution. In many areas of the country, the present distribution system is old and in some places, overloaded. Nationally, almost 25 percent of the power produced is lost in transmission in the form of heat into the atmosphere. Research and development in nanotube conductors seems to offer a long-term solution, as some of these materials have demonstrated superconducting-like properties in the laboratory. Unfortunately, realization of the nanotube solution to the problem of transmission line loss probably lies far in the future. At present, one key problem is that not all nanotubes have this property. Moreover, there is so far no way of producing nanotubes industrially, and there is no industrial way to separate out those with the desirable characteristics. At least some of the low-loss technologies involve what is now called high-temperature superconductivity. Unfortunately, this term does not mean that a material conducts at normal temperatures; it only means that superconductivity is achieved at temperatures above the freezing point of hydrogen, or -176 degrees Celsius. The power needed to maintain superconducting temperatures adds to the overall demand for power.

If not nanotubes, could HVDC offer an answer? Unlike nanotube transmission, HVDC is here and now. However, economic application of HVDC is probably limited to long-distance transmission of hundreds of miles at a minimum. This technology could become very important should solar or wind power achieve cost reductions and become more widely applied.

Reduction or elimination of this line loss would directly substitute for additional future generating capacity. In the past, reduction of line loss has been achievable through large cross-section conductors and/or shorter transmission distances, and in suitable geographical settings, pump-storage facilities. Considerations of investments in these measures are supposing a very long time horizon for the low-loss transmission technologies. At such time as low- or no-loss transmission becomes a reality, there are some interesting possibilities. First, not as much power production capacity will be needed to service a demand of given size. Second, power can be transmitted over longer distances. Power technologies whose geographic locations are dependent on specific weather conditions, such as solar and wind, are particularly dependent on future development of low- or no-loss transmission technologies, more so than conventional power technologies. A longer economic range for power transmission would allow fewer and larger central station power plants of any technology. It may be that some future combination of low- or no-loss transmission capability, an economical storage technology, and solar power will be fully competitive with today's fossil fuel thermal plants. At this point, one can only speculate as to how the advantages of low- or no-loss transmission will be utilized.

Broadband Services

As noted in Chapter 6, the regulation of the broadband service industry appears to be in a state of flux. Two technologies compete: fiber optic lines to individual residences are contending with the coaxial cable. The former technology is being pushed by local telephone companies and the latter by cable companies. The cable companies have in the past been granted exclusive franchises to serve local areas by local governments, and they have taken the position that the telephone companies should be required to have similar franchises. To this the phone companies answer that they have always had access to serving local areas without regulatory interference from local governments, and that fiber optic-based broadband service is merely an extension of the slate of products they have always offered. The implication of the cable companies' position would be to subject the telephone companies to a level of regulatory interference to which they have not heretofore been subject.

Any sort of bet placed on one or the other of these technologies must therefore be based on a guess as to how the regulatory battle will end. Another source of risk involved in the broadband future revolves around how much consumers will be willing to pay for the huge volume of entertainment services that is possible with either of the two contending technologies. The broadband case is of interest from the standpoint of the Arthur model. Both cable and fiber require substantial investment in order to be able to serve a large market. When fiber optic is laid into a residential or business neighborhood, costs have to fall with increasing numbers of subscribers, as there is a fiber infrastructure that is common to all subscribers. The same can be said about cable. Therefore, one can suspect that both technologies are characterized by falling costs as the number of subscribers increases. The random events that can ultimately determine the competitive outcome between the two technologies have already appeared in the form of the regulatory battle now in progress, which is being fought in the legal and political arena as well as in the economic arena. Moreover, there could be other influences on the ultimate decision not presently foreseen. In this setting, there appears to be potential for lock-in ultimately with no guarantee that the outcome will be the most desirable of the options.

Computing Power

Expansion of computing speed and power in continuing accordance with Moore's law rests on one of several technologies now the topics of research: optical, quantum, and nanotube-based microprocessors. As of 2005, none of these lines of development has produced more than the barest demonstration of feasibility. Any basis for preferring one of these over the others has to await at least two basic uncertainties. The first is the pace of development, which could be tied to the rate of funding for research toward the development of the three options. The second is the timing of any dramatic breakthroughs that may affect any of the options. From the standpoint of the present, the several layers of uncertainty as to how these options develop suggest that random events as in the Arthur model could play a very large role. As for the usefulness of an order-of-magnitude increase in the speed of computing that could result from any of the three options, one has to resort to history, for there has never been any lack of ideas of how to use more power on the part of the broad areas of future users: business and defense.

Automotive Power

The era of concern regarding greenhouse gas emissions has produced a contest between internal combustion and electric power as prime movers for automobiles. The reader will remember that this same contest occurred in the earliest days of the automobile industry, and its revival may seem eerily familiar (see Chapter 3 above). Even the hybrid auto had an early twentieth-century precedent in the form of Ferdinand Porsche's gas-electric car of 1900. Today, the issue is between the hybrid and the pure electric car, and the contestants are major automakers. The pure electric car's practicality awaits further improvement of the lithium-ion battery technology. While this issue is being settled, the automakers continue to attempt to achieve profitability with traditionally powered vehicles.

Complexity and the Adoption of Technology

Using the first two decades of the twentieth century as an example, the typical technology that came out of the era was not a simple application of a single thread of technical development; it was a combination of threads. Combinations are not typically in the minds of prototype inventors or in the minds of stage-one innovators. Combinations arise in a trial-and-error process. In technology, the more prototypes that emerge from stage one innovation within a time period too short for any one to be incorporated fully into the economy other than in the presence of the others, the more possibilities there are not only for combinations among the newly-emerged technologies, but also between these and existing technologies. From an inventor's concept of a heat engine that is thermally more efficient than anything available for powering a line-shaft to the standard power source for terrestrial and maritime freight transportation was a path of countless experiments involving combinations. It is this complexity that contributes to the sometime long evolution of a prototype invention to wide usefulness.

Many of the new technologies of the early twentieth century were fully emerged from stage-one innovation. This meant that combinations among them could be tried at full industrial scale and compared without ambiguity as to costs. In contrast, many of the technologies that are presently discussed are not fully emerged from stage one innovation. In many cases, comparisons of cost characteristics among these rest on pure conjecture (for example, when cost characteristics are subject to some yet-unrealized technology breakthrough) or at best on results from pilot-scale plants. In spite of this apparent difference, however, there is a basic similarity in how the market is reacting to the possibilities in recent years and in the early twentieth century. High energy costs and increasingly stringent environmental laws are inducing proposals for combinations of underdeveloped technology possibilities. What is different are the bases of comparison of different technology combinations. Today, many of the comparisons have a hypothetical quality, whereas cost comparisons in the early twentieth century could be made on the basis of working industrial experience.

Notes

1. Arthur, W. Brian, *Increasing Returns and Path Dependence in the Economy*, (Ann Arbor: University of Michigan Press, 1994), pp. 13-26.
2. The most famous of the steam cars in America was the Stanley. It had a simple firetube boiler and no condenser initially. Because of customer problems with finding refills of suitable-quality boiler water, Stanley equipped cars of 1914 and after with a condenser, which increased both thermal efficiency and water range. Stanley went out of business in 1924. By 1923, Abner Doble had equipped the Doble car with a high-pressure flash/watertube boiler, which converted a smaller volume of water to steam on an as-needed basis. This reduced the safety problems associated with the firetube boiler. This, plus automatic controls, enabled the Doble car to be started cold and operated within seconds, thus overcoming the slow cold start handicap of

the Stanley design. A much-improved condenser design dramatically extended the range between water refills. Abner Doble outlined these improvements in a presentation before the Cleveland Section of the Society of Automotive Engineers on Oct. 20, 1916. The Doble was produced in small quantities for an upscale market, and its production ceased after 1930. The Doble, for all its performance virtues, was a luxury car. Steam had no performance equivalent to the Model T.

3.   Arthur, p. 25.
4.   Pool, Robert, *Beyond Engineering*, (New York: Oxford University Press, 1997), pp. 200-201.
5.   See, for example, Hubbert, M.K., "Energy Resources," in *Resources and Man*, (Committee on Resources and Man, National Academy of Sciences—National Research Council, 1969), pp. 157-242.
6.   Nehring, Richard, "Hubbert's Unreliability," *Oil & Gas Journal*, (April 3, 17 & 24, 2006). The conclusions of the study appear on pp. 43-51 in the Apr. 24 edition.
7.   Conventional production means reliance only on natural forces to move the oil through the formation to the production well. Natural forces include gas pressure and natural water drive situations.
8.   Nehring, (Apr. 24, 2006), pp. 46-51.
9.   Production in the Weyburn Field actually dates to 1954. A field can be "unitized" when there is sufficient borehole data that the extent and physical characteristics of the reservoir are known. At this point, the field can be operated as a unit for the purpose of maximizing output. Establishment of a unit requires agreement from all ownership interests involved.
10. Conventional production methods typically can recover a maximum of 15 percent of the oil originally in place in a reservoir. The reason that enhanced recovery methods have often been highly profitable is that they tap into the remaining 85 percent of the oil in the reservoir. With enhanced recovery technology, recovery expectations can rise to over 50 percent.
11. This technique has brought about an interesting revival of production in some of the oldest of the Permian's fields in recent decades.
12. Nehring, (Apr. 24), p. 46.
13. Arthur, p. 25. The dominance of other-than-market influences in the innovation process can lead to a market sub-optimal solution. This is consistent with Arthur's broad point even though his formal analysis did not deal with the kind of technological uncertainty that exists in the stage-one innovation process.
14. Many factors influenced the conversion from steam to diesel on the railroads. If there was single tipping event pushing this conversion, it was the lengthy coal miner strike of 1949 which forced railroad companies to restrict steam operations to what was possible with existing coal inventories and to rely heavily on such diesels as they had then.
15. One such cost was that by the time the N & W adopted the diesel, the commercial builders of steam engines and their standardized appliances, such as air compressors, heat exchangers, etc., either had gone out of business or, if the manufacturer survived, were no longer supporting steam engines. The railroad was increasingly forced to operate its own manufacturing facility to support a very small population of steam locomotives.
16. For example, the Great Plains Gasification Plant of Beulah, ND, produces large quantities of CO_2, which is piped to a tertiary recovery project in the Weyburn oil field of southeastern Saskatchewan, over a distance of about 180 miles.
17. Another possibility is a continuation in the decline in the cost of industrial oxygen, which affects the operating costs of the IGCC system. This cost has been falling

steadily for the past thirty years, but continuation of this trend faces some uncertainty.

18. In several of Brian Arthur's examples, a standard was imposed on a choice situation by an extra-market influence, and his point was that this can result in a less-than-optimal choice among the technical alternatives. In the network case, such as that of the railroad coupler, the market results in no progress and the imposition of a standard from outside the market is the only way a superior technology can be adopted.

19. An example situation comes from the Atlanta area construction industry in the years before the 2000 recession. In order to cope with an area shortage of skilled electricians, a large electrical contractor inaugurated a two-year training program consisting of on-job supervised work combined with classroom instruction for two evenings each week.

20. In the early days of the light-water power reactor, the best-trained corps of operators might well have been navy veterans; but unfortunately, the influence of such people was not the decisive element in how the utilities ran their reactors.

21. Other countries which have committed heavily to nuclear power (e.g., France) routinely recycle their spent nuclear fuel. In almost five decades of experience, no one has stolen material for the purpose of making a bomb, suggesting that the probability of such an event is miniscule. This remote possibility was used to justify cessation of the US reprocessing effort in the 1970s and is today at the heart of the US problem of disposal of spent nuclear fuel.

22. Problems of storage of nuclear power plant wastes are not confined to the US. In France, approximately 80 percent of whose power production is from nuclear, the problem of waste disposal has been studied for the last twenty-five years. Indigenous green parties have been so successful at blocking proposals that the government has recently made a decision to study the problem some more.

15

Conclusions and Policy Implications

The Long Technology Cycle

The logical sequence behind economic development advanced herein is this: technological advance in the form of prototype inventions foments investment opportunity. Prototype inventions have wider implications for ensuing investment opportunities than was likely to have been contemplated by their inventors or discoverers. Active investment supports a period of economic prosperity, and an investment "boom" can support an episode of exceptional prosperity and growth. When the slate of high-return investment opportunities growing out of a group of prototype inventions has been fully exploited, investment activity diminishes and difficulties in terms of general economic prosperity often arise. In the absence of a basis in improved technology, investment alone is not capable of fomenting substantial economic growth. This appears to have been true historically whether the investment originates in the private sector or the public sector. Economic progress historically has rested mostly on human ingenuity in producing new technology that makes production of goods and services less costly.

It is not foregone that even a prototype invention will always lead ultimately to an observable rise in the rate of productivity increase over the entire economy above the background tendency for productivity to increase. Such an invention will probably lead to a limited increase in the rate of productivity in at least a limited number of production areas. However, when a substantial number of technologies emerge from stage-one innovation in a comparatively short time interval, especially when the new technologies' growth-inducing power is reinforced by enabling combinations among them and with existing technologies, then a visible impact on the overall rate of growth of productivity becomes likely. There were two such episodes in the twentieth century: an acceleration in productivity growth in the 1910s based on the technology legacy of the late nineteenth century, and the productivity acceleration of the 1990s and early twenty-first century based on the "big four" postwar prototype inventions (the

335

transistor, the integrated circuit, the laser, and optical fiber). It is the argument here that these "bunchings" in the appearance of prototype inventions and their emergence from stage-one innovation are behind Schumpeter's observation that entrepreneurial activity occurs in bunches.

In thinking over the foregoing, one question that comes to mind is what is the future of the long cycle first viewed by Kondratief and which was of such interest to the cycle theorists, especially, but not confined to, the Austrians? The long cycle was once said to have an average period of fifty years. Some of the writers who attempted to explain this cycle worked with the idea that the upturn phases of these cycles were associated with bunchings of entre-preneurial activity that were themselves responses to, among other things, opportunities presented by the emergence of new technology. Certainly, there was an exceptional period of entrepreneurial opportunity based on the technology that emerged in the last two decades of the nineteenth century. By the fifty-year norm, the collapse of 1929-1930 was approximately on this schedule. Another serious setback as a result of a major investment failure was due about 1980. Indeed, there was a serious recession in 1982, attributed to the Federal Reserve's efforts to bring inflation under control. This reces-sion has commonly been attributed to the financial side effects of bringing inflation under control and is not generally remembered as a result of an investment failure. However, it did follow a time of "stagflation" in which relatively high inflation coincided with low overall growth and a reduced rate of productivity increase. These conditions collectively imply low returns to overall investment, which indicate the lack of underlying technology-fueled investment opportunities that would have stimulated productivity growth had they been present. The 1970s can be seen as a lull in the technology-based growth that had emerged from the World War II era and the ensuing expansion based on the development of information technology. The decade of the 1980s was the time of the buildup of the effects of the exploitation of the integrated circuit in the wider economy.

Even though the pattern of maturity and investment failure following a surge in innovation activity as in the first four decades of the twentieth century has not so far repeated itself since the late 1970s, it is probably premature to dismiss the long cycle of technology origin as a thing of the past. One indication that could signify the exhaustion of the technology boost to growth that has stemmed from the "big four" postwar prototypes is the sharp diminution of demand for a wide slate of products that are the technological lineal descendants of the prototypes whose innovation started the boom in the first place.

One problem that might inhibit the force of today's technological possibili-ties on future rates of growth is the existence of slow-developing technologies, such as electric power storage and low- or no-resistance power distribution. One can think of these as needed enabling technologies in the development of such things as pollution-free methods of producing and distributing electric power, for example. There is a real possibility that continued slowness in the development these enablers will materially delay the realization of some of the more interesting possibilities of today's recent discoveries and inventions. The timing of technological breakthroughs that would hasten the general availability of some of the futuristic technologies already described is completely uncertain, but the *possibility* of such breakthroughs allows the possibility that a cluster of them could materialize and trigger a major upswing of innovation-induced investment activity.

At least one aspect of the innovation surge of the early twentieth century seems unlikely to be repeated. That process created new industrial combinations that together became much larger in relation to the total economy than the result of anything that happened in the latter half of the twentieth century. The resulting industrial narrowness of the pre-1930 prosperity and its role in the collapse after 1929 seems unlikely to be a source of trouble in the late twentieth and early twenty-first centuries if for no other reason than the overall economy of today is much more broadly based than it was before 1930. For the future, it seems unlikely that even major breakthroughs in any technological arena, such as the production and distribution of power, could trigger investment streams large enough relative to the total economy such that their failure could threaten the entire economy on the scale of 1929-1933.

It is worth asking when the investment possibilities of the postwar "big four" prototype inventions will be exhausted and what happens then. Even though the technology is radically different from that of 1900, certain patterns may be repeating. Take, for example, the integrated circuit, whose early manifestations date from around 1960. The investment boom on the basis of this invention really got under way in the 1980s, and at the present time, applications of increasingly powerful ICs continue to appear. Counting from the early microprocessors, say in around 1980, the old fifty-year standard puts the end of the boom at around 2030. At some time before then, the pattern is that the field of highly profitable investment opportunities based on the IC will have been exhausted, and there will be an economic downturn of some magnitude. Will further new technology then be at hand to create a new host of investment possibilities? The best answer possible is "maybe." At that time, the future may emerge from some of today's new technologies that are now working their way through stage-one innovation. Modern research and development financing from all sources is keeping these technologies alive today. There was no equivalent effort in stage-one innovation in the 1920s.

What can public policy do to promote the continued prosperity that is ultimately founded on ever-increasing productivity based on new technology? Prototype inventions come from a variety of origins, many of which are at best weakly related to strictly economic motivations. Growth, however, comes from innovation of prototypes, especially in the second stage of innovation. Therefore, public policy should aim toward maintaining conditions favorable to stage-two innovation. These are many, but the most important include easy access to capital for startup firms, a court system that will enforce private property (including intellectual property) rights, and a modest tax load. These and other measures will not absolutely guarantee against lapses in the growth process, but they will assure that the full benefit of such clusters of prototype inventions as do emerge can be realized.

The opening years of the twenty-first century have seen proposals calling for long-term commitments of public funds to energy development projects.[1] If the technology experience of the last two hundred years means anything, it is that such commitments are foolish public policy. The reason is twofold. First, all long-range proposals of these kinds are based on known or "within sight" technology. It should be noted that the actual evolution of technology tends to be highly complex to the point where knowledge of how even known technologies can be combined to achieve a desired result is highly imperfect. Second, all such prognostications are mostly or completely blind to possibilities that can arise from future prototype inventions.

Supply Curves: How Quickly Do They Shift?

One-Time Events

Many of the economists employed in both industry and government have the function of making educated predictions regarding how various interest rate statistics and NIPA data will evolve over the next several quarters. Of course, the notion of a "one-time event" came straight from any number of economics textbooks at various levels of instruction. It should be clear from the foregoing material of this book that while a disruptive new technology or a radical change in the regulatory climate is in some sense a "one time" event, it is not an *instantaneous* event in terms of its effects. The repercussions from such a change in the technology or regulatory background can be present in some cases for many years. What happens when you introduce a disruptive prototype invention or a radical change in government regulation? A supply curve shifts to achieve a new equilibrium. This analysis is timeless. How quickly does it do this? Somehow, this question almost never gets much attention in the usual textbook discussion. This is unfortunate, for the shift in the aggregate supply curve in response to one or a group of prototype inventions takes place during the stage-two innovations of the technologies embodied in the inventions. The duration of the shift may coincide with an entire cyclic episode. Indeed, it is difficult to closely review the history of the Great Depression without being struck by the possibility that this economic calamity, at least in part, reflected the waning of investment opportunities created by the technology legacy of the late nineteenth century. Nor is it easy to ignore the technological underpinnings of the postwar prosperities.

Institutional Inertia

Institutional inertia often is a large factor in preventing large and established organizations, both inside and outside of government, from adapting quickly and efficiently to changed circumstances. *Corporate culture* is another name for institutional inertia and is often at the heart of slowness to adapt. The examples of Xerox and Eastman Kodak offered above were very much on this point. Sometimes, large organizations do manage eventually to adapt to radically changed circumstances of their existence, but not always. Much of new technology is brought to usefulness by new organizations, and this is a central reason why public policy has to maintain favorable conditions for startups if its objective is to foment economic growth.

Measurement

How can the impact of prototype inventions and their follow-on innovations be measured? All things considered, the statistics on investment in the National Income and Product Accounts (NIPA) are not much help. To see why, consider a recent example: the personal computer. US businesses of all kinds buy these computers in great numbers, and the expenditures for them are on capital accounts and are treated as investment in the NIPA. In some cases, the investment in the facilities to manufacture these machines is foreign, and therefore not in the US NIPA. Personal computers today are largely generic products and are assemblages of components such as central processing units (CPU), memory chips, and many others. It is only when the components are of domestic manufacture that the investment in the manufacturing facilities to make them is recognized as investment in the NIPA.

If one looks back to the origins of the modern computer in the 1940s, it becomes clear that the powerful and compact computers of today resulted from a huge amount of innovation, including many innovation inventions, in the course of their development. A gathering surge in innovation activity—stage one and two—would not be reflected in the investment account in the NIPA because the research and development expenditures that are integral to innovation are not recognized as investment. Recognition would come only at such time as the results of an innovation stream engendered sufficient confidence among business investors to induce the building of facilities to produce systems based on the new technology, and at such time as business in general commenced buying the resulting systems. About the best the NIPA Accounts can do is to reflect the unmistakable fingerprints of technology-driven growth in periods when such growth dominates all the other factors producing the growth. It seems like an obvious point that a device or system that is useful to business could not have come into existence without, in some cases, decades of innovation and very substantial expenditures. All such expenditures are as future-oriented as expenditures for productive hardware which does qualify as investment in the national accounts. Recognition of research and development expenditures as investment by business would mitigate a problem that grows from offshore outsourcing of manufacturing. A fairly common arrangement, especially in the field of high technology, is for a product to be developed in the US and manufactured offshore. Recognition of the research and development basis of such expenditures would credit US investment with the stage-one development work behind an important new product.

Perhaps the best reflection of a technology boom to be found in the official statistics lies in the productivity ratios. These statistics are based on ratios of some measure of output, such as Gross Domestic Product, to the number of worker-hours that went into producing the output. Of course, labor is not the only input to the overall production function, but to date there has not been developed a truly satisfactory measure of capital input for the economy as a whole. As shown above, these data do show a productivity surge during the 1910s and during the 1990s. Circumstantially, these are relatable, respectively, to tooling for the automobile and electrification of industry in the earlier period and the impact of the computer in the latter period. However, the productivity statistics show only the *results* of stage-one and -two innovation after it has taken place, but do not give much indication of the *buildup* of innovative activity. Use of NIPA-based productivity statistics reveals the fingerprints of invention and innovation after the fact of it has become widely obvious in many other ways. This is unfortunate, for it would be difficult to specify an engine of economic growth that is more powerful and universal than the advance of technology. The received statistics force us into a not-fully-satisfactory attempt to sort this effect on growth from all the other influences on growth.

One could argue that if the market is aware of the technology-based buildup, then that is all that is needed. Indeed, the very fact of the buildup is the result of market actors pursuing opportunities that grow from the new technology. However, the whole purpose of a system of accounts, whether for a private enterprise or for the macro-economy of a great republic, is to monitor the current condition of the enterprise. If the account system fails to achieve this in a significant way, its overall value is questionable. The free market is such an efficient transmitter of information that actors within it generally do not even have to understand in detail why they should act in certain ways; they simply recognize opportunity from the information that the market provides. Admittedly, capturing something like the capital investment nature of research and development involves some statistical nightmares, but this is the nature of the information economy.

A serious measurement problem that is related to the failure of investment statistics to recognize the kind of research activity that precedes the emergence of commercially useful versions of prototype inventions is the inability to measure capital in place. This is a measurement problem with which economic statisticians, both private and government, have long struggled. What has emerged from these efforts typically involves gross investment data and some assumptions regarding the average rate of depreciation appropriate to the capital stock. So far, no one has devised a means of measuring the rate of obsolescence as contrasted with the depreciation due to wear and tear in use. This is another way of saying that there is no recognition of the destruction of the value of physical capital that is being replaced by capital embodying superior technology—that is, creative destruction. One can conjecture that were this admittedly knotty problem overcome, the resulting capital stock data would be somewhat more volatile than has been the results of estimation processes that do not recognize obsolescence. Indeed, it can be wondered whether Professor Solow would have found such a large unexplained residual in his growth model of the 1950s had he had the use of capital-in-place data that reflected creative destruction.

War and Technological Change

Readers of the preceding pages will not have missed the impression that war and preparation for it have played a large role in technological innovation from the very early years of the American republic. In the nineteenth century, the role of the army contract was often of the nature of preparation for war, as in the contracts won by Whitney, Colt, Hall, and the prosecution of war during serious conflicts, especially the Civil War. These contracts financed very important improvements in machine tools, especially as better materials became available. The ability to produce finely machined parts *en masse* was crucial to the technology of the private sector in the opening decades of the twentieth century. In the twentieth century, the simple purchase of weaponry as a driver of technology expanded into innovation of weaponry in order to stay abreast or ahead of enemy capabilities, especially in World War II and ever since. The Cold War period has been highly prolific from the standpoint of widely used consumer technology that had its origin in war preparation. Well-known examples include the Internet and the global positioning system. The old distinction between guns and butter has become highly blurred. Technologies that started out as defense-related have expanded and enriched the entire economy.

Apart from the massive civilian economy developments with defense-related heritage, how technology has affected warfare itself is of interest. This history is very long, but the arguably most important recent development has been the perfected ability to hit a target. From the time that gunpowder was first introduced into warfare until well past World War II, most of the physical mass that has been thrown at human and inanimate targets has missed. As recently as the 1950s, there was a rule of thumb to the effect that it takes a ton of metal to kill someone. During the conflict in Afghanistan, by contrast, it was possible for frontline troops to call for an air strike on a single house with realistic expectation that the target would be hit on the first try. This capability grew out of ingenious marriages of microprocessing capability, the global positioning system, and other elements, including a relativistic correction. All of this technology was brought together based on inventions and innovations that took place mostly after 1970. In other words, such precision as is now possible in warfare emerged in a handful of decades, a mere instant in the long history of warfare. Needless to say, military theorists the world over are fully occupied with the effort to anticipate what this means for the future of warfare. The equally cogent question concerns what this technology has meant and means for standards of living, for the pattern that the technology of war preparation and prosecution has always had important implications for peaceful activities is not likely to disappear.

Not all who read this will be pleased with this thought, but there is a large amount of modern technology that grew out of warfare preparation. Given that much of war-related research is shrouded in secrecy, one is confined to guessing how omnipresent this effort is at supporting the scientists and engineers engaged in the requisite research at the present time. While it doesn't seem likely, one can wonder if the rapid technological flux characteristic of the last four decades could be maintained if the world were to become truly peaceful. The possibility is not totally without precedent. The defense element in technology was not that strongly present in the years from the end of the American Civil War through the eve of World War II. Even in this span of years, however, the government's hand was not entirely absent. Could all those prototype inventions that emerged from stage-one innovation after 1880 have done so without the improvement of machine tooling that grew out of all those nineteenth-century army contracts? Moreover, the ubiquity of the defense motive since 1940 suggests that a peaceful world sometime in the twenty-first century could force some serious, perhaps painful, adjustments.

Inasmuch as defense-motivated research and development might be confused with a form of industrial policy, a distinction should be made. For the most part, the US government's defense-related invention and innovation effort has had the limited purpose of improving the country's state of defense. This has been true across the entire history of the republic. This is what has made certain kinds of defense procurement and innovation so successful as a support for technology that has wide benefits for the economy as a whole. As a general rule, the defense effort has *not* included formal efforts to foresee the impact of the defense-related work on the civilian economy. This extension of the government's work has been left to the private market, and this relation has proved to be an efficient provider of the benefits of technological improvement to the economy as a whole.

Of course, defense-related research and development is far from the entire story of the government's role in the development of technology. Many of today's medications have been based on NIH-financed scientific research.

Also, environmental regulation has been a very important goad to technological development in the past forty years. In some ways, however, the technological response to regulation such as under the Clean Air Act and amendments has not followed the pattern of defense-related technology. We are familiar with the pattern in which the defense effort throws off ideas that the private sector develops, often with the result of richer living for people in general and of reduced costs of living. So far, environmental regulation has resulted in new kinds of hardware that results in apparent *increased* costs to the consumer. Examples include such ubiquitous items as catalytic converter in motor vehicles and most of the alternative technologies for producing electricity that are currently under discussion.

To a large extent, this new pattern is more apparent than real for the reason that results that are mostly confined to a cleaner atmosphere have little or no market price and, therefore, cannot be valued by conventional accounting means. The European Community has recognized this problem and has begun programs such as tradable pollution rights in an attempt to place a cost on pollution externalities. At least they have the idea even if their initial attempts at imposing such costs have been somewhat faltering. The US has so far shied away from formal measures to impose explicit costs on polluting, and, not surprisingly, the market has not responded with the solution to the air and water quality except in the form of responses to specific regulatory measures. As matters now stand, it does not pay an individual business enterprise to be aggressive in reducing externalities other than in direct response to specific regulation. The problem is vaguely reminiscent of the problem of adoption of the safety coupler in the nineteenth century. It pays no one to adopt cleaner but more costly methods if competitors fail to act simultaneously and adopt also. Then there is the modern problem of which technology to adopt.

Commoditization

A group of prototype inventions can lead to a substantial impetus to economic growth through innovation and combinations of innovations. One of the repeating patterns that the technological history of the twentieth century brings out is that goods and services that are the progeny of prototype inventions tend to become commodities. This is not obvious when products resulting from a prototype invention first appear, for at that time they are truly novelties. However, the new technologies and their products become commonplace with the passage of time. Witness the automobile and the personal computer. In opposition to the commoditization tendency is the effort to achieve product differentiation, or at least the illusion of it. As noted earlier in these pages, the progress of a technology can be characterized as an ongoing tendency for all resulting products to look alike and the effort by different producers to obscure this process. Progress in this or that line of products takes the form of increasingly small increments of change.

Is Creative Destruction Always Creative?

As noted in Chapter 13 above, creative destruction refers to the idea that a new technological combination competes with an established one, proves superior to it, and causes its obsolescence and abandonment. The final result of this process is that people supposedly live better. Creative destruction is, thus, an integral part of economic growth. For the most part, a reading of Western economic history seems to uphold this idea. If, however, the choice among competing systems is not always determined by objective economic and/or technical criteria, as W. Brian Arthur has pointed out, than there appears to be room for some long-term mistakes, in which what is chosen is really no better than what is abandoned, or maybe even worse than. In other words, is creative destruction always that creative? Moreover, if an abandoned technology later appears to have had some utility after all, under what conditions can it be revived?

There is at least one example that bears on this. Consider the electric street railway. These systems directly replaced horse-drawn omnibuses. As noted in Chapter 2, the street railways appeared after 1885 and represented a very early application of the electric motor. The lines were extended widely, not just in large cities, but also in places of modest size. Collectively, they afforded a clean and safe public transportation means to people of all economic classes. Street railways expanded until about 1920, and then went into what proved to be a long and fatal decline. Depression conditions led to the financial failure of many local systems. The systems that survived the Depression survived through World War II with the aid of gasoline rationing, but were on borrowed time. Once wartime gasoline rationing was lifted, people returned to their cars. With a few exceptions, the last of these had been retired by the mid-1950s. The street railways had been abandoned in favor of a universal dependence on the automobile. Such public transportation as survived generally was in the form of the rubber-tired bus systems, and, as a rule, getting anywhere on a local bus was so slow that no one used the service except people with no automobile alternative.

What were the characteristics of these two technologies? The street railway, in common with other forms of rail transportation, had some perceived weaknesses. Among these was the inability of streetcars to maneuver in traffic. Inasmuch as rail services ran on fixed schedules, their use required accommodation of personal schedules to the streetcar's timetable. A more important characteristic was that the economics of rail operation at any level dictate that stops to discharge or receive freight or passengers have to be limited in number and at publically recognized locations. When communities depended on rail transport, either for local transport or for contact with the rest of the world, they tended to develop such that people lived and worked conveniently to the railway station or the streetcar line. Before the automobile began to dominate local personal transportation, the "conveniently" here often had to mean walking distance. Rail-based transportation at any level was a powerful *concentrator* of economic activity.

By contrast, the automobile had the ability to stop anywhere for the convenience of its passengers. It lacked the railway's ability to serve large numbers of passengers at the same time, and because of this, a community growing around the automobile had no incentive to compactness. An automobile owner could go at any time he or she pleased and for any distance. With the development of the network of paved roads, the distance potential of auto travel expanded to the whole country and beyond. The automobile had the exact opposite effect on general development from that of the streetcar or train. It not only did not encourage concentration of economic activity, *it enabled urban sprawl.* Ownership of a car conferred a sense of freedom and power, which many today would argue is illusory. In a sense, the country's rush to the rubber tire rested on emotion more than cold calculation on the part of consumers.[2] One would think that the ideal system would take the form of autos and busses serving rail-stops, but this did not happen at first. The automobile simply took over, and rail-based public transportation was abandoned. In the 1960s, there was said to exist something called the "rubber tire lobby," a creature of automakers and large oil companies. One of the effects of this or something like it was an action of the California Legislature, giving the right-of-way to the automobile where before it had resided with the streetcar. Among other effects of this was the killing of the passenger operations of the Pacific Electric Railway, which had previously served the entire Los Angeles Basin.

A handful of cities have avoided the full commitment to the automobile. Generally speaking, these are larger places that attained their size and commercial importance before the days of the automobile. The list includes New York, Philadelphia, Chicago, and San Francisco. Philadelphia is a very interesting case. In the late nineteenth century, high officials of the Pennsylvania Railroad, headquartered in downtown Philadelphia, settled in a number of communities along the railroad's main line: Wayne, Paoli, Ardmore, Bryn Mawr, and others—the famous "main line." These people were in position to create commuter train schedules convenient for commuting to downtown offices. Their presence in the suburbs encouraged others to join them and also to use the rail commuter services. A visitor to these communities today sees neighborhoods near the rail stations that have freestanding houses on small lots. These are old houses whose values have been maintained and that have been physically modernized. The visitor has to be impressed by the physical compactness of these neighborhoods in contrast to the spread-out character of bedroom communities in other areas that have gotten their growth in the context of the automobile. As for local transportation in the central city, Philadelphia retained the use of its original street railway by incrementally digging it under the most congested downtown streets. The history of Philadelphia's local transportation system suggests that the wisdom of abandoning early local transit systems in so many places was questionable.

Was the abandonment of rail passenger service at all levels an example of creative destruction or merely destruction? A hint of the answer to this comes from simple observation of events of the last several decades. Cities with serious traffic problems have increasingly turned to "light-rail" systems in an attempt to divert drivers from their personal vehicles. To go back several decades, there was an attempt to build "heavy-rail" systems of public people movers. Light rail and heavy rail differ in that the latter embodies complete separation of rail and rubber-tired traffic. Heavy-rail systems were built in Atlanta, San Francisco Bay Area (BART), Washington, DC, and Los Angeles. The combination of heavy rail and feeder-bus lines seemed to be a modern solution to traffic systems at first, for it afforded commuters and others a fast service unimpeded by street traffic, and it itself is not an impediment to street traffic. The Atlanta system's experience may be indicative of what happens when a nineteenth-century solution is imposed on a city with late twentieth-century population sprawl. As the system expanded, the comfort and speed it afforded did attract many from cars, especially commuters and travelers to and from the airport. The pre-existing bus system was completely rerouted to feed passengers into the rail system. Unfortunately, this ridership resulted in an all-too familiar pattern of rush at the ends of the workday with only light traffic at other hours. The capital cost of the heavy-rail rail systems has proved prohibitive—$100 million per mile and up. Fares do not come close to covering operating costs, and the deficit requires a heavy subsidy from local and other public sources. One of the major problems with heavy-rail systems is that their required large investment buys far more people-moving capacity than is needed in the many American cities that have grown up around the automobile. These places simply do not have the population density that can remotely justify such capacity. As a result, heavy-rail systems' utility is effectively limited to connecting places, such as downtowns, big airports, big malls, and edge cities, which are nodes of higher than average population density. This leaves much of the city un-served.

Light rail is at street level. It is difficult to distinguish modern light-rail systems from the streetcar lines that were junked before 1960 in so many places except that the modern cars are quieter and more comfortable for the rider. Light rail is much cheaper than heavy rail, and therefore it is more attainable. Light rail is less dependent on nodes of high population density than are heavy-rail systems. Cities which have built light-rail systems in recent decades and which also junked their early street railways back in and before the fifties include Denver, San Diego, Dallas, Houston, Salt Lake City, and Charlotte. Where are the light-rail systems being built? They are being built in the larger cities. What of the middling-sized places that once had street systems? If there seems to be a return to a nineteenth-century technology to relieve twenty-first-century auto traffic problems, it is incomplete. The sad fact is that none of this turning back of the calendar benefits the smaller places. It is difficult to find a municipality of any size that has a road net adequate to the traffic load imposed on it.

The larger places are spending tens of millions to rebuild nineteenth-century systems. Creative indeed!

Notes

1. For example, see Zweibel, Ken; Mason, James; and Fthenakis, Vasilis, "A Solar Grand Plan," *Scientific American,* (January 2008), pp. 64-73. This article proposes concentration of US electric power production in the desert southwest. It includes several improvements over earlier proposals along this general line, such as long-distance distribution via high-voltage direct current (this avoids the uncertainty of the timing of commercial superconductor availability) and use of solar energy to heat water rather than photoelectric production of power (avoids the cost of massive quantities of pure silicon). It represents an advance in this general concept from earlier proposals. Nevertheless, the estimated time of completion is 2050, and the estimated price tag is well over $600 billion.
2. This is not really a condemnation of the consumer. It took a number of years for traffic problems to develop, and, until these became acute, individual motorists did enjoy considerable personal freedom based on their automobiles.

Bibliography

Books

Arthur, W. Brian, *Increasing Returns and Path Dependence in the Economy,* (Ann Arbor: The University of Michigan Press, 1994).

Atkins, P.W., *The Second Law,* (New York: Scientific American Books, 1984).

Babcock & Wilcox Corp., *Steam: Its Generation and Use (35th. ed.),* (New York, 1913).

------, *Steam: Its Generation and Use (39th. ed.),* (New York, 1978).

Batelle Memorial Institute, *Cobalt Monograph,* (Brussels: Centre D'Information du Cobalt, 1960).

Bosworth, Barry, and Triplett, Jack, *Is the 21st Century Productivity Expansion Still in Services? And What Should be Done About it?* (New York: National Bureau of Economic Research: 2006).

Burdick, Donald L. and Leffler, William L., *Petrochemicals in Nontechnical Language,* (Tulsa: Pennwell Books, 1990).

Burke, James, *Connections,* (Boston: Little Brown and Company, 1978).

Cardwell, Donald, *The Norton History of Technology,* (New York: W.W. Norton & Co., 1995).

Chas. A. Strelinger & Co., *A Book of Tools, Machinery and Supplies Lindsay,* (Detroit, 1895. Reprinted by Lindsay Publishing Co., 1991).

Collins, Douglas, *The Story of Kodak,* (New York: Harry N. Abrams, Inc., 1990).

Cummins, Lyle, *Diesel's Engine,* (Wilsonville, OR: Carnot Press, 1993).

------, *The Diesel Odyssey of Clessie Cummins,* (Wilsonville, OR: Carnot Press, 1998).

Drucker, Peter F., *Management: Tasks, Responsibilities, Practices,* (New York: Harper & Row, 1974).

Eames, Charles, and Eames, Ray, *A Computer Perspecitve,* (Cambridge, MA: Harvard University Press, 1973).

Fogel, Robert, *Railroads and American Economic Growth,* (Baltimore, MD: The Johns Hopkins University Press, 1964).

Friedman, Milton, and Schwartz, Anna J., *A Monetary History of the United States, 1867-1960,* (Princeton, NJ: Princeton University Press, 1963).

Gary, James H. and Handwerk, Glenn E., *Petroleum Refining: Technology and Economics,* (New York: Marcel Decker, Inc., 1975).

Haberler, Gottfried, *Prosperity and Depression,* (Lake Success, NY: United Nations, 1946).

Hawke, David Freeman, *Nuts & Bolts of the Past,* (New York: Harper & Row, 1988).

Hawkins, N., *The New Catechism of the Steam Engine,* (New York: Theo. Audel & Co., 1904. Reprint by Lindsay Publications, Inc., 1987).

Hayek, F. A., *The Road to Serfdom,* (Chicago: University of Chicago Press, 1944).

Hicks, J.R., *Value and Capital (Second ed.),* (London: Oxford University Press, 1965).

-------, *Capital and Growth,* (New York: Oxford University Press, 1965).

-------, *A Contribution to the Theory of the Trade Cycle,* (Oxford: Oxford University Press, 1950).

Hubbert, M.K., "Energy Resources," in *Resources and Man,* (Committee on Resources and Man. National Academy of Sciences—National Research Council), pp.157- 242.

Hughes, Thomas P., *American Genesis,* (New York: Viking Press, 1989).

Keynes, J.M., *The General Theory of Employment, Interest, and Money,* (New York, 1936).

Kuhn, Thomas S., *The Structure of Scientific Revolutions,* (Chicago: University of Chicago Press, 1996).

Leslie, Stuart W., *Boss Kettering,* (New York: Columbia University Press, 1983).

Lipsey, Richard G.; Carlew, Kenneth L.; and Bekar, Clifford, *Economic Transformations,* (Oxford: Oxford University Press, 2005).

Mansfield, Edwin, *Technological Change: An Introduction to a Vital Area of Economics,* (New York: W.W. Norton & Co., 1971).

Maples, Robert E., *Petroleum Refinery Process Economics,* (Tulsa: Pennwell Books, 1993).

MacMillan, Harold, *Blast of War,* (New York: Carroll & Graf Publishers, Inc. 1967), p. 68.

Mowery, David C. and Rosenberg, Nathan, *Paths of Innovation: Technological Change in 20th Century America,* (Cambridge: Cambridge University Press, 1998).

Pacey, Arnold, *Technology in World Civilization,* (Cambridge: The MIT Press, 1990).

Pasour, E.C., Jr., *Agriculture and the State: Market Processes and Bureaucracy,* (Oakland, CA: The Independent Institute, 1990).

-------, and Randal R. Rucker, *Plowshares and Pork Barrels: The Political Economy of Agriculture,* (Oakland, CA: The Independent Institute, 2005).

Pool, Robert, *Beyond Engineering: How Society Shapes Technology,* (New York: Oxford University Press, 1997).

Porter, Roy, *Blood & Guts: A Short History of Medicine,* (New York: W.W. Norton & Co., 2002).

Porter, Roy (ed.), *The Cambridge Illustrated History of Medicine,* (New York: The Cambridge University Press, 1996).

Rosenberg, Nathan, *Inside the Black Box: Technology and Economics,* (Cambridge: Cambridge University Press, 1982).

-------, *Exploring the Black Box: Technology, Economics, and History,* (Cambridge: Cambridge University Press, 1994).

Schlaes, Amity, *The Forgotten Man,* (New York: Harper Collins, 2007).

Schumpeter, Joseph A., *Capitalism, Socialism, and Democracy,* (New York: Harper Perennial, 1950).

-------, "The Analysis of Economic Change," from *The Review of Economic Statistics, 1935,* as reprinted in *AEA Readings in Business Cycle Theory,* (Homewood, IL: Richard D. Irwin, Inc., 1951).

-------, *The Theory of Economic Development* (reprint of the 1934 edition), (New Brunswick, NJ: Transaction Publishers, 1983).

Schwingwe, Julian, *Einstein's Legacy,* (New York: Scientific American Books, 1986).

Sharkey, William W., *The Theory of Natural Monopoly,* (New York: Cambridge University Press, 1986).

Skousen, Mark, *Vienna and Chicago: Friends or Foes,* (Washington: Capital Press, 2005).

Smith, Douglas K. and Alexander, Robert C., *Fumbling the Future: How Xerox Invented, Then Ignored, The First Personal Computer,* (New York: William Morrow and Co., 1988).

Smith, Vernon L., *Investment and Production,* (Cambridge: Harvard University Press, 1966).

-------, *Papers in Experimental Economics,* (Cambridge: Cambridge University Press, 1991).

Stewart, Harry L., *Hydraulic and Pneumatic Power for Production* (4th. ed.), (New York: Industrial Press, Inc., 1976).

Tussing, Arlon and Tippee, Bob, *The Natural Gas Industry: Evolution, Structure, and Economics,* (Tulsa: Pennwell Books, 1995).

U.S. Census of Manufactures, *Power and Machinery Employed in Manufactures,* (Washington: U.S.G.P.O., 1888; Reprinted by Lindsay Publishing Co., Inc., 1994).

United States Steel Co., *The Making, Shaping, and Treating of Steel* (8th. ed.), (Pittsburgh, PA, 1964).

U.S.D.A. and Tennessee Valley Authority, *Superphosphate: Its History, Chemistry, and Manufacture,* (Washington: U.S.G.P.O., 1964).

Usher, Abbott Payson, *A History of Mechanical Inventions,* (Boston: Beacon Press, 1929).

Walton, Gary M. and Rockoff, Hugh, *History of the American Economy (6th. ed.),* (New York: Harcourt, Brace Jovanovich Publishers, 1990).

Waterson, Michael, *Economic Theory of the Industry,* (New York: Cambridge University Press, 1984).

White, John H., *The American Railroad Passenger Car,* (Baltimore: Johns Hopkins Press, 1978).

Williamson, Oliver H., *Markets and Hierarchies: Analysis and Antitrust Implications,* (New York: The Free Press, 1975).

Articles in Periodicals

Anderson, Howard, "Why Big Companies Can't Invent," *Technology Review,* Vol. 107 (May 2004), pp. 56-58.

Bell, Trudy E., "The Victorian Global Positioning System," *The Bent of Tau Beta Pi* (Spring 2002), pp. 14-21.

Benderley, Beryl Lee, "From Poison Gas to Wonder Drug," *Invention and Technology,* Vol. 18 (Summer 2002), pp. 48-54.

Bernstein, Marc, "Thomas Midgley and the Law of Unintended Consequences," *Invention and Technology,* Vol. 17 (Spring 2002), pp. 38-45.

Beyer, David S., "The Mechanics of Saving Life-II," *Scientific American Supplement* (Sept. 17, 1910), 184-185.

Canfield, Paul C. and Bud'ko, Sergey, "Low-Temperature Superconductivity is Warming Up," *Scientific American,* Vol. 292 (April 2005), pp. 81-87.

Churchill, Arthur, "Edison and His Early Work," *Scientific American Supplement,* (April 1, 1905), pp. 2451-2452.

Coase, R.H., "The Nature of the Firm," *Economica,* New Series, IV (1937), pp. 386-403. As reprinted in *AEA Readings in Price Theory,* (Chicago: Richard D. Irwin, Inc., 1952), pp. 331-351.

De Muralt, C.L., "Heavy Electric Locomotives," *Scientific American Supplement,* (April 2, 1910), pp. 318.

Fairly, Peter, "Solar on the Cheap," *Technology Review,* Vol. 105, (Jan/Feb 2002), pp. 48-53.

Fitzgerald, Michael, "To Fight, Verizon Switches," *Technology Review, Vol. 107* (Dec. 2004), pp. 46-52.

Freedman, David H. "Fuel Cells vs. The Grind," *Technology Review,* Vol. 105, (Jan/Feb 2002), pp. 40-47.

Fridel, Robert, "New Light on Edison's Light," *Invention and Technology,* (Great Inventions Supplement), pp. 26-31.

Genera, Charles, "Nature vs. Nature," *Region Focus,* Vol. 9 (Fall 2005).

Gibbs, W. Wayt, "Computing at the Speed of Light," *Scientific American,* Vol. 291, (Nov. 2004), pp. 40-48.

Golembeski, Dean J., "Struggling to Become and Inventor," *Invention and Technology,* (Great Inventions Supplement), pp. 40-48.

Gradenwitz, Alfred, "A Gigantic Shipyard Crane," *Scientific American Supplement,* Vol. LIX, (Feb. 1, 1905), pp. 24333-24334.

Hamilton, Linda, "Liquid Crystals," *Invention and Technology* Vol. 17, (Spring 2002), pp. 20-29.

Hamley, David H. and Corley, Raymond, Raymond E., "How to Control and Engine of Limited Power," *Trains,* Vol. 34, (January 1974), pp. 26-28.

Hannon, Wm. H., Marsh, Gerald E., and Stanford, George S., "Smarter Use of Nuclear Waste," *Scientific American,* Vol. 293, (Dec. 2005), pp. 84-91.

Heppenheimer, T.A., "Nuclear Power: What Went Wrong?" *Invention and Technology,* Vol. 18 (Fall 2002), pp. 46-56.

Howard, Webster E., "Better Displays with Organic Films," *Scientific American,* Vol. 290, (Feb. 2004), pp. 18-27.

Judson, Horace Freeland. "The Glimmering Promise of Gene Therapy," *Technology Review*, Vol. 109, no. 5, (Nov.-Dec. 2006), pp. 40-47.

Kelly, Jack, "The Most Perfect Weapon," *Invention and Technology*, Vol. 20 (Fall 2004), pp. 18-27.

King, Jim, "The Lucrative Elution," *Technology Review*, Vol. 108 (Oct. 2005), pp 32-34.

Lenatti, Chuck, "Nanotech's First Breakthrough," *Technology Review*, Vol. 107 (March 2004), pp. 46-52.

Lok, Corrie, "Life Vest," *Technology Review*, Vol. 108 (Feb. 2005), pp. 88.

Lowenstein, Roger, "The Integrator," *Technology Review*, Vol. 108 (Oct. 2005), pp. 80.

Manning, Willard; Joseph Newhouse; Naihua Duan; Emmett Keeler; Arleen Leibowitz; and Susan Marquis, "Health Insurance and the Demand for Medical Care: Evidence from a Randomized Experiment," *American Economic Review*, Vol. 77, No. 3, (1987), pp. 283-295.

Martin, Jean, "Mule to MARTA," *The Atlanta Historical Bulletin*, XX, (Winter 1976), pp. 1-208.

Nehring, Richard, "Hubbert's Unreliability," *Oil & Gas Journal*, 104.16, (Apr. 24, 2006), pp. 43-51.

Paluka, Tim, "Holography: The Whole Picture," *Invention and Technology*, Vol. 18 ,(Winter 2003), pp. 12-21.

Paniccia, Mario, *et. al.*, "A High-Speed Silicon Optical Modulator Based on a Metal-Dioxide-Semiconductor-Capacitor," *Nature*, Vol. 427, (Feb. 12, 2004), pp. 615-618.

Perlin, John, "Solar Power: The Slow Revolution," *Invention and Technology*, Vol. 18, (Summer 2002), pp. 20-25.

Phillips, Don, "Digital Railroad," *Technology Review*, Vol. 105, (March 2002), pp. 75-78.

Quinn, Jim, "Hall of Fame Interview: Thomas Fogarty," *Invention and Technology*, Vol.19, (Winter 2004), pp. 60-63.

-------, "Hall of Fame Interview: Stephanie Kwalek," *Invention and Technology*, Vol.18, (Winter 2003), pp. 60-63.

Rhodes, Anne K., "Kansas Refinery Starts Up Coke Gasification Unit," *Oil & Gas Journal*, (Aug. 5, 1995).

Robinson, Ken K. and Tatterson, David E., "Fischer-Tropsch Oil-From-Coal Promising as Transport Fuel," *Oil & Gas Journal*, Vol. 105.8, (Feb. 26, 2007), pp. 20-31.

Rosenberg, Nathan and Steinmueller, W. Edward, "Why are Americans such Poor Imitators?" *American Economic Review*, Vol. 78 (May 1988), pp. 229-234.

Scanlon, Lisa, "The Birth of Cool," *Technology Review*, Vol. 107, (Mar. 2004), pp. 84.

Service, Robert, "Intel's Breakthrough," *Technology Review*, Vol. 108, (July 2005), pp. 62-65.

Shagam, Janet Yagoda, "Designing Drugs," *Invetion and Technology*, Vol. 21, (Fall 2005), pp. 18-27.

Simcoe, Robert J., "The Revolution in your Pocket," *Invention and Technology*, Vol. 20 (Fall 2004), pp. 18-27.

Solow, Robert M., "Technical Change and the Aggregate Production Function." *Review of Economics and Statistics*, 39, (1957).

Soojung-Kim Pang, Alex, "The Making of the Mouse," *Invention and Technology*, Vol. 17, (Winter 2002), pp. 48-54.

Talbot, David, "LEDs vs. the Lightbulb," *Technology Review*, Vol. 106, (May 2003), pp. 30-36.

Westinghouse, George, "The Electrification of Railways-I," *Scientific American Supplement*, (July 9, 1910), pp. 22-23.

-------, "The Electrification of Railways-II," *Scientific American Supplement*, (July 16, 1910), pp. 38-39.

Wood, Lamont, "Copper for Fiber," *Scientific American*, Vol. 293, (July 2005), p. 24.

Yafa, Stephen, "The Man Who Made Cotton King," *Invention and Technology*, Vol. 20, (Winter 2005).

Zweibel, Ken; Mason, James; and Fthenakis, Vasilis, "A Solar Grand Plan," *Scientific American*, Vol. 294, (January 2008), pp. 64-73.

Presented Papers, Working Papers, etc.

Amick, et. al., "A Large Coal IGCC Plant," Presented at Nineteenth Annual Pittsburgh C Coal Conference, (Sept. 23-27, 2002).

Finkelstein, Amy N., "The Aggregate Effects of Health Insurance: Evidence from the Introduction of Medicare," (April 2006), www.nber.org-afinkels/papers/Finkelstein_Medicare_Medicare_April06.pdf

Fulgheri, Paolo, and Sevlir, Merih, "The Ownership and Financing of Innovation in R & D Races," EOGI Finance Working Paper No. 18, (2003), http://ssrn.com/abstract=397500

Holt, Neville, "IGCC Power Plants—EPRI Design and Cost Studies," EPRI/GTC Gasification Technologies Conference, San Francisco, CA, (Oct. 16, 1998).

Klepper, Steven, "The Evolution of the U.S. Automobile Industry and Detroit as its Capital," Pittsburgh, Carnegie Mellon Univ., mimeo, (2001).

Oliver, Richard A., "Application of BGL Gasification of Solid Hydrocarbons for IGCC Power Generation," Paper presented at the Gasification Technologies Conference,   San Francisco, CA, (2000).

Petrofund Energy Trust, *Annual Report, 2005*.

Rieffel, Eleanor and Wolfgang Polak, "An Introduction to Quantum Computing for Non-Physicists," FX Palo Alto Laboratory, (2000).

Schrijver, Alexander, "On the History of Combinatorial Optimization (till 1960)," See Alexander Schrijver under Google Wikipedia.

Index

For Product Safety Concerns and Information please contact our EU
representative GPSR@taylorandfrancis.com Taylor & Francis Verlag GmbH,
Kaufingerstraße 24, 80331 München, Germany

Batch number: 08153774

Printed by Printforce, the Netherlands